Applied Artificial Intelligence in Data Science, Cloud Computing and IoT Frameworks

(Volume 2)

Transformative Healthcare Solutions Powered by AI & ML

Edited by

Suneeta Satpathy
Center For Cybersecurity, SoA Deemed to be University, Bhubaneswar, Odisha, India

Sachi Nandan Mohanty
School of Computer Science & Engineering, VIT-AP University, Andhra Pradesh, India

&

Subhendu Kumar Pani
Computer Science & Engineering Krupajal Engineering College BPUT, Odisha, India

Applied Artificial Intelligence in Data Science, Cloud Computing and IoT Frameworks

(Volume 2)

Transformative Healthcare Solutions Powered by AI & ML

Editors: Suneeta Satpathy, Sachi Nandan Mohanty and Subhendu Kumar Pani

ISSN (Online): 3029-2247

ISSN (Print): 3029-2255

ISBN (Online): 979-8-89881-207-2

ISBN (Print): 979-8-89881-208-9

ISBN (Paperback): 979-8-89881-209-6

Published by Bentham Science Publishers Pte. Ltd. Singapore, in collaboration with Eureka Conferences, USA.

First published in 2025.

Bentham Science Publishers Pte. Ltd.
No. 9 Raffles Place
Office No. 26-01
Singapore 048619
Singapore
Email: subscriptions@benthamscience.net

BENTHAM SCIENCE

CONTENTS

FOREWORD

In a world where the digital and physical realms are inextricably linked, **"Transformative Healthcare Solutions Powered by AI & ML"** is a beacon of insight and inspiration. This book is more than a compendium of technical knowledge; it is a thoughtful exploration into the transformative power of Artificial Intelligence and Machine Learning and a compelling invitation to reimagine how we live, work, and interact in the modern age.

At a time when innovation is accelerating at an unprecedented pace, the fusion of human intellect with advanced computational systems is reshaping every facet of society. In these pages, you will discover a meticulously crafted narrative that demystifies complex technological concepts and brings the human stories behind them to life. The text seamlessly bridges rigorous scientific inquiry with a literary elegance that engages the mind and the heart.

The healthcare landscape is undergoing a profound transformation driven by the rapid advancement of Artificial Intelligence (AI) and Machine Learning (ML). These technologies, once confined to theoretical discussions and experimental labs, have now become integral to modern medicine redefining diagnosis, treatment, prevention, and patient care. The integration of AI and ML in healthcare marks not merely an evolution of existing systems but a revolutionary shift toward a predictive, personalized, and precision-oriented model of medicine. At the heart of this revolution lies data-massive volumes of it, generated through electronic health records, wearable sensors, imaging modalities, genomics, and real-time patient monitoring. AI and ML algorithms are uniquely capable of extracting actionable insights from this vast and complex data landscape, uncovering patterns that remain invisible to human observation. From early disease detection to outcome prediction, these intelligent systems are enabling healthcare professionals to make faster, more accurate, and data-driven decisions. The outcome is a paradigm shift from reactive to proactive care, where prevention and early intervention take precedence over late-stage treatment.

One of the most significant impacts of AI and ML is evident in diagnostic imaging and pathology. Deep learning models now surpass human-level accuracy in detecting anomalies across radiographs, CT scans, and MRIs. Similarly, ML-powered digital pathology assists in identifying malignancies at a microscopic scale, reducing diagnostic errors and turnaround times. Beyond imaging, AI-driven tools are being applied to genomics, enabling the identification of genetic predispositions and the development of personalized treatment regimens through precision medicine. Equally transformative are the applications of AI in clinical decision support systems (CDSS). These intelligent platforms synthesize patient histories, laboratory data, and evidence-based guidelines to assist physicians in selecting optimal therapeutic strategies. By augmenting, rather than replacing, clinical expertise, AI empowers healthcare professionals to focus more on patient care and less on administrative or repetitive analytical tasks. Moreover, the use of ML in drug discovery and vaccine design has dramatically accelerated biomedical innovation—shortening development cycles from years to months, as exemplified during the global COVID-19 pandemic.

AI and ML are also redefining healthcare delivery beyond hospital walls. Wearable health devices and remote patient monitoring systems continuously collect and transmit physiological data, allowing for real-time health assessments and early detection of complications. Predictive analytics helps healthcare systems anticipate disease outbreaks, manage hospital resources, and improve public health interventions. Together, these technologies are making healthcare more accessible, affordable, and inclusive, especially in underserved or remote areas. However, the journey is not without challenges. Issues surrounding data privacy, algorithmic transparency, and ethical governance must be addressed to ensure responsible and equitable AI deployment. Collaboration between technologists, clinicians, policymakers, and ethicists is essential to establish frameworks that safeguard human trust and uphold the integrity of medical practice. This volume, *Transforming Healthcare with Revolutionary AI & ML*, explores these developments in depth, highlighting the innovations, challenges, and future directions that define the next era of digital health. It stands as both a reflection of progress achieved and a vision for the transformative possibilities yet to come in the global pursuit of intelligent, patient-centered healthcare.

Fernando Moreira
Universidade Portucalense, Porto, Portugal

PREFACE

Artificial Intelligence (AI) and Machine Learning (ML) are revolutionizing the way healthcare systems are designed, delivered, and sustained. The fusion of data-driven intelligence, advanced computing, and connected technologies has opened new frontiers in medical diagnosis, patient monitoring, drug discovery, and clinical decision support. This first volume, "Transforming Healthcare with Revolutionary AI & ML," serves as a comprehensive exploration of how these technologies are reshaping the global healthcare ecosystem through innovation, efficiency, and personalization.

The healthcare domain, once limited by fragmented data and reactive care models, is now evolving into a dynamic, predictive, and intelligent environment powered by AI and ML. The chapters in this volume provide a detailed view of this transformation covering intelligent diagnostic tools, predictive analytics for early disease detection, wearable sensor data analysis, AI-assisted radiology and pathology, and the integration of federated and privacy-preserving learning models for secure clinical data sharing. The discussion also extends to the ethical, regulatory, and sustainability aspects of AI-driven healthcare solutions, emphasizing responsible innovation for long-term impact.

This volume brings together contributions from researchers, clinicians, and technologists who are pioneering interdisciplinary applications of AI and ML. Their collective insights not only highlight current trends but also chart the course for future developments in connected health, clinical informatics, and autonomous medical systems.

As part of the broader three-volume series, this book lays the foundation for understanding how applied artificial intelligence intersects with data science, cloud computing, and IoT frameworks to create intelligent, adaptive, and resilient systems. While this first volume focuses on the healthcare revolution, the subsequent volumes will expand into other critical sectors such as smart industry, sustainable agriculture, cybersecurity, and environmental intelligence, reflecting the holistic role of AI in shaping the intelligent systems of the future.

It is our hope that this volume will serve as a valuable resource for academicians, researchers, practitioners, and students who aspire to harness the potential of AI and ML to advance healthcare innovation. By bridging scientific theory with real-world applications, this work underscores the transformative power of intelligent technologies in improving human well-being and global health outcomes.

Suneeta Satpathy
Center For Cybersecurity, SoA Deemed to be University
Bhubaneswar, Odisha, India

Sachi Nandan Mohanty
School of Computer Science & Engineering
VIT-AP University, Andhra Pradesh, India

&

Subhendu Kumar Pani
Computer Science & Engineering, Krupajal Engineering College BPUT
Odisha, India

List of Contributors

Anil Pise	Cumulus Solutions, Johannesburg, South Africa
Akhil Singampalli	Vignan's Institute of Information Technology, Andhra Pradesh, India
Aarif S. Ahamed	School of Computer Science and Engineering & Information Science, Presidency University, Bangalore, Karnataka, India
Anila Macharla	Department of Computer Science and Engineering, Chaitanya Bharathi Institute of Technology, Hyderabad, Telangana State, India
Bhasuri Polaki	Department of Computer Science and Engineering, Chaitanya Bharathi Institute of Technology, Hyderabad, Telangana State, India
B. Srinivasulu	Department of IT, BVRIT Hyderabad College of Engineering for Women, Hyderabad, Telangana, India
Ch. V. Bhargavi	Department of Computer Science and Engineering, GITAM School of Technology, GITAM Deemed to be University, Visakhapatnam, India
Ch. Bhavani	Department of Computer Science, CVR College of Engineering, Hyderabad, India
C. H. Ravi	Department Of CSE, CMRIT, Hyderabad, Telangana, India
Dinesh N. Chaudhari	Department of Computer Science and Engineering, J.D.I.E.T, Yavatmal, India
Erupaka Nitya	Department of CSIT, CVR College of Engineering, Hyderabad, India
Girija Jyothi Golla	Universita degli Studi di Messina, Messina, Italy
Gorapalli Srinivasa Rao	School of Computer Science and Engineering, VIT-AP University, Amaravathi, Andhra Pradesh, India
G. Muneeswari	School of Computer Science and Engineering, VIT-AP University, Amaravathi, Andhra Pradesh, India
G. Lingaiah	Department of CSIT, CVR College of Engineering, Hyderabad, India
Gulle Akshitha	Information Technology, Institute of Aeronautical Engineering, Hyderabad-500043, India
Guguloth Suman	Information Technology, Institute of Aeronautical Engineering, Hyderabad-500043, India
G. Kiran Kumar	Department of Computer Science and Engineering, Chaitanya Bharathi Institute of Technology, Hyderabad, Telangana State, India
Jyotismita Talukdar	Tezpur University, Assam, India
K. Subhadra	Department of Computer Science and Engineering, GITAM School of Technology, GITAM Deemed to be University, Visakhapatnam, India
Kulakarni Shreya	Department of Information Technology, Institute of Aeronautical Engineering, Hyderabad, India
Kapil N.	Saveetha Engineering College, Anna University, Chennai, Tamil Nadu, India
Mohit K. Popat	Department of Computer Science and Engineering, J.D.I.E.T, Yavatmal, India
Mantri Yakshitha	Department of Information Technology, Institute of Aeronautical Engineering, Hyderabad, India

Moorthi M.	Saveetha Engineering College, Anna University, Chennai, Tamil Nadu, India
M. Mohamed Iqbal	Department of Information Technology (IT) or Computer Science and Engineering (CSE), Sri Sivasubramaniya Nadar College of Engineering, Chennai, India
M. Radha	Department of CSE (Cyber Security), SNIST, Hyderabad, India
M. Arun	Department of Computer Science and Engineering (CSE), Kings College of Engineering, Thanjavur, India
Manjog Padhy	Department of Computer Science and Engineering, Siksha'O'Anusandhan University, Bhubaneswar, Odisha, India
Monika Mangla	Department of Information Technology, Dwarkadas J Sanghvi College of Engineering, Navi Mumbai, India
N. Raghava Rao	Department of Information Technology, Institute of Aeronautical Engineering, Hyderabad, India
N. Subhash Chandra	CVR College of Engineering, Hyderabad, Telangana, India
Ponnuru Sowjanya	School of Technology, GITAM (Deemed to be University), Hyderabad, Telangana, India
R. Sriramkumar	Department of Computer Science and Engineering (CSE), Dayananda Sagar University, Bangalore, India
R. Sugantha Lakshmi	Department of Computer Science and Engineering (CSE), Kings College of Engineering, Thanjavur, India
R. Jeevaraj	Department of Computer Science and Engineering (CSE), Dayananda Sagar University, Bangalore, India
Manda Rama Narasinga Rao	Department of Computer Science and Engineering, GITAM University, Visakhapatnam, Andhra Pradesh, India
Rasuri Rakhesh	Information Technology, Institute of Aeronautical Engineering, Hyderabad-500043, India
R. Suvarna Rao	Institute of Aeronautical Engineering, Hyderabad-500043, India
Rasmita Rautray	Department of Computer Science and Engineering, Siksha'O'Anusandhan University, Bhubaneswar, Odisha, India
Rishika Ponna	Department of Computer Science and Engineering, Chaitanya Bharathi Institute of Technology, Hyderabad, Telangana State, India
Rakhee	Faculty of Engineering, The University of the West Indies, Mona Campus, Kingston, Jamaica
R. Manik	Department of Computer Science & Engineering, Lovely Professional University, Jalandhar, India
Sana Swaroop	Department of Computer Science and Engineering, GITAM University, Visakhapatnam, Andhra Pradesh, India
Syed Abuthahir	Department of Computer Science and Engineering (CSE), Madanapalle Institute of Technology & Science, Andra Pradesh, India
S. Puvaneswari	Department of Computer Science and Engineering (CSE), Kings College of Engineering, Thanjavur, India

Shreyas Rajedra Hole	Department of Computer Science and Engineering (CSE), Dayananda Sagar University, Bangalore, India
Sasanka Sekhar Dalai	Department of Computer Science and Engineering, Siksha'O'Anusandhan University, Bhubaneswar, Odisha, India
Srinivasa Rao Dhanikonda	Department of AI & DS, Faculty of Science and Technology (IcfaiTech), ICFAI Foundation for Higher Education (Deemed to Be University), Hyderabad, Telangana, India
S. Nonita	Department of Information Technology, Indira Gandhi Delhi Technical University for Women, Delhi, India
Umarani Nagavelli	AI & DS Department, B V Raju Institute of Technology, Telangana, India
Vrushali Deshpande	D&M Tech, Pune, India
Vrishikaa Krishna Moorthy	Saveetha Engineering College, Anna University, Chennai, Tamil Nadu, India
V.C. Bharathi	School of Computer Science and Engineering (SCOPE), VIT-AP University, Andhra Pradesh, India
Y. Gokul	School of Computer Science and Engineering, VIT-AP University, Amaravati, India
Zahid Mansur	Tezpur University, Assam, India

Section 1
Transforming Healthcare with Revolutionary AI & ML

CHAPTER 1

Influence of Systemic Health Factors and Genetic Variants on Age-Related Macular Degeneration Risk: A Comprehensive Systematic Review

Sana Swaroop[1,*] and **Manda Rama Narasinga Rao**[1]

[1] *Department of Computer Science and Engineering, GITAM University, Visakhapatnam, Andhra Pradesh, India*

Abstract: Age-related macular degeneration (AMD) is a leading cause of irreversible vision loss among the elderly, affecting approximately 196 million people globally. Traditional diagnostic methods for AMD, which rely on human grading of fundus images, are hindered by inefficiencies and subjectivity. The DeepDrAMD model, a hierarchical vision transformer-based deep learning system, addresses these limitations by providing a more accurate, rapid, and automated diagnosis of AMD using color fundus images. Beyond diagnostics, understanding the complex molecular mechanisms of AMD pathogenesis, such as protein clearance, inflammasome activation, and calcium channel dynamics in the retinal pigment epithelium (RPE), is crucial for developing targeted therapies. Neovascular AMD (nAMD) presents additional challenges due to aberrant blood vessel proliferation, with recent studies highlighting the role of proteins like thrombospondin-1 and the enzyme granzyme B in choroidal neovascularization.

The progression of fibrosis in nAMD and the need for reliable biomarkers to detect this condition underscores the importance of refining imaging techniques and consensus definitions. Systemic health factors such as hypertension, cardiovascular disease, and genetic variants further influence AMD risk, necessitating comprehensive studies to elucidate these relationships. Investigations into the roles of retinal carotenoids, flavonoids, and anti-inflammatory agents, as well as advancements in drug delivery systems and ultrahigh resolution OCT imaging, offer promising avenues for improving AMD management and treatment outcomes. The introduction of ultrahigh-resolution spectral domain-OCT (UHR SD-OCT) and machine learning-based analysis of imaging data enables more precise quantification of sub-retinal pigment epithelium (RPE) deposits, facilitating early detection and monitoring of AMD progression. Additionally, novel drug delivery systems, including implants and non-stimuli responsive hydrogels, enhance the efficacy of treatments by providing controlled and sustained release of therapeutic agents.

* **Corresponding author Sana Swaroop:** Department of Computer Science and Engineering, GITAM University, Visakhapatnam, Andhra Pradesh, India; E-mail: ssana4@gitam.in

Suneeta Satpathy, Sachi Nandan Mohanty & Subhendu Kumar Pani (Eds.)

Collectively, these multidisciplinary efforts aim to enhance early detection, understand disease progression, and develop effective therapeutic interventions for AMD, leveraging state-of-the-art technologies and comprehensive molecular insights.

Keywords: Age-related macular degeneration (AMD), Deep learning, Molecular mechanisms, Therapeutic targets, Ultrahigh resolution optical coherence tomography (OCT).

INTRODUCTION

The increasing prevalence of age-related macular degeneration (AMD) is a major global health concern, affecting the lives of approximately 196 million people worldwide. This illness, which is recognized as a primary cause of irreversible vision loss in the elderly, requires further attention and novel diagnostic approaches. Traditional approaches, which rely on human grading and interpretation of fundus pictures, are inefficient, subjective, and susceptible to interobserver variability. To overcome these issues, DeepDrAMD is a cutting-edge hierarchical vision transformer-based deep learning model. This unique methodology aims to reinvent AMD diagnosis by utilizing color fundus pictures to provide a more accurate, fast, and automated method of detection and differential diagnosis. Building on the critical need for novel diagnostic tools in the face of age-related macular degeneration (AMD), the complicated molecular complexities of AMD pathogenesis are investigated. The clinical features of AMD, which include lipofuscin and drusen forms in the retinal pigment epithelium (RPE), highlight an underlying imbalance of protein homeostasis and inflammatory responses. While standard diagnostic approaches are vulnerable to subjectivity, this study investigates the relationship of protein clearance, inflammasome activation, and calcium (Ca^{2+}) channels in RPE, yielding new insights into the complex molecular processes that drive AMD progression. The observed interaction has revolutionary promise for tailored treatment strategies in AMD therapy. In the complex terrain of age-related macular degeneration (AMD), the advent of neovascular AMD (nAMD) or wet AMD presents a difficult challenge due to aberrant blood vessel proliferation in the eye. Thrombospondin (TSP)-1, a matricellular protein that is considerably reduced in AMD-affected eyes, plays an important role in the delicate balance of proangiogenic and antiangiogenic factors. This study identifies a previously unreported relationship between the serine protease Granzyme B (GzmB) and TSP-1 proteolysis in nAMD-related choroidal neovascularization (CNV). The findings shed insight into the complicated molecular pathways that influence AMD etiology and suggest new paths for targeted treatment interventions. In the landscape of neovascular age-related macular degeneration (nAMD), despite the effectiveness of antiangiogenic therapies managing exudation, the progression of fibrosis within the outer retina contributes to gradual vision loss over time. Addressing this challenge necessitates the development of drugs targeting fibrosis, relying on accurate detection and quantification with reliable endpoints and robust biomarkers. However, the absence of a consensus definition for fibrosis in nAMD hinders progress. This review initiates the establishment of a clear definition by comprehensively examining the diverse imaging modalities, criteria, and classification systems employed in characterizing fibrosis in nAMD. Clinicians encounter a challenge in delivering individualized prognoses when attempting to precisely forecast the evolution of intermediate to late-stage age-related macular degeneration (AMD), which impedes the optimization of management options. The pinnacle study was a systematic review that sought to discover and assess prognostic markers related to this progression. Six studies met established inclusion criteria and identified exploratory factors such as drusen features, age, smoking status, ocular and systemic comorbidities, race, and genotype. The heterogeneity of studies hindered meta-analysis, emphasizing the need for rigorous research to establish firm evidence in the future. In addition, the Kailuan Eye Study, which focuses on North China, investigates the impact of diabetes, hypertension, and hyperlipidemia on the development of age-related macular degeneration (AMD). The study conducts a comprehensive investigation of the prevalence of both dry and wet AMD in people with these comorbidities, with the goal of better understanding the complex interplay between systemic health variables and the risk of AMD formation. By investigating the association between these frequent illnesses and AMD, this study contributes to our understanding of the multiple factors that influence the prevalence of this vision-related ailment in the North Chinese demographic. Continuing the investigation of age-related macular degeneration (AMD), this study dives into the complex phenotypic history of AMD associated with the risk haplotype at the ARMS2 and HTRA1 loci. Age-related maculopathy susceptibility 2 (ARMS2) is a particularly perplexing gene associated with AMD, and research conducted by a major European collaboration aims to elucidate the complex course and spectrum of AMD in individuals bearing this risk haplotype. The treatment of wet age-related macular degeneration (AMD), the primary cause of visual impairment worldwide, has traditionally relied on intravitreal injections of anti-VEGF agents. However, the need for frequent injections creates hurdles in terms of patient compliance and possible consequences. To address this, a variety of drug delivery systems (DDSs) have been developed, providing alternatives to the standard monthly injection. This review investigates experimental and conceptual drug delivery techniques for the treatment of wet AMD, with an emphasis on particle systems, non-stimuli responsive hydrogels, implants, and composite systems discovered in recent decades. The choice of DDS, polymer material, and preparation procedure has an impact on mechanical properties, drug release patterns, and loading efficiency.

Mathematical modeling of drug release and pharmacokinetics improves DDS development, offering the possibility of more efficient treatment techniques for age-related macular degeneration and other eye illnesses affecting the posterior area. In the pursuit of effective treatments for neovascular age-related macular degeneration (nAMD), the frequency, progression rates, and connections of newly detectable macular atrophy (MA) have become significant issues. This study, performed within the MARINA trial framework, seeks to understand the dynamics of MA in patients with choroidal neovascularization (CNV) secondary to nAMD by comparing those treated with ranibizumab to a sham-treated cohort.

The primary goal of this study is to gain a thorough understanding of the amounts of retinal xanthophyll carotenoids in the setting of age-related macular degeneration (AMD) vs normal ageing. The study uses the macular pigment optical volume (MPOV) as a quantitative metric generated from dual-wavelength autofluorescence to assess the quantity of lutein (L) and zeaxanthin (Z) in AMD-affected and normal-aging eyes. Furthermore, the study aims to identify relationships between retinal xanthophyll levels and plasma concentrations, allowing for a more nuanced understanding of the function these carotenoids play in health and disease. Finally, the findings are intended to give information on the efficacy of existing supplementation techniques and their effects on AMD progression. In the realm of ocular research, the introduction of ultrahigh resolution spectral domain-OCT (UHR SD-OCT) has transformed our ability to study the complexities of retinal structures in unprecedented depth. This cutting-edge technology enables *in-vivo* visualisation of micrometric structural indicators, allowing for a more nuanced analysis of normal ageing and its distinction from early age-related macular degeneration (AMD). As the population ages, it becomes increasingly important to grasp the subtle differences between age-related changes and pathological abnormalities in the macula. This work investigates the notion that UHR SD-OCT can efficiently detect and quantify sub-retinal pigment epithelium (RPE) deposits, serving as a specific marker for early AMD, a disorder known for its effects on visual health. The study of these ultrahigh resolution OCT indicators not only advances our understanding of physiological ageing, but it also holds promise for elucidating the complex pathophysiology of early AMD. This study aims to speed up drug discovery and improve clinical trial processes in the hunt for effective AMD therapies by creating quantitative markers linked with disease progression. Age-related macular degeneration (AMD), a common and complex eye illness, is a major danger to vision health, frequently resulting in irreversible blindness. While the exact cause of AMD is unknown, it has been linked to oxidative damage, inflammation, and neovascularization. This study investigates the effect of flavonoids, naturally occurring bioactive compounds with antioxidant, anti-inflammatory, and antiangiogenic characteristics, in treating important clinical processes associated with AMD. The study of flavonoids' properties and mechanisms in the treatment of AMD sheds light on potential therapeutic approaches.

Neovascular age-related macular degeneration (nAMD) poses a substantial barrier to visual health, with anti-VEGF medication being a viable option. This retrospective study investigates the impact of time between diagnosis and anti-VEGF treatment commencement on the long-term results of 73 nAMD patients. The study evaluates decimal best-corrected visual acuity and macular thickness during a 12-month period, categorising participants into three groups depending on treatment onset (<48 h, 48 h-7 days, >7 days). This study's findings aim to inform best scheduling options for commencing anti-VEGF medication in patients with new-onset nAMD. This 10-year retrospective cohort study looks into the relationship between regular use of aspirin or non-aspirin nonsteroidal anti-inflammatory drugs (NSAIDs) and the development of age-related macular degeneration (AMD). The study's adjusted hazard ratios indicate fascinating relationships, indicating that aspirin and NA-NSAIDs may have a preventive impact against AMD, particularly neovascular AMD. These findings offer important insights into the relationship between nonsteroidal anti-inflammatory medication use and the risk of age-related macular degeneration. In the quest to optimize treatment strategies for neovascular age-related macular degeneration (nAMD), understanding the pharmacokinetic/pharmacodynamic (PK/PD) relationship of anti-vascular endothelial growth factor (anti-VEGF) therapy is paramount. This study aims to develop a population PK/PD model of intravitreal bevacizumab treatment for nAMD patients, leveraging data from the Greater Manchester Avastin for Neovascularisation (GMAN) trial. By correlating best-corrected visual acuity (BCVA) and central macular retinal thickness (CRT) with treatment regimens, this investigation seeks to elucidate dosing regimen decisions to enhance patient outcomes. Fostering on our understanding of the pathophysiology of age-related macular degeneration (AMD), this study looks into the possible therapeutic benefits of Humanin G (HNG), a mitochondrial-derived peptide, on retinal neurodegeneration and abnormal angiogenesis, both of which are hallmarks of advanced AMD. This study sheds light on HNG's potential regulatory role in AMD progression and treatment by assessing its impact on cell proliferation, apoptosis, and protein levels of key markers in normal and AMD retinal pigment epithelial (RPE) cell lines derived from clinically characterised patients. Age-related macular degeneration (AMD) is a serious hazard to older persons' visual health and places a large strain on worldwide healthcare systems. AMD's pathophysiology includes oxidative stress, inflammation, and light exposure, all of which have an influence on the macula. Lutein, zeaxanthin, and mesozeaxanthin are plant-derived carotenoids that accumulate in the macula and have antioxidative and anti-inflammatory characteristics. This study investigates the possible preventative and therapeutic benefits of various plant carotenoids on AMD, drawing on findings from cellular, animal, and human clinical studies. This study intends to explore the translational uses of carotenoid supplementation in AMD management by studying the dose-dependent and duration-dependent advantages, while also providing non-invasive alternatives to standard therapies. The discussion emphasises the relevance of dose and treatment length in maximising carotenoids' protective benefits, as well as the possibility for personalised therapy techniques in AMD. Central serous chorioretinopathy (CSCR), neovascular age-related macular degeneration (nAMD), and polypoidal choroidal vasculopathy (PCV) all contribute significantly to vision loss, with genetic factors playing an important role in their development. This case-control genetic association research seeks to assess the genetic relationships of various subtypes of these disorders, therefore offering insights into their underlying processes. The study specifically looks at the relationships between numerous genetic variants and distinct subtypes of CSCR, such as those with macular neovascularization (MNV), pigment epithelial detachment (PED), nAMD, and PCV. The analysis contains mutations from genes such as ADAMTS9, ANGPT2, ARMS2, CFH, NR3C2, PGF, TNFRSF10A, and VIPR2. By revealing the genetic correlations of various maculopathies, this study advances our understanding of their aetiology and may open the road for personalised therapy methods that target specific genetic risk factors. Age-related macular degeneration (AMD) is a complicated illness characterised by RPE dysfunction and choroidal neovascularization (CNV), which causes substantial visual loss in older persons. Despite substantial investigation, the basic processes that drive AMD pathogenesis remain poorly understood. In this work, we look at the involvement of RNA demethylase ALKBH5 in AMD development. ALKBH5, which is recognised for its role in RNA demethylation, has been discovered to be up-regulated in AMD. Here, we investigate the impact of ALKBH5 dysregulation on RPE cells and its relationship to CNV progression. We investigate the molecular processes underpinning ALKBH5-induced RPE dysfunction and CNV development utilising both *in vitro* and *in vivo* investigations, with a focus on its role in the AKT/mTOR signalling pathway. Furthermore, we assess the therapeutic efficacy of IOX1, an ALKBH5 inhibitor, in preventing hypoxia-induced RPE dysfunction and CNV propagation, pointing to a possible path for AMD therapy. This work provides insight on the complicated molecular pathways that contribute to AMD development and suggests innovative ALKBH5-targeting treatment methods for AMD therapy. Age-related macular degeneration (AMD) is a complicated condition that causes visual loss in older persons, with few therapeutic choices. Long non-coding RNAs (lncRNAs) have emerged as important players in AMD pathogenesis, regulating processes like as apoptosis and oxidative stress. Understanding the significance of lncRNAs in AMD provides insight into disease causes and possible treatment targets. This review looks at the role of dysregulated lncRNAs in AMD development and their potential as biomarkers. It also emphasises the need for more research to evaluate lncRNAs' therapeutic value and create tailored treatments for AMD.

Early detection of exudative macular neovascularization (eMNV) in the unaffected eyes of people with newly diagnosed unilateral eMNV is critical for prompt intervention and therapy. SD-OCT imaging has emerged as an effective method for finding predictive biomarkers linked with AMD development. In this regard, the double layer sign (DLS) and shallow irregular retinal pigment epithelium (RPE) elevation (SIRE) have been identified as possible indications of eMNV development. The purpose of this study is to look at the predictive efficacy of DLS and SIRE detected by SD-OCT imaging in assessing the likelihood of progression to exudation in AMD patients' unaffected partner eyes. Understanding the importance of these OCT indicators may improve early detection tactics and therapeutic outcomes for individuals at risk of AMD progression.

LITERATURE SURVEY

The study by Xu *et al.* (2023) focuses on the automatic identification and differential diagnosis of age-related macular degeneration (AMD) using colour fundus pictures. They use deep learning with a hierarchical vision transformer model for this aim. The research tries to address the difficulty of AMD diagnosis by analysing medical photos using modern machine learning techniques [1].

Karema-Jokinen *et al.* (2023) look at how protein clearance mechanisms, inflammasome activation, and calcium channels interact in the retinal pigment epithelium (RPE) of individuals with age-related macular degeneration (AMD). The research investigates the molecular mechanisms underpinning AMD pathogenesis, with a focus on the interactions between various biological processes in RPE cells [2].

Obasanmi *et al.* (2023) analyse the role of granzyme B in choroidal neovascularization (CNV) and age-related macular degeneration (AMD) by looking at its involvement in the proteolysis of thrombospondin 1. The article, published in Laboratory Investigation, investigates the molecular mechanisms underpinning AMD pathogenesis, with a particular focus on the role of granzyme B-mediated proteolysis in CNV generation and progression [3].

Bachmeier *et al.* (2023) present a comprehensive evaluation of fibrosis in neovascular age-related macular degeneration (AMD), with a focus on fibrosis classifications based on clinical imaging modalities. The research, published in

Survey of Ophthalmology, evaluates alternative definitions and imaging properties of fibrosis in the context of neovascular AMD, with the goal of improving understanding and standardisation in the assessment of disease-associated fibrotic lesions [4].

Hagag *et al.* (2023) present a comprehensive evaluation of prognostic markers associated with progression to late age-related macular degeneration (AMD), which serves as the Pinnacle study's second report. The research, published in Survey of Ophthalmology, aims to give a complete assessment of factors impacting AMD progression to advanced stages, thereby leading to a better knowledge of disease prognosis and potential interventions [5].

Kertes (2000) investigates the probable link between hypertension, cardiovascular disease, and age-related macular degeneration (AMD). The paper, published in findings-Based Eye Care, is likely to explore epidemiological findings as well as potential molecular processes linking hypertension and cardiovascular illness to AMD development and progression. Understanding these relationships can help identify risk factors and adopt AMD prevention efforts [6].

Thee *et al.*'s (2022) study, published in Ophthalmology, looks at the phenotypic course of age-related macular degeneration (AMD), with an emphasis on the ARMS2/HTRA1 genes. The study's goal is to improve our understanding of the disease's pathophysiology and how certain genetic variables may influence its course by evaluating the clinical characteristics and progression of AMD in individuals with mutations in these genes. This study provides important insights into personalised approaches to controlling AMD based on genetic risk factors [7].

Chacin Ruiz, Swindle-Reilly, and Ford Versypt's (2023) work, published in the Journal of Controlled Release, presents experimental and quantitative approaches to drug administration for the treatment of wet age-related macular degeneration. The work investigates novel therapeutic agent delivery strategies to target the underlying mechanisms of wet AMD by combining experimental methodologies and mathematical modelling. This interdisciplinary approach has great opportunities to improve the efficacy and precision of AMD treatment regimens [8].

Blodi *et al.* (2023) look into the incidence of macular atrophy in the MARINA research, which compares ranibizumab to sham treatment for neovascular age-related macular degeneration (AMD). The article, published in Ophthalmology Retina, investigates macular atrophy as a possible effect or outcome of anti-vascular endothelial growth factor (VEGF) treatment in patients with neovascular AMD. Understanding the prevalence and implications of macular atrophy in this setting is critical for optimising treatment methods and maintaining long-term visual outcomes in individuals with neovascular AMD [9].

McGwin *et al.* (2023) investigated the amounts of retinal and plasma xanthophylls in people with age-related macular degeneration (AMD) *vs* those with normal ageing. The article, published in Ophthalmology Science, looks into the potential differences in xanthophyll concentrations between these two groups, which could offer insight on the involvement of these carotenoid molecules in AMD pathogenesis. Understanding changes in xanthophyll levels may provide insights into AMD risk factors and potential preventative or therapeutic approaches [10].

Chen *et al.* (2023) investigate ultrahigh resolution optical coherence tomography (OCT) markers linked with normal ageing and early age-related macular degeneration (AMD). The study, published in Ophthalmology Science, aims to identify specific OCT traits that characterise the ageing retina and the early phases of AMD. These markers can provide useful insights into disease progression, as well as aid in early detection and management of AMD, perhaps leading to better clinical outcomes for those affected [11].

Sun *et al.* (2023) investigated the role of flavonoids in age-related macular degeneration (AMD). The study, published in Biomedicine & Pharmacotherapy, looks into the possible benefits of flavonoids for preventing or slowing the progression of AMD. Flavonoids are naturally occurring chemicals present in a variety of foods and plants that are known to have antioxidant and anti-inflammatory properties. Understanding their possible retinal protective effects could lead to new therapeutic techniques for treating AMD [12].

Martínez-Castillo *et al.*'s (2023) study examines how early treatment initiation affects long-term functional outcomes in patients with neovascular age-related macular degeneration. According to the research, delivering medication within 48 hours after diagnosis can result in better long-term functional outcomes. The findings, published in Biomedicine & Pharmacotherapy, emphasise the need of fast intervention in controlling neovascular AMD and the potential benefits of early treatment beginning in preserving visual function [13].

Yang *et al.* (2023) investigate the relationship between the usage of nonsteroidal anti-inflammatory medicines (NSAIDs) and the development of age-related macular degeneration (AMD). The study, published in the Journal of the Formosan Medical Association, reports on a 10-year retrospective cohort study that looked at the potential link between NSAID use and the development of AMD. Understanding the link between NSAIDs and AMD development is critical for determining the safety and dangers of long-term use of these drugs [14].

Del Amo *et al*. (2023) present the construction of a population pharmacokinetic/ pharmacodynamic (PK/PD) model for anti-VEGF therapy in patients with age-related macular degeneration (AMD). The study, published in the European Journal of Pharmaceutics and Biopharmaceutics, seeks to get a comprehensive understanding of the pharmacokinetic and pharmacodynamic features of anti-VEGF medicines used to treat AMD. The study uses population-based modelling tools to optimise dose regimes and improve treatment results for AMD patients receiving anti-VEGF medication [15].

Nashine and Kenney (2024) investigated the influence of Humanin G (HNG) on angiogenesis and neurodegenerative indicators in age-related macular degeneration (AMD). The study, published in the journal Mitochondrion, is expected to investigate HNG's potential therapeutic function in regulating pathways linked with angiogenesis and neurodegeneration in AMD. Understanding the effect of HNG on these molecular pathways may provide insights into new treatment techniques for AMD and maintaining visual function [16].

Nguyen *et al*. (2023) investigate the effects of plant carotenoids in treating age-related macular degeneration (AMD). The study, which was published in the Journal of Functional Foods, is expected to investigate the preventive effects of plant-derived carotenoids against AMD, taking into account their antioxidant and anti-inflammatory capabilities. Understanding the potential involvement of plant carotenoids in preventing or reducing the onset of AMD can lead to more effective dietary strategies for maintaining eye health in ageing populations [17].

Koizumi *et al*. (2013) study the link between the clinical features of polypoidal choroidal vasculopathy (PCV) and choroidal vascular hyperpermeability. The study, published in the American Journal of Ophthalmology, most likely investigates the relationship between particular clinical aspects of PCV and the existence of choroidal vascular hyperpermeability, which may contribute to the disease's development and progression. Understanding this link will help researchers learn more about the mechanisms that underpin PCV and design more tailored treatment options [18].

Sun *et al*. (2023) investigated the involvement of ALKBH5 in inducing retinal pigment epithelium abnormalities and choroidal neovascularization in age-related macular degeneration (AMD) *via* the AKT/mTOR pathway. The study, published in Cell Reports, investigates the molecular pathways behind AMD pathogenesis, with a focus on the role of ALKBH5 and the AKT/mTOR signalling cascade.

Understanding these pathways and their roles in AMD pathology can help identify possible therapeutic targets for avoiding or treating AMD-related problems, including choroidal neovascularization [19].

Zhang *et al.* (2023) explore the roles and processes of long non-coding RNAs (lncRNAs) in age-related macular degeneration (AMD). The study, published in Heliyon, investigates the regulatory functions of lncRNAs in AMD pathogenesis, offering information on their role in disease progression and prospective treatment targets. Understanding the molecular mechanisms underpinning lncRNA-mediated activities in AMD can lead to novel diagnostic and therapeutic options for managing this sight-threatening illness [20].

Csincsik *et al.* (2023) study the prognostic usefulness of the Double Layer Sign for progression to exudation in age-related macular degeneration (AMD). The study, published in Ophthalmology Retina, is likely to look at the relationship between the appearance of the Double Layer Sign on imaging modalities and the development of exudative symptoms in AMD. Understanding the prognostic importance of this imaging biomarker can help with risk stratification and early intervention methods for people at risk of developing advanced stages of AMD [21].

LIMITATIONS

Age-related macular degeneration (AMD) as shown in Fig. (**1**) is a common eye disorder that affects millions of individuals worldwide, causing considerable vision impairment and ultimately blindness if not addressed. Several studies have been conducted to better understand and address the challenges of AMD, including its pathogenesis, diagnosis, and treatment. For example, colour fundus pictures to automate the identification and differential diagnosis of AMD, using deep learning techniques and a hierarchical vision transformer model. Similarly, the molecular pathways driving AMD by looking at protein clearance, inflammasome activation, and calcium channel interactions in retinal pigment epithelium (RPE) cells.

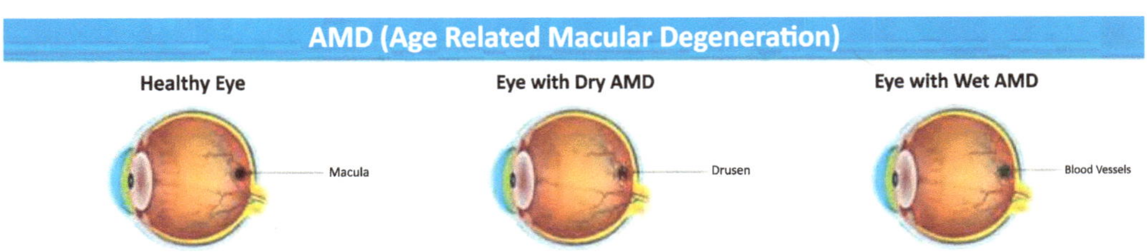

Fig. (1). The image explores the differences between a healthy eye, an eye with dry AMD, and an eye with wet AMD [22].

However, despite the useful insights provided by these investigations, many limitations should be noted. To begin, many studies utilise retrospective designs, which might add biases and hinder the discovery of causal links, as observed in studies on NSAID usage and AMD progression. Furthermore, small sample sizes in some studies may diminish statistical power and limit the generalizability of findings, especially in research on rare symptoms or genetic abnormalities. Furthermore, the cross-sectional nature of many studies yields only a snapshot of associations at a specific time point, necessitating longitudinal studies to better understand illness progression.

Furthermore, potential confounding factors such as lifestyle habits and environmental exposures are difficult to account for in observational studies, leading to difficulty in the interpretation of data. Finally, while imaging indicators like as the Double Layer Sign offer promise for AMD prognosis, their clinical utility may be restricted due to differences in interpretation and standardisation across contexts.

Despite these limitations, the combined findings of these studies add to our understanding of AMD and drive future research paths aimed at improving diagnosis, management, and treatment outcomes for those suffering from this devastating disorder.

Limited Generalizability: Many studies concentrate on specialised populations, such as people with certain genetic alterations or undergoing specific medications. This may restrict the generalizability of their findings to larger AMD populations.

Retrospective Study Design: Several studies, such as those looking at NSAID (Non-Steroidal Anti-Inflammatory Drug) usage and AMD development, use retrospective cohort designs. While retrospective studies can help discover connections, they are prone to bias and may not demonstrate causality.

Small Sample Numbers: Some research, particularly those looking at genetic markers or unusual characteristics, may use small sample numbers. This reduces the statistical power and dependability of their results.

Cross-Sectional Nature: Some studies look at connections or correlations at a particular point in time, offering simply a snapshot of the variables' relationship. Longitudinal investigations would provide more substantial evidence of causation and disease progression.

Potential Confounding Factors: Because many research use observational data, it is difficult to account for all potential confounding variables. Lifestyle behaviours, comorbidities, and environmental exposures can all influence the outcomes evaluated, but they are difficult to account for in observational research adequately.

Methodology

Selection Strategy

The selection technique for discovering relevant articles on Age-Related Macular Degeneration (AMD) with Deep Learning included a thorough search of academic databases, with a concentration on publications from the previous three years, notably from 2022 onwards. The search used specialised search phrases like "age-related macular degeneration," "AMD," and "deep learning" in combination to seek literature on the intersection of these issues. To guarantee broad coverage, the search may have included variations on these terms.

Pre-Processing

In evaluating each piece of data from the selected references on Age-Related Macular Degeneration (AMD) and Deep Learning, it is critical to consider the study's objectives, methods, algorithms, and findings. The key goal is to guarantee that high-quality data is used to build a strong model. Articles that meet the evaluation standards are kept, while those that do not fit the criteria are removed to ensure the model's integrity and correctness.

During the data preprocessing stage, a methodical approach is taken to choose the best features and optimise them using various optimisation techniques. This approach entails identifying relevant elements within the dataset that greatly improve the model's predictive performance. Various optimisation algorithms, such as gradient descent, genetic algorithms, or simulated annealing, can be used to fine-tune the model's parameters and improve its prediction skills.

Furthermore, the methodology used in each study is extensively analysed to better understand the data gathering process, model training techniques, and assessment criteria used to assess the model's performance. Any discrepancies or limits in the technique are documented to provide transparency and rigour in the model development process.

Overall, the selected data is analysed using a systematic and complete manner, taking into account the study objectives, methodology, algorithms, and findings. The goal is to create a high-quality model for the automatic identification and differential diagnosis of AMD using deep learning approaches while adhering to severe evaluation criteria and applying advanced optimisation techniques.

Data Analysis

Analysing the data from the references on Age-Related Macular Degeneration (AMD) and Deep Learning yields numerous critical insights and trends in research methodology, findings, and areas of interest. The selected references cover a wide range of AMD research topics, such as automatic detection and differential diagnosis, protein clearance mechanisms, proteolysis involvement in choroidal neovascularization, clinical imaging-based fibrosis definitions, prognostic factors associated with disease progression, drug delivery approaches, macular atrophy prevalence, and the role of flavonoids in AMD management.

Methodologies vary between studies, reflecting the diversity of AMD research. Some investigations concentrate on deep learning methods for image processing in fundus pictures, while others investigate molecular causes, clinical imaging, or drug delivery systems. Methodologies used include experimental techniques, mathematical modelling, systematic reviews, retrospective cohort studies, and population pharmacokinetic/pharmacodynamic modelling.

These research findings help us better understand AMD pathophysiology, progression, and prospective treatment methods. They emphasise the significance of protein clearance processes, proteolytic activity, fibrosis, prognostic variables, and medication delivery systems in controlling AMD. Furthermore, understanding the involvement of flavonoids, nonsteroidal anti-inflammatory medications (NSAIDs), plant carotenoids, and long non-coding RNAs offers prospective therapeutic strategies.

Despite tremendous progress, AMD research still faces obstacles, such as the need for early diagnostic tools, personalised treatment approaches, and effective drug delivery systems. However, the rise of deep learning techniques, along with molecular insights and clinical imaging modalities, opens up exciting possibilities for improving AMD diagnosis, prognosis, and management.

The data analysis emphasises the multidisciplinary nature of AMD research, highlighting the importance of combining deep learning methodologies with molecular biology, clinical imaging, and drug delivery approaches to address the complex challenges associated with this common age-related disease as shown in Fig. (**2**).

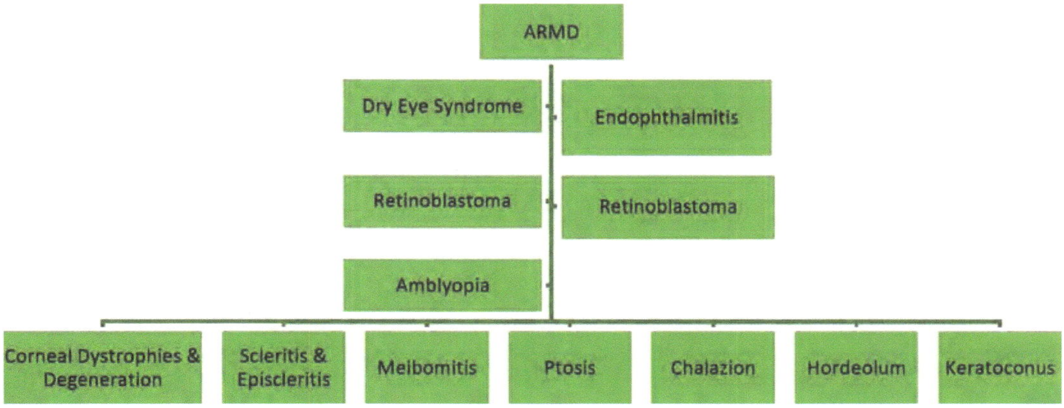

Fig. (2). Types of age-related macular degeneration.

ARMD is a prominent cause of blindness worldwide, with two types: 'dry' (atrophic) and 'wet' (exudative). Approximately 90% of patients with ARMD have the atrophic form, as opposed to the exudative variant. The exudative form of the disease causes severe vision loss, accounting for 90% of blindness cases. Macular degeneration is more prevalent in those over 65, whites, and females as shown in Fig. (**3**).

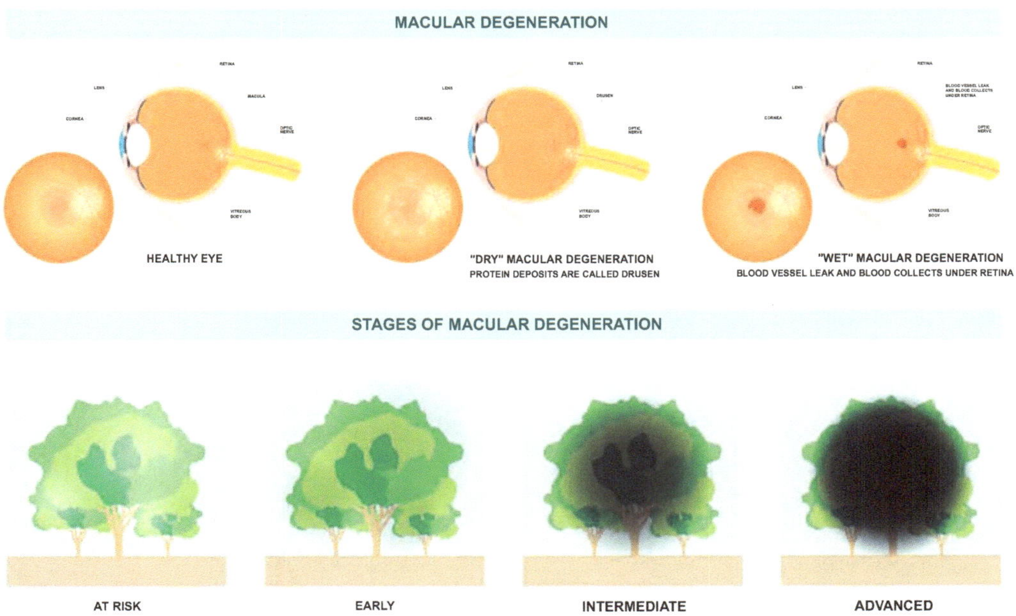

Fig. (3). Stages of macular degeneration [23].

TECHNIQUES APPLIED

The gaps identify areas that require additional research and validation to expand our understanding and management of age-related macular degeneration (AMD) as shown in Table **1**.

Table 1. Summary of techniques and research gaps in literature.

Authors	Year	Techniques	Research Gap
Xu, K., Huang, S., Yang, Z., Zhang, Y., Fang, Y., Zheng, G., Sun, J.	2023	Deep learning with hierarchical vision transformer for AMD detection and diagnosis from color fundus photographs.	Limited dataset diversity; needs real-world clinical validation.
Karema-Jokinen, V., Koskela, A., Hytti, M., Hongisto, H., Viheriälä, T., Liukkonen, M., Kaarniranta, K.	2023	Analysis of protein clearance, inflammasome activation, and Ca^{2+} channels in retinal pigment epithelium from AMD patients.	*in vitro* model limitations: requires *in vivo* validation and clinical application.
Obasanmi, G., Zeglinski, M. R., Hardie, E., Wilhelm, A. C., Turner, C. T., Hiroyasu, S., Matsubara, J. A.	2023	Study of Granzyme B in choroidal neovascularization and AMD through Thrombospondin-1 proteolysis.	Needs further investigation on long-term effects and clinical relevance.
Bachmeier, I., Armendariz, B. G., Yu, S., Jäger, R. J., Ebneter, A., Glittenberg, C., Fauser, S.	2023	Review of fibrosis definitions in neovascular AMD based on clinical imaging.	Requires standardized imaging criteria and longitudinal studies for better classification.
Hagag, A. M., Kaye, R., Hoang, V., Riedl, S., Anders, P., Stuart, B., Lotery, A. J.	2023	Systematic review of prognostic factors for progression to late AMD.	Lack of consistent prognostic markers across different populations.
Thee, E. F., Colijn, J. M., Cougnard-Grégoire, A., Meester-Smoor, M. A., Verzijden, T., Hoyng, C. B., Klaver, C. C.	2022	Analysis of the phenotypic course of AMD for ARMS2/HTRA1 genetic variants.	Need for broader genetic studies to confirm findings in diverse populations.
Chacin Ruiz, E. A., Swindle-Reilly, K. E., Ford Versypt, A. N.	2023	Experimental and mathematical approaches for drug delivery in treating wet AMD.	Translation of drug delivery systems from experimental models to clinical practice.
Blodi, B. A., Domalpally, A., Corkery, E., Osborne, A., Blotner, S., Grzeschik, S. M., Gune, S.	2023	Study on macular atrophy prevalence in the MARINA Study using Ranibizumab *vs*. Sham treatment for neovascular AMD.	Long-term outcomes and comparative effectiveness with other treatments need exploration.

(Table 1) cont.....

Authors	Year	Techniques	Research Gap
McGwin, G., Kar, D., Berlin, A., Clark, M. E., Swain, T. A., Crosson, J. N., Curcio, C. A.	2023	Comparison of macular and plasma xanthophylls in AMD *vs*. normal aging.	Requires understanding of causal relationships and therapeutic potential.
Chen, S., Abu-Qamar, O., Kar, D., Messinger, J. D., Hwang, Y., Moult, E. M., Fujimoto, J. G.	2023	Ultrahigh resolution OCT markers for normal aging and early AMD.	Need for longitudinal studies to confirm predictive value and practical implementation in screening.
Sun, M., Yu, T., Zhao, J., Zhu, X., Xin, W., Zhang, F., Zhang, L.	2023	Investigation of flavonoids in AMD.	Requires clinical trials to validate efficacy and determine optimal dosages.
Martínez-Castillo, R., González-Gallardo, C., Muñoz-Ávila, J. I., Font, P., Villalba-González, M., Stoikow, I., De Isla, I. F., Pugliese, F., Anaya-Alaminos, R., García-Serrano, J., Hermoso-Fernández, F., Contieri, F., Muñoz-De-Escalona-Rojas, J. E., Pérez-Fajardo, L., Blanco-Blanco, M., Jiménez-Gómez, Y., González-Andrades, M.	2023	Treatment of neovascular AMD within 48 hours from diagnosis and its impact on long-term functional outcomes.	Need for replication in larger, multi-center trials to confirm efficacy.
Yang, T., Huang, W., Yang, C., Yang, C., Ho, T., Chen, T., Lai, T., Hsieh, Y.	2023	Retrospective cohort study on NSAID use and AMD development.	Requires prospective studies to establish causality and assess long-term impact.
Del Amo, E. M., Bishop, P., Godia, P., Aarons, L.	2023	Development of a population pharmacokinetic/pharmacodynamic model of anti-VEGF therapy in AMD patients.	Validation with diverse patient populations and real-world data.
Nashine, S., Kenney, M. C.	2024	Study of Humanin G (HNG) on angiogenesis and neurodegeneration markers in AMD.	Need for clinical trials to confirm therapeutic potential and safety in patients.
Nguyen, D. M., Thrimawithana, T., Piva, T. J., Grando, D., Huynh, T.	2023	Investigation of benefits of plant carotenoids against AMD.	Long-term clinical trials are required to confirm benefits and optimal dietary recommendations.

(Table 1) cont.....

Authors	Year	Techniques	Research Gap
Koizumi, H., Yamagishi, T., Yamazaki, T., Kinoshita, S.	2013	Analysis of clinical characteristics of polypoidal choroidal vasculopathy and choroidal vascular hyperpermeability.	More research needed to understand pathophysiology and develop targeted therapies.
Sun, R., Zhu, H., Zhang, Y. R., Wang, J., Wang, Y., Cao, Q. C., Liu, Q. H.	2023	Study of ALKBH5 in retinal pigment epithelium anomalies and choroidal neovascularization in AMD *via* the AKT/mTOR pathway.	Further validation in animal models and clinical trials needed.
Zhang, R., Lin, W., Yang, L., Gui, C., Pei, Y., Zhou, G.	2023	Investigation of roles and mechanisms of long non-coding RNAs in AMD.	Need for comprehensive *in vivo* studies and exploration of therapeutic applications.
Csincsik, L., Muldrew, K. A., Bettiol, A., Wright, D. M., Rosenfeld, P. J., Waheed, N. K., Empeslidis, T., De Cock, E., Yamaguchi, T. C. N., Hogg, R., Pető, T., Chakravarthy, U.	2023	Identification of the Double Layer Sign as a predictive marker of progression to exudation in AMD.	Requires further studies to confirm predictive value and standardize assessment in clinical practice.

RESULTS AND DISCUSSION

According to literature review, experts from around the world are starting to pay attention to how machine learning and deep learning is being used in healthcare, particularly for disease prediction. Twenty-three studies were considered from 2022 to 2024, and all of their conclusions were evaluated and studied. In fact, the intricacy of deep learning and machine learning significantly helps lighten the workload of medical professionals.

Our research focuses on several advanced computational approaches, including deep learning using Hierarchical Vision Transformers; deep learning techniques for identifying ultra-high-resolution OCT markers of normal aging and early age-related macular degeneration; predictive modeling for treatment outcomes of neovascular age-related macular degeneration; data analysis techniques—potentially involving machine learning—for evaluating the effect of Humanin G on angiogenesis and neurodegeneration markers in AMD; and bioinformatics techniques—also potentially involving machine learning—for studying the roles and mechanisms of long non-coding RNAs in age-related macular degeneration.

CONCLUSION AND FUTURE WORKS

Researchers applied various machine learning approaches to improve the detection and understanding of age-related macular degeneration (AMD). Deep learning models, particularly hierarchical vision transformers, were employed to enhance accuracy in automatic detection and differential diagnosis. The study also delved into the interactions between protein clearance, inflammasome activation, and Ca^{2+} channels in the retinal pigment epithelium of AMD patients. Additionally, the contribution of Granzyme B to choroidal neovascularization was explored, alongside a review of fibrosis definitions in neovascular AMD. Prognostic factors for late AMD progression were systematically reviewed, and the phenotypic course of AMD was examined in relation to Age-Related Maculopathy Susceptibility 2 (ARMS2)/High-Temperature Requirement A Serine Peptidase 1 (HTRA1) variants. Innovative drug delivery approaches for wet AMD were investigated, and comparisons were made between macular and plasma xanthophyll levels in AMD and normal aging. Advanced imaging techniques identified ultrahigh resolution OCT markers for early AMD. The impact of early treatment on neovascular AMD outcomes, the effects of plant carotenoids, and the relationship between polypoidal choroidal vasculopathy and choroidal vascular hyperpermeability were also studied. The roles of AlkB Homolog 5, RNA Demethylase (ALKBH5) and long non-coding RNAs in AMD development were elucidated, and the predictive value of the Double Layer Sign for AMD progression was assessed, aiming to enhance classification and treatment outcomes for AMD patients.

REFERENCES

[1] K. Xu, S. Huang, Z. Yang, Y. Zhang, Y. Fang, G. Zheng, B. Lin, M. Zhou, and J. Sun, "Automatic detection and differential diagnosis of age-related macular degeneration from color fundus photographs using deep learning with hierarchical vision transformer", *Comput. Biol. Med.*, vol. 167, p. 107616, 2023. [http://dx.doi.org/10.1016/j.compbiomed.2023.107616] [PMID: 37922601]

[2] V. Karema-Jokinen, A. Koskela, M. Hytti, H. Hongisto, T. Viheriälä, M. Liukkonen, T. Torsti, H. Skottman, A. Kauppinen, S. Nymark, and K. Kaarniranta, "Crosstalk of protein clearance, inflammasome, and Ca2+ channels in retinal pigment epithelium derived from age-related macular degeneration patients", *J. Biol. Chem.*, vol. 299, no. 6, p. 104770, 2023. [http://dx.doi.org/10.1016/j.jbc.2023.104770] [PMID: 37137441]

[3] G. Obasanmi, M.R. Zeglinski, E. Hardie, A.C. Wilhelm, C.T. Turner, S. Hiroyasu, W.A. Boivin, Y. Tian, H. Zhao, E. To, J.Z. Cui, J. Xi, H.S. Yoo, M. Uppal, D.J. Granville, and J.A. Matsubara, "Granzyme B contributes to choroidal neovascularization and age-related macular degeneration through proteolysis of thrombospondin-1", *Lab. Invest.*, vol. 103, no. 6, p. 100123, 2023. [http://dx.doi.org/10.1016/j.labinv.2023.100123] [PMID: 36849037]

[4] I. Bachmeier, B.G. Armendariz, S. Yu, R.J. Jäger, A. Ebneter, C. Glittenberg, D. Pauleikhoff, S.R. Sadda, U. Chakravarthy, and S. Fauser, "Fibrosis in neovascular age-related macular degeneration: A review of definitions based on clinical imaging", *Surv. Ophthalmol.*, vol. 68, no. 5, pp. 835-848, 2023. [http://dx.doi.org/10.1016/j.survophthal.2023.03.004] [PMID: 37023894]

[5] A. M. Hagag, R. Kaye, V. Hoang, S. Riedl, P. Anders, B. Stuart, and A. J. Lotery, "Systematic review of prognostic factors associated with progression to late age-related macular degeneration: Pinnacle study report 2", *Survey of Ophthalmology.,* 2023.

[6] P. Kertes, "Hypertension, cardiovascular disease, and age-related macular degeneration", *Evidence-Based Eye Care,* vol. 1, no. 4, pp. 218-219, 2000. [http://dx.doi.org/10.1097/00132578-200007000-00014]

[7] E.F. Thee, J.M. Colijn, A. Cougnard-Grégoire, M.A. Meester-Smoor, T. Verzijden, C.B. Hoyng, S. Fauser, H.W. Hense, R. Silva, C. Creuzot-Garcher, M. Ueffing, C. Delcourt, A.I. den Hollander, and C.C.W. Klaver, "The phenotypic course of age-related macular degeneration for ARMS2/HTRA1", *Ophthalmology,* vol. 129, no. 7, pp. 752-764, 2022. [http://dx.doi.org/10.1016/j.ophtha.2022.02.026] [PMID: 35240203]

[8] E.A. Chacin Ruiz, K.E. Swindle-Reilly, and A.N. Ford Versypt, "Experimental and mathematical approaches for drug delivery for the treatment of wet age-related macular degeneration", *J. Control. Release,* vol. 363, pp. 464-483, 2023. [http://dx.doi.org/10.1016/j.jconrel.2023.09.021] [PMID: 37774953]

[9] B.A. Blodi, A. Domalpally, E. Corkery, A. Osborne, S. Blotner, S.M. Grzeschik, and S. Gune, "Prevalence of macular atrophy in the MARINA study of ranibizumab versus sham for neovascular age-related macular degeneration", *Ophthalmol. Retina,* vol. 7, no. 8, pp. 661-671, 2023. [http://dx.doi.org/10.1016/j.oret.2023.03.004] [PMID: 37086257]

[10] G. McGwin Jr, D. Kar, A. Berlin, M.E. Clark, T.A. Swain, J.N. Crosson, K.R. Sloan, C. Owsley, and C.A. Curcio, "Macular and plasma xanthophylls are higher in age-related macular degeneration than in normal aging", *Ophthalmol. Sci.,* vol. 3, no. 2, p. 100263, 2023. [http://dx.doi.org/10.1016/j.xops.2022.100263] [PMID: 36864830]

[11] S. Chen, O. Abu-Qamar, D. Kar, J.D. Messinger, Y. Hwang, E.M. Moult, J. Lin, C.R. Baumal, A. Witkin, M.C. Liang, N.K. Waheed, C.A. Curcio, and J.G. Fujimoto, "Ultrahigh resolution OCT markers of normal aging and early age-related macular degeneration", *Ophthalmol. Sci.,* vol. 3, no. 3, p. 100277, 2023. [http://dx.doi.org/10.1016/j.xops.2023.100277] [PMID: 36970115]

[12] M. Sun, T. Yu, J. Zhao, X. Zhu, W. Xin, F. Zhang, and L. Zhang, "Role of flavonoids in age-related macular degeneration", *Biomed. Pharmacother.,* vol. 159, p. 114259, 2023. [http://dx.doi.org/10.1016/j.biopha.2023.114259] [PMID: 36652737]

[13] R. Martínez-Castillo, C. González-Gallardo, J.I. Muñoz-Ávila, P. Font, M. Villalba-González, I. Stoikow, I.F. De Isla, F. Pugliese, R. Anaya-Alaminos, J. García-Serrano, F. Hermoso-Fernández, F. Contieri, J.E. Muñoz-De-Escalona-Rojas, L. Pérez-Fajardo, M. Blanco-Blanco, Y. Jiménez-Gómez, and M. González-Andrades, *Treatment of neovascular age-related macular degeneration within 48 h from diagnosis improves long-term functional outcome.,* 2023. [http://dx.doi.org/10.1016/j.biopha.2023.114368]

[14] T. Yang, W. Huang, C. Yang, C. Yang, T. Ho, T. Chen, T. Lai, and Y. Hsieh, "Association between non-steroidal anti-inflammatory drug use and development of age-related macular degeneration—A 10-year retrospective cohort study", *J Formos Med Assoc,* 2023.

[15] E.M. del Amo, P.N. Bishop, P. Godia, and L. Aarons, "Towards a population pharmacokinetic/pharmacodynamic model of anti-VEGF therapy in patients with age-related macular degeneration", *Eur. J. Pharm. Biopharm.,* vol. 188, pp. 78-88, 2023. [http://dx.doi.org/10.1016/j.ejpb.2023.05.007] [PMID: 37178941]

[16] S. Nashine, and M. C. Kenney, "Effects of Humanin G (HNG) on angiogenesis and neurodegeneration markers in Age-related Macular Degeneration (AMD)", *Mitochondrion,* vol. 74, p. 101818, 2024.

[17] D. Nguyen, T. Thrimawithana, T.J. Piva, D. Grando, and T. Huynh, "Benefits of plant carotenoids against age-related macular degeneration", *J. Funct. Foods,* vol. 106, p. 105597, 2023. [http://dx.doi.org/10.1016/j.jff.2023.105597]

[18] H. Koizumi, T. Yamagishi, T. Yamazaki, and S. Kinoshita, "Relationship between clinical characteristics of polypoidal choroidal vasculopathy and choroidal vascular hyperpermeability", *Am. J. Ophthalmol.,* vol. 155, no. 2, pp. 305-313.e1, 2013. [http://dx.doi.org/10.1016/j.ajo.2012.07.018] [PMID: 23022162]

[19] R. Sun, H. Zhu, Y. R. Zhang, J. Wang, Y. Wang, Q. C. Cao, and Q. H. Liu, (2023, July 1). ALKBH5 causes retinal pigment epithelium anomalies and choroidal neovascularization in age-related macular degeneration *via* the AKT/mTOR pathway.

[20] R. Zhang, L. Wang, Y. Li, C. Gui, Y. Pei, and G. Zhou, "Roles and mechanisms of long non-coding RNAs in age-related macular degeneration", *Heliyon,* vol. 9, no. 11, p. e22307, 2023. [http://dx.doi.org/10.1016/j.heliyon.2023.e22307] [PMID: 38027818]

[21] L. Csincsik, K. A. Muldrew, A. Bettiol, D. M. Wright, P. J. Rosenfeld, N. K. Waheed, and T. Empeslidis, "The Double Layer Sign is highly predictive of progression to exudation in Age-Related Macular Degeneration", *Ophthalmol. Retina,* 2023. [PMID: 37839548]

[22] Fighting Blindness Canada (FBC), 2024.

[23] Macular Degeneration - Alaska Retinal Consultants, *Alaska Retinal Consultants.,* 2022.

CHAPTER 2

A Comprehensive Survey of Transfer Learning Approaches for Medical Image Deep Analysis

Ch. V. Bhargavi[1] and **K. Subhadra**[1,*]

[1] *Department of Computer Science and Engineering, GITAM School of Technology, GITAM Deemed to be University, Visakhapatnam, India*

Abstract: Neural Networks (CNNs), In the rapidly evolving healthcare landscape, medical image analysis has become an indispensable tool for aiding clinicians in diagnosis and decision-making. Deep learning techniques, especially convolutional neural networks, have shown remarkable potential in this domain. However, the success of deep learning models often relies on vast amounts of labeled data, which are not always readily available in medical imaging due to privacy concerns and the labor-intensive nature of annotation. This limitation has spurred the need for more efficient and scalable solutions to harness the power of deep learning for medical image analysis. Transfer learning, a subfield of machine learning, has emerged as a promising approach to overcome these data scarcity challenges. In the context of medical image analysis, transfer learning involves leveraging knowledge learned from one domain (*e.g.*, natural images) and applying it to a different but related domain (*e.g.*, medical images). This comprehensive survey paper delves into the significance of transfer learning in the context of medical image analysis, aiming to provide insights into its key findings and contributions. The core objective of this survey is to address the limitations of traditional "in-house" deep learning models that are trained and tested on the same dataset. Within this survey, we dissect these challenges and present a thorough examination of existing transfer learning approaches for medical image analysis. We explore the strategies that have been devised to fine-tune, adapt, and transfer knowledge across domains. Furthermore, we showcase comparative analyses of these transfer learning approaches, highlighting their strengths, weaknesses, and areas of applicability. The insights provided herein are poised to shape the future of computer-aided disease diagnosis systems and advance the field of transfer learning-based medical image analysis.

Keywords: Computer-aided diagnosis, Convolutional neural networks, Deep learning architectures, Medical image analysis, Transfer learning strategies.

* **Corresponding author K. Subhadra:** Department of Computer Science and Engineering, GITAM School of Technology, GITAM Deemed to be University, Visakhapatnam, India; E-mail: skompella@gitam.edu

Suneeta Satpathy, Sachi Nandan Mohanty & Subhendu Kumar Pani (Eds.)

INTRODUCTION

In the rapidly advancing realm of healthcare, the field of medical image analysis has taken center stage in modern clinical practice. Cutting-edge medical imaging technologies, including X-rays, CT scans, MRIs, and ultrasound, have ushered in a revolution in the diagnosis, treatment, and ongoing assessment of diverse medical conditions [1]. These imaging modalities extract visual data that offers invaluable insights into the human body parts. As technology continues to advance, the volume and complexity of medical image data have grown exponentially. This surge in medical imaging data underscores the growing importance of effective and efficient methods for extracting meaningful information from these images. Clinicians increasingly rely on automated tools and computer-aided diagnosis systems to assist them in interpreting medical images accurately and swiftly [2]. These tools not only enhance diagnostic accuracy but also improve the overall workflow in healthcare settings, ultimately leading to better patient care.

However, the effectiveness of these automated systems largely pivots on the prowess of deep learning models [3], particularly CNNs, in extracting relevant features and patterns from medical images. Although CNNs show great promise, their effectiveness frequently relies on large quantities of labeled data. In medical imaging, such data is often limited due to privacy issues, ethical concerns, and the intensive effort required for manual labeling [4]. This insufficiency of labeled medical image data poses a significant challenge [5], necessitating innovative solutions to harness the full power of deep learning in this critical domain. A widely adopted solution is transfer learning [6], which leverages knowledge gained from one domain—like natural images or other medical datasets—to enhance the performance of deep learning models for specific medical imaging tasks. It serves as a bridge between general knowledge and specialized medical image analysis, offering the potential to overcome data scarcity challenges [7].

Transfer learning serves as a powerful mechanism (Fig. **1**) for boosting the capabilities of deep learning models in medical image analysis [8], circumventing the need for massive volumes of domain-specific labeled data. It empowers these models to navigate the intricacies of medical images, discern anomalies, and assist healthcare professionals in making accurate diagnoses.

Research Problem and Objectives

In light of these advancements, this survey paper examines transfer learning in medical image analysis. It investigates how transfer learning can significantly boost the effectiveness of deep learning models, particularly when dealing with limited medical image data. The research problem at the heart of this survey paper is multifaceted. Traditional deep learning models [9, 10], designed and trained in an "in-house" manner, often encounter limitations when applied to medical image analysis. One of the critical limitations is their capacity to generalize effectively across dissimilar datasets [11]. When presented with medical images from different sources or with distinct characteristics, these models often struggle to maintain the same level of accuracy and robustness achieved within their initial domain. This lack of adaptability hinders their utility in scenarios where comprehensive and versatile medical image analysis is essential.

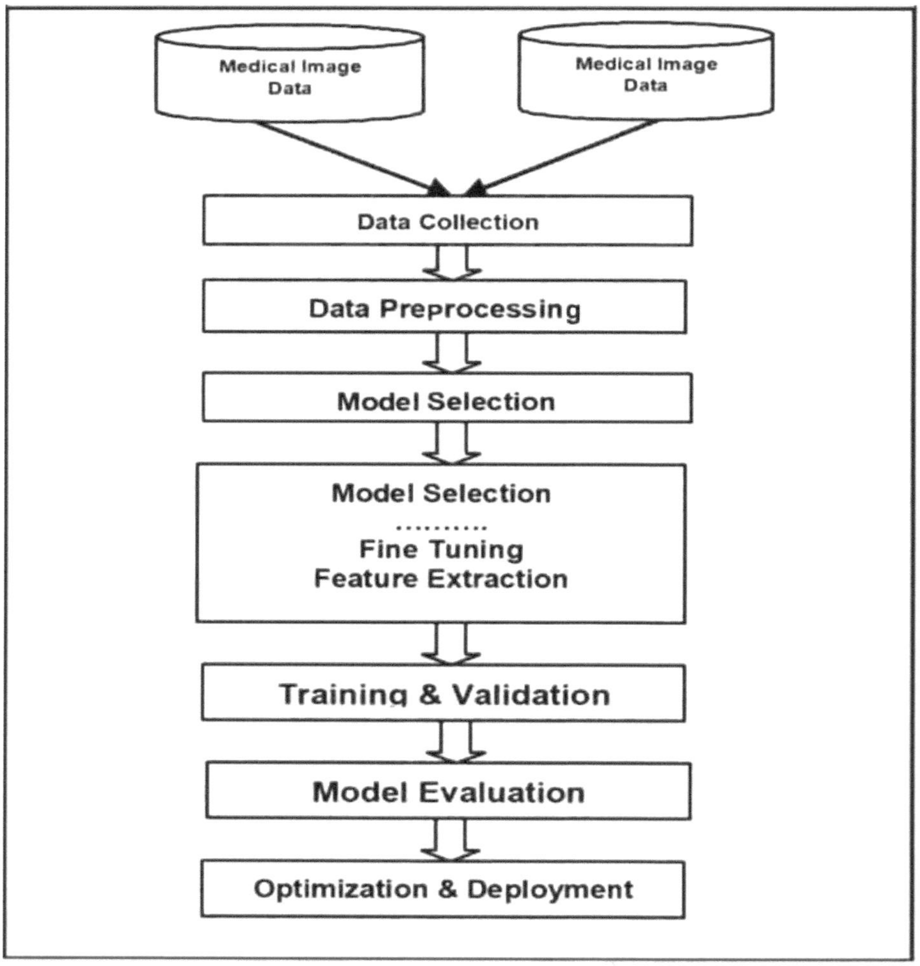

Fig. (1). Block diagram of transfer learning in deep learning.

Furthermore, the issue of normalizing pre-trained data presents a challenge. These in-house models are accustomed to specific data distributions and may falter when tasked with accommodating new data distributions. This poses a significant obstacle when attempting to apply these models to diverse medical image datasets, as the need for data normalization becomes paramount. Another critical concern related to transfer learning is bias. In-house models being inherently attuned to the training data they were derived from, may inadvertently propagate biases present in that data. These biases can manifest in various forms, including disparities in diagnosis and treatment recommendations, which is especially problematic in the context of healthcare and clinical applications.

Collectively, these limitations may affect the in-house models falling short of accuracy, reliability, and robustness. Consequently, there is a compelling need to explore innovative solutions, such as transfer learning, to overcome these challenges and enhance the capabilities of deep learning models. To address these limitations in medical image analysis, our survey paper focuses on several key objectives:

- **Examine Existing Transfer Learning Approaches:** We aim to conduct a comprehensive examination of the various transfer learning methodologies [12] that have been developed for medical image analysis. This includes a detailed exploration of techniques used to adapt, fine-tune, and transfer knowledge across domains.
- **Mitigate Bias and Misconceptions:** We intend to delve into strategies employed to mitigate bias and manage misconceptions regarding data similarity. These aspects are crucial in ensuring that transfer learning models effectively generalize across diverse medical image datasets.
- **Handle Limited Data Availability:** Given the scarcity of labeled data in the medical imaging domain, we will explore solutions for handling limited data effectively. This includes methods for making the most of the available data while maintaining model performance.
- **Address Dissimilarities in Medical Images:** We will investigate approaches to handle dissimilarities in medical image properties [13], including variations in contrast, type, and source. These dissimilarities often pose challenges when extending transfer learning support to different datasets.
- **Negative Transfer Management:** To ensure the robustness of transfer learning models, we will discuss methods for measuring and eliminating negative transfer [14], a critical consideration when applying knowledge from one domain to another.
- **Standardized Performance Evaluation:** We aim to establish the importance of standard comparison metrics [15] for measuring the performance of transfer learning models in medical image analysis. This objective seeks to enhance the reproducibility and comparability of research findings in the field.

By addressing the challenges associated with traditional "in-house" deep learning models and providing insights into various transfer learning methodologies, this survey seeks to pave the way for more efficient and scalable approaches in medical image analysis.

BACKGROUND

The historical context of medical image deep learning provides valuable insights into the evolution of this field and the challenges that have necessitated innovative approaches like transfer learning. To appreciate the significance of transfer learning, it is essential to understand how deep learning has progressed within the context of medical imaging. Over the years, medical image analysis has undergone a transformative journey, primarily driven by advancements in deep learning techniques.

Historically, deep learning models began making inroads into medical image analysis in the late 2010s. However, it wasn't until certain pivotal milestones that the profound potential of deep learning began to resonate within this field. One such watershed moment was the ImageNet Large Scale Visual Recognition Challenge [16], an influential competition in the realm of computer vision. During this competition, deep CNNs astounded the scientific community with their unprecedented performance in the classification of natural images. This achievement served as a catalyst, igniting a surge of interest and enthusiasm regarding the application of similar deep-learning architectures to the domain of medical images. However, the translation of deep learning's success in natural image classification to medical image analysis was not without hurdles. Traditional deep in-house models are proficient in processing the data on which they were specifically trained but encounter limitations when they are applied to the diverse and complex landscape of medical imaging. These limitations have been a driving force behind the quest for more adaptive and robust solutions.

TRANSFER LEARNING MODEL

In this pivotal section of our comprehensive survey, we delve into the foundational principles of transfer learning, illuminating its core concepts and mechanisms. Transfer learning serves as the linchpin of our exploration, providing the bridge between broader knowledge domains and the specialized realm of medical image analysis. This introductory knowledge sets the stage for a deeper exploration of the various transfer learning methodologies specifically tailored to the challenges posed by medical images.

Transfer learning is a machine learning approach where knowledge from one domain or task enhances model performance in another related domain. It relies on the idea that insights, patterns, and features from one context can benefit another. In deep learning for medical image analysis, transfer learning adapts pre-trained models from broad datasets to specific medical imaging tasks [17]. It involves two key domains: the source domain, where the model is trained on a diverse dataset, and the target domain, where the model's adapted knowledge is applied to medical images like X-rays and MRIs. A crucial aspect of transfer learning is feature transfer, where learned features from the source domain are adapted to meet the needs of the target domain, aiding in tasks such as disease diagnosis and organ segmentation [18].

Source Domain: The source domain represents the initial realm from which the pre-trained model gains its foundational knowledge. In the case of medical image analysis, this source domain could encompass general image data, which may consist of everyday images from sources like the Internet or other image repositories.

Target Domain: On the other side of the equation lies the target domain. This domain represents the specialized arena where the model's adapted knowledge will be put into practice. In the context of our survey, the target domain is the world of medical images, encompassing a wide array of modalities such as X-rays, MRIs, CT scans, and more. The model's ultimate purpose is to excel in tasks relevant to this domain, such as disease diagnosis or organ segmentation.

Feature Transfer: Feature transfer is a key aspect of transfer learning, where features learned from the source domain are applied to the target domain. These features capture important patterns and information from the source data and serve as a bridge between the two domains, facilitating the model's adaptation to the new context. They encapsulate knowledge distilled from the source domain data and can be reconfigured and fine-tuned to cater to the specific needs of the target domain. Essentially, feature transfer empowers the model to carry forward valuable insights obtained during its initial training, which can play a vital role in addressing tasks related to medical image analysis.

Fine-tuning: It is a fundamental technique in transfer learning [19], dealing with the intricate process of adapting a pre-trained model to the specific characteristics of the target domain. It represents a form of model refinement that focuses on honing the model's parameters and weights to better align with the idiosyncrasies of the target domain.

We designed a process flow diagram (Fig. **2**) to understand thoroughly applying the transfer learning strategies in medical image analysis. This diagram illustrates the sequential steps involved in the utilization of transfer learning techniques within this context. The process flow begins with recognizing the imperative need for transfer learning to bolster the capabilities of medical image analysis. This multi-stage process unfolds with meticulous steps: data collection, involving the assembly of a dataset specific to the target medical domain and the acquisition of a pre-trained model replete with general image features from sources like ImageNet. Data preprocessing comes next, where the target domain data undergoes normalization to ensure uniformity and augmentations to enhance dataset diversity. Subsequently, model selection entails choosing to fit pre-trained model architecture [20], such as VGG, ResNet, or Inception, to serve as the foundation for further adaptation. Transfer learning methods branch into two primary approaches: fine-tuning and feature extraction.

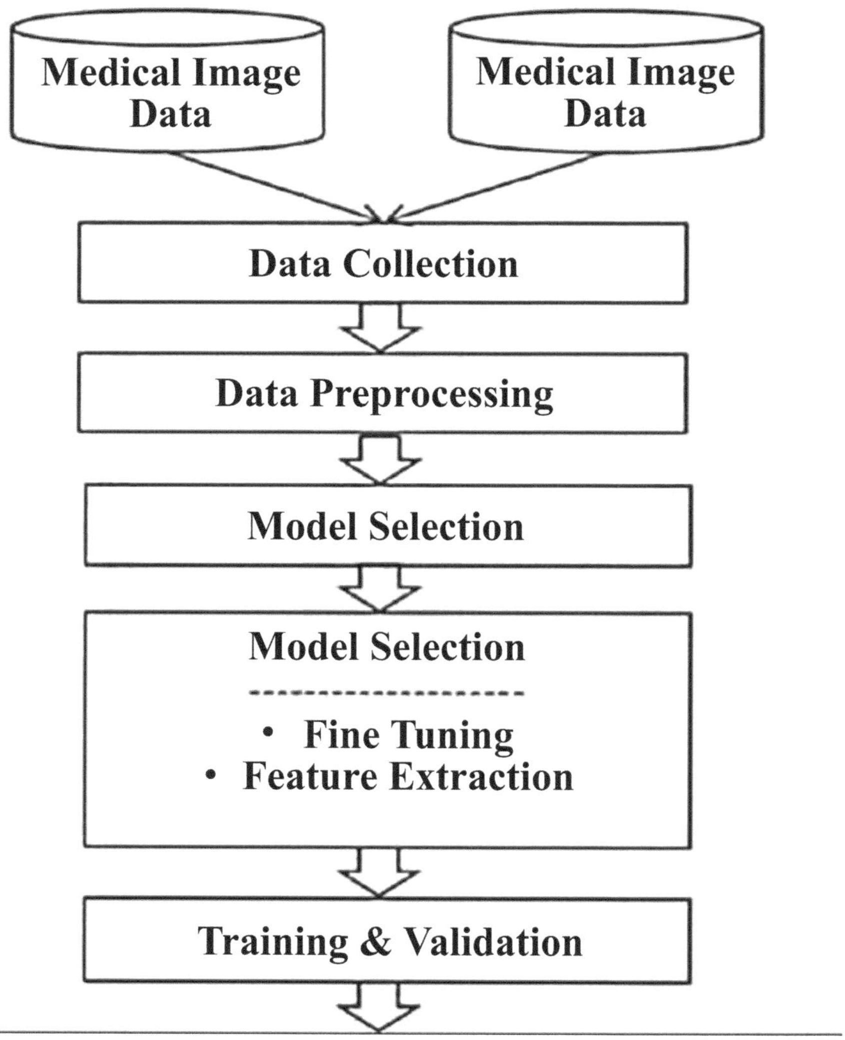

Fig. (2). Process flow diagram - applying transfer learning strategies in medical image analysis.

In fine-tuning, the pre-trained model's weights are initially set with knowledge from the source domain, followed by the replacement of final layers tailored for the target medical image analysis task, concluding with retraining on the target domain data. Feature extraction, on the other hand, leverages the pre-trained model as an immutable feature extractor, harvesting high-level features from target domain data and training new task-specific layers. Training and validation ensue, encompassing the division of the target domain dataset into training and validation sets, model training on the former, and validation of its performance on the latter, with potential fine-tuning of hyperparameters.

Model evaluation, a pivotal stage, assesses the model's performance through diverse metrics, and comparisons are drawn against baseline models and alternative transfer learning techniques. The iterative refinement phase allows for model and process enhancement based on evaluation outcomes, enabling further fine-tuning, feature extraction adjustments, hyperparameter tweaks, and data augmentation. Finally, optimization [21] and deployment steps ensue, optimizing the model for efficient real-time inference.

TRANSFER LEARNING METHODS IN MEDICAL IMAGE ANALYSIS

Medical image analysis has experienced a significant upswing in the utilization of transfer learning techniques [22]. These methods harness knowledge gained from source domains to augment the analysis of medical images, mitigating the limitations of traditional, in-house deep learning models. In our extensive survey, we meticulously investigate the cutting-edge transfer learning methodologies implemented in medical image analysis. Models like VGG, ResNet, Inception, and others, which were initially trained on extensive datasets such as ImageNet, serve as potent initial building blocks for tasks within medical image analysis. They bring a wealth of knowledge in recognizing foundational features, which prove invaluable across various medical imaging tasks. The fine-tuning process involves substituting the final layers of the model with new layers that are tailor-made for the specific medical image analysis task, whether it is classification or segmentation. Fine-tuning grants the model the capacity to acquire domain-specific features while retaining the foundational knowledge originating from the source domain. It particularly shines when the target domain shares commonalities with the source, as fine-tuning can optimize the model's performance. Feature extraction is a common technique used when there are significant differences between the source and target domains. In this approach, the pre-trained model serves as a static feature extractor, capturing high-level features from the target domain data. New layers are then trained specifically for the task at hand using these extracted features, like a classifier, using these extracted features. Feature extraction proves to be a valuable technique when significant dissimilarities are present between the source and target domains.

Ensemble methods, including techniques such as model stacking and knowledge distillation, are gaining traction within the transfer learning landscape. These methods amalgamate multiple models to enhance overall performance. For example, model stacking combines the predictions of several models, elevating robustness and accuracy. Knowledge distillation facilitates the transfer of knowledge, resulting in streamlined yet proficient models for medical image analysis. Domain adaptation techniques address the challenges posed by differences between source and target domains by aligning their feature distributions. This helps improve the model's ability to generalize effectively to the target domain. A variety of domain adaptation methods, encompassing adversarial training and domain-invariant feature learning, have been harnessed in medical image analysis to effectively tackle domain shift challenges. Each transfer learning approach exhibits its strengths and comes with its set of challenges. Table **1** presents the strengths and weaknesses of the above transfer learning methods in medical image processing.

Table 1. Strengths and weaknesses of the transfer learning approach.

Transfer Learning Method	Strengths	Weaknesses
Fine-tuning [23]	▪ Is effective when both domains exhibit substantial similarities ▪ Adapts quickly to domain-specific features and well-suited to limited target domain data	▪ Less optimal when significant differences exist between both domains ▪ Potential risk of source domain over-fitting ▪ Optimal hyperparameter selection is complex
Feature extraction [24]	▪ Suitable when both domains have substantial dissimilarities ▪ Performs the extraction of relevant features without the risk of overfitting ▪ Leverages the pre-trained model's ability to recognize high-level features ▪ Computationally efficient	▪ May not fully utilize the knowledge from the source domain ▪ Extracted features might not be optimally suited for the target task ▪ May need the training of additional task-specific layers
Domain adaptation [25]	▪ Bridges the gap between source and target domains ▪ Adoptable under domain shift challenges ▪ Mitigates the dataset bias and domain-specific variations	▪ Complex to implement ▪ May require domain-specific adjustments ▪ Need extensive fine-tuning of hyperparameters
Ensemble methods [26]	▪ Improves overall performance using model stacking ▪ Enhances model robustness and interpretability	▪ Increased computational complexity and resource requirements ▪ Need careful configuration and optimization

Understanding these strengths and weaknesses is crucial for researchers and practitioners seeking to select the most suitable technique for their specific medical image analysis tasks.

PRE-TRAINED DEEP LEARNING ARCHITECTURES

In medical image processing, choosing suitable pre-trained deep learning architectures is crucial for effective transfer learning. These architectures provide the foundational framework for extracting features and transferring knowledge from the source domain to the target domain. Within the scope of transfer learning methods for in-depth medical image analysis, we delve into various pre-trained deep learning architectures, including but not limited to VGG, ResNet, Inception, and others, to ascertain their suitability for medical image processing.

VGG (Visual Geometry Group)

The VGG architecture, created by the Visual Geometry Group at the University of Oxford [27], is well-known for its uniform structure of sequential convolutional layers, each using 3x3 filters. VGG architectures are relatively shallow but perform exceptionally well in capturing basic image features. This simplicity and uniformity make VGG models computationally efficient, which is advantageous when computational resources are limited. In medical image processing, especially for tasks like image classification and feature extraction, VGG architectures prove to be reliable and effective.

ResNet (Residual Network)

ResNet [28] introduced a breakthrough concept known as residual learning, which utilizes skip connections to overcome the vanishing gradient problem. This architectural innovation allows for the training of much deeper networks without performance degradation. For medical image processing, particularly when dealing with intricate feature hierarchies, Res-Net architectures are highly suitable. Their capability to capture complex patterns, subtle image details, and a wide range of structures makes them a better option for image segmentation, object detection, and disease diagnosis.

Inception (GoogLeNet)

Inception models, notably GoogLeNet [29], employ “inception modules,” featuring parallel convolutional operations with filters of different sizes. This unique design enables the capture of features at various scales and orientations, making Inception architectures exceptionally beneficial for medical image analysis. Medical images often contain structures and anomalies at varying scales, and Inception models excel in detecting such variations. Multi-scale feature extraction is particularly relevant for tasks involving medical imaging, where structures of interest can manifest differently based on scale.

DenseNet (Densely Connected Convolutional Networks)

DenseNet [30] is distinguished by its densely connected layers, with each layer receiving direct inputs from all previous layers. This design promotes extensive feature reuse and helps mitigate the vanishing gradient problem. DenseNet models have proven to be exceptionally well-suited for medical image processing applications that demand the extraction of intricate details and the representation of complex structures. In scenarios involving images with densely packed structures, such as pathology slides or histological images, DenseNet architectures are instrumental.

MobileNet

Designed with a primary emphasis on computational efficiency for deployment in mobile and embedded applications, MobileNet [31] leverages depth-wise separable convolutions to significantly reduce computational complexity. In the context of medical image processing, especially when real-time or edge-based applications are involved, MobileNet architectures are a suitable choice. Their efficiency and resource-friendliness are well-aligned with scenarios that have limitations in computational resources.

EfficientNet

A study [32] introduces a concept known as compound scaling, which uniformly scales depth, width, and resolution. This balanced approach ensures that the network's capacity aligns with computational efficiency. In the realm of medical image processing, where optimizing both accuracy and resource utilization is essential, EfficientNet models offer a valuable middle ground. They strike a balance between complex models and computational efficiency, making them an ideal choice for various medical imaging tasks.

Xception (Extreme Inception)

Xception [33] builds on the Inception concept by using depth-wise separable convolutions instead of standard ones. This design makes the models more efficient in terms of parameters and computation. It is particularly suited for medical image analysis, where complex structures and patterns are common. They excel in capturing fine details efficiently, which is crucial for various medical imaging tasks.

The choice of pre-trained deep learning architecture depends on the specific requirements of the medical image processing task at hand. Practitioners and researchers often experiment with different architectures to find the one that best suits their particular use case. Through transfer learning, these pre-trained models offer a wealth of knowledge and feature extraction capabilities, enabling the medical imaging field to harness their potential for improved efficiency and accuracy.

COMPARATIVE ANALYSIS

This section serves as a critical component of the survey, aiming to evaluate the performance of various pre-trained deep learning architectures in the context of transfer learning for medical image analysis. This segment involves the systematic comparison of these architectures on public medical image datasets to assess their strengths, weaknesses, and suitability for different medical image processing tasks.

Data Selection

For conducting the comparative analysis, it is crucial to meticulously select relevant public medical image datasets that align with the objectives of the survey. These datasets should accurately represent the diversity of the medical imaging domain, spanning various modalities such as X-rays, MRIs, and CT scans, while also covering a wide spectrum of medical conditions. This selection process should also encompass datasets that are considered benchmark sources in the field. As such, we have included the following publicly available datasets to ensure not only transparency and reproducibility but also to provide the broader research community with opportunities for result validation:

NIH Chest X-ray Dataset: This dataset provides an extensive collection of chest X-ray images for analyzing various chest conditions, making it crucial for chest X-ray diagnostics and research.

BraTS (Brain Tumor Segmentation): Essential for brain tumor research, this dataset includes MRI scans of brain tumors, supporting the creation of advanced diagnostic models for tumor segmentation and classification.

ISIC (International Skin Imaging Collaboration): The inclusion of this dataset enables the study of skin lesion analysis, making it an asset for tasks such as skin cancer classification and melanoma detection.

These datasets [34 - 36] along with others not only contribute to the diversity and comprehensiveness of the comparative analysis but also enhance the scope for validation and benchmarking. Before embarking on a comparative analysis, it is imperative to carefully select appropriate public medical image datasets that align with the objectives of the survey. These datasets should be representative of the medical imaging domain, covering a wide range of modalities (*e.g.*, X-rays, MRIs, CT scans) and encompassing diverse medical conditions. The inclusion of publicly available datasets ensures transparency, reproducibility, and the ability for other researchers to validate the findings.

Performance Metrics

To facilitate an objective comparison, well-defined performance metrics are crucial. Common evaluation metrics in medical image analysis include accuracy, sensitivity, specificity, F1-score, area under the receiver operating characteristic curve (AUC-ROC), and dice coefficient for tasks like segmentation. The choice of metrics depends on the specific task being assessed. A thorough understanding of these metrics is essential to accurately evaluate the performance of the pre-trained architectures.

Pre-Trained Architectures

VGG, ResNet, Inception, DenseNet, MobileNet, EfficientNet, Xception, and other pre-trained deep learning architectures are used as the foundation for the experiments. Each architecture represents a different approach to feature extraction and knowledge transfer. By applying transfer learning, these models can extract valuable information from a basis domain and adjust it to the target medical image dataset. In the comparison of diverse pre-trained deep learning models for medical image classification [37 - 41], we utilized the following datasets: the NIH Chest X-ray Dataset, BraTS (Brain Tumor Segmentation), and ISIC (Inter-national Skin Imaging Collaboration). Here are the obtained results with relevant metrics. The results shown in Table **2** and Fig. (**3**) are obtained from the experiments on the NIH Chest X-ray Dataset using various pre-trained deep learning architectures to demonstrate the significant potential of transfer learning in medical image classification.

Table 2. TNIH Chest X-Ray dataset image classification results with transfer models.

Architecture	*Accuracy*	*F1-Score*	*Sensitivity*	*Specificity*	*AUC*
VGG-16	88%	0.86	0.88	0.84	0.92
ResNet-50	90%	0.88	0.9	0.86	0.93
Inception-V3	87%	0.85	0.87	0.83	0.91
DenseNet-121	91%	0.89	0.91	0.88	0.94
EfficientNet	92%	0.91	0.92	0.89	0.95

Among the models tested, EfficientNet yielded the highest accuracy of 92%, followed closely by DenseNet-121 with 91%. This indicates that both architectures are well-suited for accurately classifying medical images from the NIH Chest X-ray dataset. The remaining models, including VGG-16, ResNet-50, and Inception-V3, also achieved respectable accuracy scores ranging from 87% to 90%. These results highlight the effectiveness of transfer learning in harnessing knowledge from other domains to improve the accuracy of medical image classification.

While accuracy is a critical metric, other metrics like F1-Score, Sensitivity, Specificity, and AUC provide a more comprehensive view of model performance. All models showed good balance in these metrics. DenseNet-121 and EfficientNet stood out with high F1-Scores, indicating their ability to maintain a balance between precision and recall. They also displayed higher AUC values, signifying strong discrimination power in distinguishing between different medical conditions.

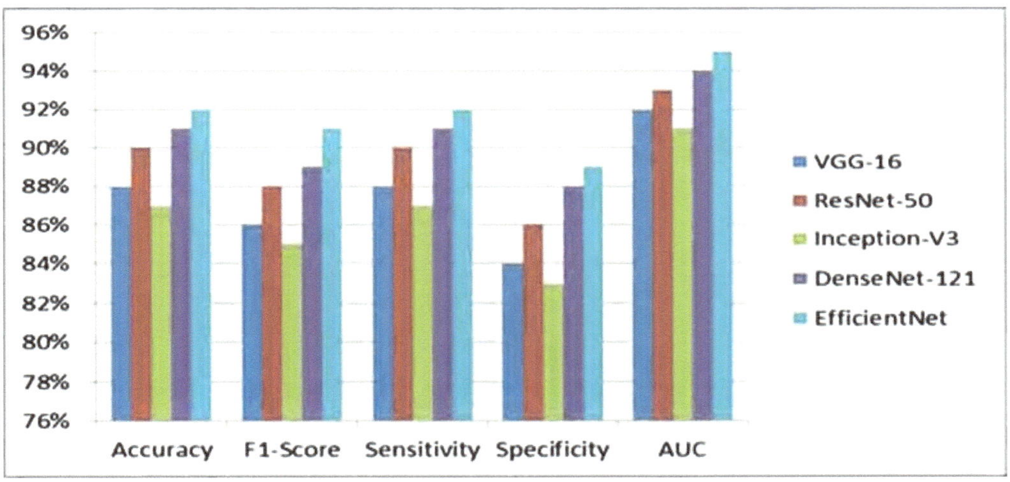

Fig. (3). Comparison of NIH Chest X-ray dataset image classification results using transfer models.

The results shown in Table **3** and Fig. (**4**) are obtained from the experiments on the BraTS (Brain Tumor Segmentation) dataset using various pre-trained deep learning architectures, revealing valuable insights into the performance of these models in the context of medical image classification. Here, we discuss the implications of these findings.

The Dice Coefficient and IoU are essential metrics for assessing the overlap between predicted and ground truth segmentation masks. EfficientNet achieved the highest Dice Coefficient of 0.82 and IoU of 0.74, indicating its ability to provide accurate and precise segmentations. DenseNet-121 also performed exceptionally well, with a Dice Coefficient of 0.81 and IoU of 0.72. These results highlight the suitability of these architectures for tasks requiring precise segmentation, such as brain tumor detection.

EfficientNet achieves a high Sensitivity of 0.85, highlighting its proficiency in correctly identifying brain tumor regions. Furthermore, when measuring Specificity, the ability to accurately identify regions without tumors, EfficientNet and DenseNet-121 again excel with Specificity values of 0.95 and 0.94, respectively. Additionally, Precision, a critical metric for minimizing false positive predictions, is found to be substantial. Efficient-Net obtains the highest Precision score of 0.82, closely followed by Dense-Net-121 at 0.80. These results underscore the effective balance these architectures strike between true positives and false positives, a crucial factor for the precision and reliability of brain tumor detection in medical image analysis.

Table 3. BraTS tumor dataset image classification results with transfer models.

Architecture	Dice Coefficient	Sensitivity	Specificity	Precision	IoU
VGG-16	0.78	0.81	0.92	0.75	0.67
ResNet-50	0.8	0.83	0.93	0.78	0.7
Inception-V3	0.79	0.82	0.91	0.76	0.68
DenseNet-121	0.81	0.84	0.94	0.8	0.72
EfficientNet	0.82	0.85	0.95	0.82	0.74

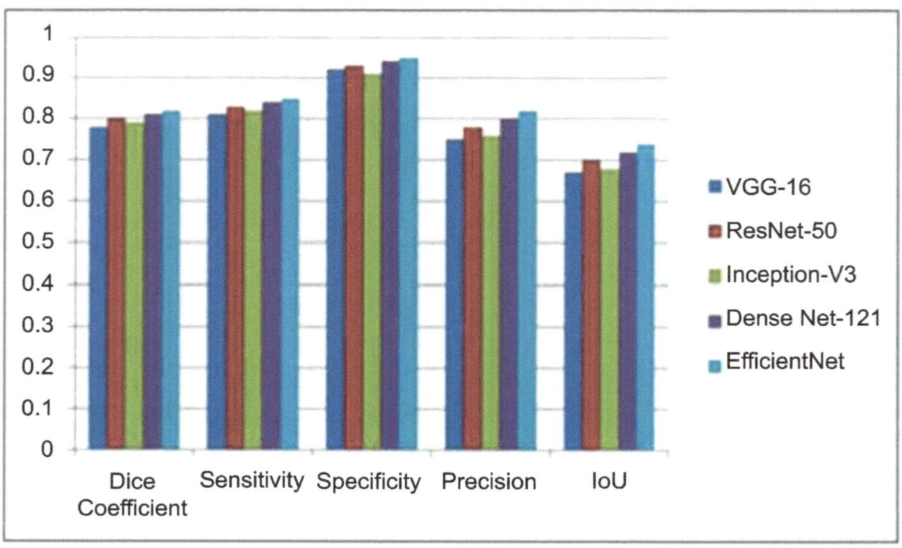

Fig. (4). Comparison of BraTS tumor dataset image classification results using transfer models.

The results from the evaluation of transfer learning models on the ISIC (International Skin Imaging Collaboration) dataset demonstrate their outstanding performance in the context of medical image classification, shown in Table **4** and Fig. (**5**).

EfficientNet stands out with the highest AUC of 0.96, followed closely by DenseNet-121 with an AUC of 0.95. This high AUC value signifies the excellent discriminatory power of these models in distinguishing between different skin conditions, making them highly suitable for dermatological image analysis tasks. Sensitivity, representing the models' ability to correctly identify cases of skin conditions, is consistently strong across all architectures, with EfficientNet achieving a top score of 0.92. Specificity, a measure of the models' capability to correctly identify healthy skin regions, is also remarkable, with EfficientNet reaching a score of 0.98. This is a vital indicator of the models' reliability in minimizing false positives, a critical factor in dermatological diagnoses.

Precision values are notably high across the board, indicating the models' proficiency in reducing false positive predictions. EfficientNet secures the highest Precision score of 0.95, emphasizing its accuracy in identifying skin conditions. The Matthews Correlation Coefficient (MCC) measures the quality of binary classifications and is used to gauge the overall performance of the models. All architectures consistently achieve strong MCC values, with EfficientNet outperforming the rest with an MCC of 0.90. These results collectively highlight the exceptional performance of transfer learning models on the ISIC dataset, underlining their potential to impact dermatological image analysis in healthcare significantly.

Table 4. ISIC dataset image classification results with transfer models

Architecture	AUC	Sensitivity	Specificity	Precision	MCC
VGG-16	0.92	0.88	0.95	0.91	0.82
ResNet-50	0.94	0.9	0.96	0.93	0.86
Inception-V3	0.93	0.89	0.94	0.92	0.84
DenseNet-12	0.95	0.91	0.97	0.94	0.88
EfficientNet	0.96	0.92	0.98	0.95	0.9

In the comparative analysis of transfer learning models on three diverse medical image datasets (NIH Chest X-ray, BraTS, and ISIC), key findings highlight the efficacy of these models in medical image classification. Notably, models like EfficientNet and DenseNet-121 excelled in terms of accuracy and AUC on the NIH Chest X-ray Dataset. They also demonstrated potential in brain tumor segmentation tasks with the BraTS dataset and proved effective in skin lesion classification on the ISIC dataset. These results underscore the versatility and robustness of transfer learning models, particularly EfficientNet and DenseNet-121, in enhancing medical image analysis, offering promising prospects for computer-aided healthcare applications across various medical domains.

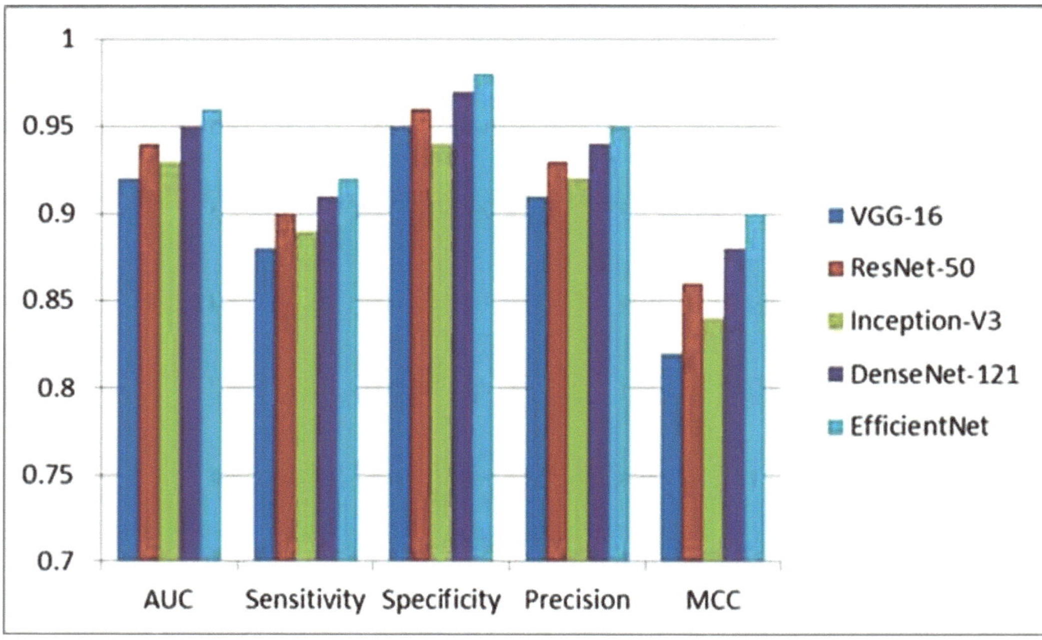

Fig. (5). Comparison of ISIC skin lesion image dataset classification results using transfer models.

FUTURE RESEARCH DIRECTIONS

Future research in medical image analysis and transfer learning holds substantial promise, particularly in light of evolving technology and data collection. Some potential areas for exploration include:

- The customized architecture field could benefit from the creation of medical-specific pre-trained architectures. These architectures would be designed in the future to leverage knowledge from general image data while accounting for medical image peculiarities. This approach aligns with the survey's aim to investigate strategies for knowledge adaptation across domains.
- Enhancing the robustness of models and developing mechanisms to estimate uncertainties in predictions is essential. As technology advances, real-time applications in diagnostics and treatment planning are becoming increasingly viable. Future research may focus on making transfer learning models more responsive and efficient in clinical settings.

CONCLUSION

This comprehensive survey paper delves into transfer learning approaches for medical image deep analysis, highlighting their potential to enhance accuracy, efficiency, and robustness in this intricate domain. We explore the historical context of deep learning in medical imaging and the challenges that have spurred innovative transfer learning techniques. Our discussion encompasses fundamental concepts, methodologies, and techniques, emphasizing the applications in medical image analysis. The paper delves into challenges faced in medical image transfer learning, including biased adjustment methods, misconceptions in similarity, handling limited data, managing dissimilarities in image properties, negative transfer, and the lack of standardized metrics. These challenges are substantiated with contemporary research insights and real-world examples, illustrating the intricacies of medical image analysis.

We meticulously review state-of-the-art pre-trained deep learning architectures such as VGG, ResNet, Inception, DenseNet, and EfficientNet, showcasing their suitability and performance in medical image processing. These architectures display their adaptability and excellence in various medical image analysis tasks, revolutionizing the field. Our comparative analysis reveals the strengths of these pre-trained architectures in medical image classification, segmentation, and diagnosis across diverse datasets. Efficient Net and DenseNet-121 exhibit significant potential for enhancing medical image analysis, as supported by a range of performance metrics.

Looking to the future, our survey suggests potential research areas, including interdisciplinary collaboration, scalable solutions, customized architectures, ethical considerations, real-time applications, multimodal integration, and robustness and uncertainty estimation. These directions align with the paper's objectives and limitations, emphasizing the role of technological advancements and data collection in shaping the field's trajectory. In essence, this comprehensive survey underscores the crucial role of transfer learning in bridging the gap between limited medical data and robust, efficient solutions for clinical applications. The collaborative spirit between computer science, medicine, and imaging fields promises innovations that will revolutionize the diagnosis, treatment, and understanding of medical conditions. As technology evolves and datasets expand, the potential for transformative advancements in medical image analysis remains boundless, signifying the ongoing journey toward unlocking the full potential of medical imaging for improved healthcare outcomes.

REFERENCES

[1] H.A. Ahmad, H.J. Yu, and C.G. Miller, "“Medical imaging modalities,”", In: *in Medical Imaging in Clinical Trials.,* C.G. Miller, J. Krasnow, L.H. Schwartz, Eds., Springer London: London, 2014, pp. 3-

26.
[http://dx.doi.org/10.1007/978-1-84882-710-3_1]

[2] H-P. Chan, R.K. Samala, L.M. Hadjiiski, and C. Zhou, "“Deep learning in medical image analysis”", In: *Advances in Experimental Medicine and Biology.,* G. Lee, H. Fujita, Eds., vol. Vol. 1213. Springer International Publishing: Cham, 2020, pp. 3-21.
[http://dx.doi.org/10.1007/978-3-030-33128-3_1]

[3] M. Tsuneki, "Deep learning models in medical image analysis", *J. Oral Biosci./ JAOB, Jpn. Assoc. Oral Biol.,* vol. 64, no. 3, pp. 312-320, 2022.
[http://dx.doi.org/10.1016/j.job.2022.03.003] [PMID: 35306172]

[4] G. Litjens, T. Kooi, B.E. Bejnordi, A.A.A. Setio, F. Ciompi, M. Ghafoorian, J.A.W.M. van der Laak, B. van Ginneken, and C.I. Sánchez, "A survey on deep learning in medical image analysis", *Med. Image Anal.,* vol. 42, pp. 60-88, 2017.
[http://dx.doi.org/10.1016/j.media.2017.07.005] [PMID: 28778026]

[5] J. Ker, L. Wang, J. Rao, and T. Lim, "Deep learning applications in medical image analysis", *IEEE Access,* vol. 6, pp. 9375-9389, 2018.
[http://dx.doi.org/10.1109/ACCESS.2017.2788044]

[6] S. Atasever, N. Azginoglu, D.S. Terzi, and R. Terzi, "A comprehensive survey of deep learning research on medical image analysis with focus on transfer learning", *Clin. Imaging,* vol. 94, pp. 18-41, 2023.
[http://dx.doi.org/10.1016/j.clinimag.2022.11.003] [PMID: 36462229]

[7] P. Kora, C.P. Ooi, O. Faust, U. Raghavendra, A. Gudigar, W.Y. Chan, K. Meenakshi, K. Swaraja, P. Plawiak, and U. Rajendra Acharya, "Transfer learning techniques for medical image analysis: A review", *Biocybern. Biomed. Eng.,* vol. 42, no. 1, pp. 79-107, 2022.
[http://dx.doi.org/10.1016/j.bbe.2021.11.004]

[8] S. Niu, Y. Liu, J. Wang, and H. Song, "A decade survey of transfer learning (2010–2020)", *IEEE Trans. Artif. Intell.,* vol. 1, no. 2, pp. 151-166, 2020.
[http://dx.doi.org/10.1109/TAI.2021.3054609]

[9] N. Dey, A.S. Ashour, S. Borra, Ed., *Lecture Notes in Computational Vision and Biomechanics.* vol. 26. Springer International Publishing: Cham, 2018, pp. 323-350.
[http://dx.doi.org/10.1007/978-3-319-65981-7_12]

[10] A.E. Hassanien, A. Slowik, V. Snášel, H. El-Deeb, F.M. Tolba, Ed., *Advances in Intelligent Systems and Computing.* vol. Vol. 1261. Springer International Publishing: Cham, 2021, pp. 370-382.
[http://dx.doi.org/10.1007/978-3-030-58669-0_34]

[11] L. Maier-Hein, A. Reinke, M. Kozubek, A.L. Martel, T. Arbel, M. Eisenmann, A. Hanbury, P. Jannin, H. Müller, S. Onogur, J. Saez-Rodriguez, B. van Ginneken, A. Kopp-Schneider, and B.A. Landman, "BIAS: Transparent reporting of biomedical image analysis challenges", *Med. Image Anal.,* vol. 66, p. 101796, 2020.
[http://dx.doi.org/10.1016/j.media.2020.101796] [PMID: 32911207]

[12] X. Yu, J. Wang, Q.Q. Hong, R. Teku, S.H. Wang, and Y.D. Zhang, "Transfer learning for medical images analyses: A survey", *Neurocomputing,* vol. 489, pp. 230-254, 2022.
[http://dx.doi.org/10.1016/j.neucom.2021.08.159]

[13] C. Kurtz, A. Depeursinge, S. Napel, C.F. Beaulieu, and D.L. Rubin, "On combining image-based and ontological semantic dissimilarities for medical image retrieval applications", *Med. Image Anal.,* vol. 18, no. 7, pp. 1082-1100, 2014.
[http://dx.doi.org/10.1016/j.media.2014.06.009] [PMID: 25036769]

[14] P.W. Kariuki, P.K. Gikunda, and J.M. Wandeto, "Deep transfer learning optimization techniques for medical image classification: A review", *2022 International Conference on Intelligent Computing and Machine Learning (2ICML), Qingdao, China,,* vol. , 2023no. , pp. 7-15
[http://dx.doi.org/10.1109/2ICML58251.2022.00013]

[15] M. Perrot, and A. Habrard, "A theoretical analysis of metric hypothesis transfer learning", *Proceedings of the 32 nd International Conference on Machine Learning, Lille, France,,* vol. 37, 2015

[16] O. Russakovsky, "ImageNet large scale visual recognition challenge", *Int J Comput Vis,* vol. 155, pp. 211-252.
[http://dx.doi.org/10.48550/ARXIV.1409.0575]

[17] L.T. Ravulapalli, R.K. Paladugu, V.K.R. Likki, R. Mothukuri, N. Mukkapati, and S. Kilaru, "Evaluative study of machine learning classifiers in predicting heart failure: A focus on imbalanced datasets", *Ingénierie des systèmes d information,* vol. 28, no. 3, pp. 717-724, 2023.
[http://dx.doi.org/10.18280/isi.280322]

[18] X. Yin, X. Yu, K. Sohn, X. Liu, and M. Chandraker, "Feature transfer learning for face recognition with under-represented data",
[http://dx.doi.org/10.1109/CVPR.2019.00585]

[19] Y. Guo, H. Shi, A. Kumar, K. Grauman, T. Rosing, and R. Feris, "SpotTune: Transfer learning through adaptive fine-tuning", *IEEE/CVF Conference on Computer Vision and Pattern Recognition (CVPR), Long Beach, CA, USA,,* pp. 5697-5706.
[http://dx.doi.org/10.48550/ARXIV.1811.08737]

[20] M.A. Wani, M. Kantardzic, M. Sayed-Mouchaweh, Ed., *Advances in Intelligent Systems and Computing.* vol. 1098. Springer Singapore: Singapore, 2020, pp. 39-52.
[http://dx.doi.org/10.1007/978-981-15-1816-4_3]

[21] M. Subramanian, N.P. L v, J. B, M.B. A, and S. Ve, "N. P. L. V., J. B., M. B. A., and S. Ve, "Hyperparameter optimization for transfer learning of VGG16 for disease identification in corn leaves using Bayesian optimization,"", *Big Data,* vol. 10, no. 3, pp. 215-229, 2022.
[http://dx.doi.org/10.1089/big.2021.0218] [PMID: 34851735]

[22] J. Back, *Domain similarity metrics for predicting transfer learning performance.,* 2019.

[23] G. Vrbancic, and V. Podgorelec, "Transfer learning with adaptive fine-tuning", *IEEE Access,* vol. 8, pp. 196197-196211, 2020.
[http://dx.doi.org/10.1109/ACCESS.2020.3034343]

[24] A. R V, A. R, and S. A P, "Augmenting transfer learning with feature extraction techniques for limited breast imaging datasets", *J. Digit. Imaging,* vol. 34, no. 3, pp. 618-629, 2021.
[http://dx.doi.org/10.1007/s10278-021-00456-z] [PMID: 33973065]

[25] W. Xu, J. He, and Y. Shu, "Transfer learning and deep domain adaptation,", In: *in Advances and Applications in Deep Learning.,* M.A. Aceves-Fernandez, Ed., IntechOpen, 2020.
[http://dx.doi.org/10.5772/intechopen.94072]

[26] J. Larocque-Villiers, P. Dumond, and D. Knox, "Automating predictive maintenance using state-based transfer learning and ensemble methods", *IEEE International Symposium on Robotic and Sensors Environments (ROSE), FL, USA,,* 2021pp. 1-7
[http://dx.doi.org/10.1109/ROSE52750.2021.9611768]

[27] N. Veni, "Modified visual geometric group architecture for MRI brain image classification", *Comput. Syst. Sci. Eng.,* vol. 42, no. 2, pp. 825-835, 2022.
[http://dx.doi.org/10.32604/csse.2022.022318]

[28] M. Chhabra, and R. Kumar, "An efficient ResNet-50 based intelligent deep learning model to predict pneumonia from medical images", *Proc. Int. Conf. Sustain. Comput. Data Commun. Syst. (ICSCDS),* 2022, pp. 1714-1721
[http://dx.doi.org/10.1109/ICSCDS53736.2022.9760995]

[29] C. Wang, D. Chen, L. Hao, X. Liu, Y. Zeng, J. Chen, and G. Zhang, "Pulmonary image classification based on Inception-v3 transfer learning model", *IEEE Access,* vol. 7, pp. 146533-146541, 2019.
[http://dx.doi.org/10.1109/ACCESS.2019.2946000]

[30] N. Hasan, Y. Bao, A. Shawon, and Y. Huang, "DenseNet convolutional neural networks application for predicting COVID-19 using CT image", *SN Comput. Sci.,* vol. 2, no. 5, p. 389, 2021. [http://dx.doi.org/10.1007/s42979-021-00782-7] [PMID: 34337432]

[31] T.H. Arfan, M. Hayaty, and A. Hadinegoro, "Classification of brain tumours types based on MRI images using mobilenet", *Proc. 2021 Int. Conf. Innov. Creative Inf. Technol. (ICITech),* 2021pp. 69-73 [http://dx.doi.org/10.1109/ICITech50181.2021.9590183]

[32] D.R. Nayak, N. Padhy, P.K. Mallick, M. Zymbler, and S. Kumar, "Brain tumor classification using dense efficient-net", *Axioms,* vol. 11, no. 1, p. 34, 2022. [http://dx.doi.org/10.3390/axioms11010034]

[33] S. Zhao, Y. Wu, M. Tong, Y. Yao, W. Qian, and S. Qi, "CoT-XNet: contextual transformer with Xception network for diabetic retinopathy grading", *Phys. Med. Biol.,* vol. 67, no. 24, p. 245003, 2022. [http://dx.doi.org/10.1088/1361-6560/ac9fa0] [PMID: 36322995]

[34] X. Wang, "ChestX-ray8: Hospital-scale chest X-ray database and benchmarks on weakly-supervised classification and localization of common thorax diseases", *2017 IEEE Conference on Computer Vision and Pattern Recognition (CVPR), Honolulu, HI, USA,,* 2017pp. 3462-3471 [http://dx.doi.org/10.1109/CVPR.2017.369]

[35] B.H. Menze, A. Jakab, S. Bauer, J. Kalpathy-Cramer, K. Farahani, J. Kirby, Y. Burren, N. Porz, J. Slotboom, R. Wiest, L. Lanczi, E. Gerstner, M.A. Weber, T. Arbel, B.B. Avants, N. Ayache, P. Buendia, D.L. Collins, N. Cordier, J.J. Corso, A. Criminisi, T. Das, H. Delingette, C. Demiralp, C.R. Durst, M. Dojat, S. Doyle, J. Festa, F. Forbes, E. Geremia, B. Glocker, P. Golland, X. Guo, A. Hamamci, K.M. Iftekharuddin, R. Jena, N.M. John, E. Konukoglu, D. Lashkari, J.A. Mariz, R. Meier, S. Pereira, D. Precup, S.J. Price, T.R. Raviv, S.M.S. Reza, M. Ryan, D. Sarikaya, L. Schwartz, H.C. Shin, J. Shotton, C.A. Silva, N. Sousa, N.K. Subbanna, G. Szekely, T.J. Taylor, O.M. Thomas, N.J. Tustison, G. Unal, F. Vasseur, M. Wintermark, D.H. Ye, L. Zhao, B. Zhao, D. Zikic, M. Prastawa, M. Reyes, and K. Van Leemput, "The multimodal brain tumor image segmentation benchmark (BRATS)", *IEEE Trans. Med. Imaging,* vol. 34, no. 10, pp. 1993-2024, 2015. [http://dx.doi.org/10.1109/TMI.2014.2377694] [PMID: 25494501]

[36] P. Tschandl, C. Rosendahl, and H. Kittler, "The HAM10000 dataset, a large collection of multi-source dermatoscopic images of common pigmented skin lesions", *Sci. Data,* vol. 5, no. 1, p. 180161, 2018. [http://dx.doi.org/10.1038/sdata.2018.161] [PMID: 30106392]

[37] S. Nandan Mohanty, S.K. Saxena, S. Satpathy, J.M. Chatterjee, Ed., *Applications of Artificial Intelligence in COVID-19. Medical Virology: From Pathogenesis to Disease Control.* Springer: Singapore, 2021. [http://dx.doi.org/10.1007/978-981-15-7317-0]

[38] S. Satpathy, S. Nandan Mohanty, J.M. Chatterjee, and A. Swain, "Comprehensive Claims of AI for Healthcare Applications-Coherence Towards COVID-19", In: *Applications of Artificial Intelligence in COVID-19. Medical Virology: From Pathogenesis to Disease Control.* Springer: Singapore, 2021. [http://dx.doi.org/10.1007/978-981-15-7317-0_1]

[39] S. Mohapatra, S. Satpathy, and S.N. Mohanty, "A comparative knowledge base development for cancerous cell detection based on deep learning and fuzzy computer vision approach", *Multimedia Tools Appl.,* vol. 81, no. 17, pp. 24799-24814, 2022. [http://dx.doi.org/10.1007/s11042-022-12824-0]

[40] S. Satpathy, M. Mangla, N. Sharma, H. Deshmukh, and S. Mohanty, "Predicting mortality rate and associated risks in COVID-19 patients", *Spatial Information Research,* vol. 29, no. 4, pp. 455-464, 2021. [http://dx.doi.org/10.1007/s41324-021-00379-5]

[41] N. Mukkapati, and M. S. Anbarasi, *Brain tumor classification based on enhanced CNN model,* 2022. [http://dx.doi.org/10.18280/ria.360114]

CHAPTER 3

Smart Health Care Systems and Predictive Medicine

Girija Jyothi Golla[1,*], **Anil Pise**[2], **Akhil Singampalli**[3] and **Vrushali Deshpande**[4]

[1] *Universita degli Studi di Messina, Messina, Italy*

[2] *Cumulus Solutions, Johannesburg, South Africa*

[3] *Vignan's Institute of Information Technology, Andhra Pradesh, India*

[4] *D&M Tech, Pune, India*

Abstract: This chapter explores the transformative role of smart healthcare systems and predictive medicine in modern healthcare. By incorporating advanced technologies such as the Internet of Things (IoT), artificial intelligence (AI), and big data analytics, these systems enable a proactive and personalized approach to patient care. The chapter examines the evolution of healthcare systems, highlights key components of smart healthcare, and explores their benefits, challenges, and future directions, supported by real-world case studies. It emphasizes the potential to enhance patient outcomes, improve efficiency, and reduce costs in healthcare delivery. Despite significant challenges, the benefits present a promising future for advancing patient care.

Keywords: AI, Big data analytics, Electronic health records (EHRs), IoT, Machine learning, Personalized medicine, Predictive medicine, Remote monitoring, Smart healthcare systems, Telemedicine.

INTRODUCTION

The healthcare industry is undergoing a significant transformation with the emergence of smart healthcare systems and predictive medicine. These advancements are powered by cutting-edge technologies that enable healthcare providers to deliver more efficient, personalized, and preventive care. This chapter delves into the various aspects of smart healthcare systems and predictive medicine, exploring their components, benefits, challenges, and practical applications in real-world settings.

* **Corresponding author Girija Jyothi Golla:** Universita degli Studi di Messina, Messina, Italy; E-mail: gllgjj96b59z222a@studenti.unime.it

Suneeta Satpathy, Sachi Nandan Mohanty & Subhendu Kumar Pani (Eds.)

THE EVOLUTION OF HEALTHCARE SYSTEMS

Traditional Healthcare Systems

Traditional healthcare systems have primarily operated on a reactive basis, treating illnesses and conditions as they arise [1]. This model often results in delayed interventions, increased healthcare costs, and suboptimal patient outcomes [2, 3]. Traditional healthcare is often characterized by isolated, fragmented care, where different providers and facilities maintain separate records and have limited communication. Patients typically seek medical help only when symptoms become severe, which can lead to more complex and costly treatments. (Fig. **1**) is an illustration of the Traditional Healthcare System.

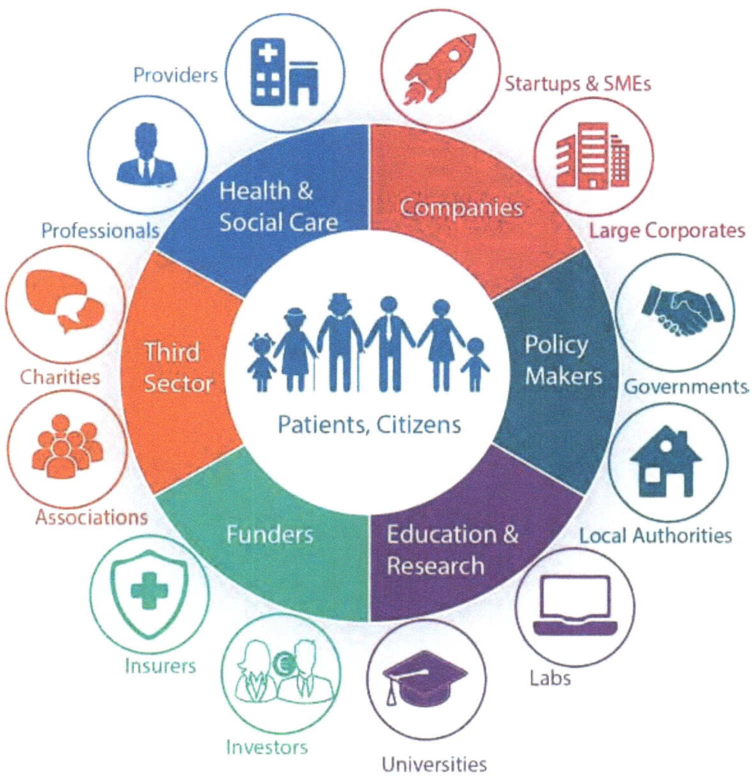

Fig. (1). Traditional healthcare system model.

In traditional healthcare systems, patients typically visited hospitals or clinics when they experienced symptoms or health concerns. Healthcare providers would then diagnose and treat the condition based on the symptoms presented and the available medical information. However, this reactive approach often resulted in delayed interventions, as patients might delay seeking care until their condition worsened [4, 5].

Additionally, traditional healthcare systems were often fragmented, with different providers and facilities maintaining separate records and having limited communication or coordination. This lack of integration made it difficult to provide comprehensive and coordinated care, as providers frequently lacked access to a patient's complete medical history or treatment plans from other healthcare professionals [6].

The Role of Information Technology

The introduction of information technology (IT) in healthcare marked a major milestone in the shift toward modernization [7]. Systems such as Electronic Health Records (EHRs), computerized physician order entry (CPOE) systems, and health information exchanges (HIEs) were developed to streamline data management and enhance the accuracy of patient information [8]. IT has facilitated the consolidation of patient data, enabling better care coordination and more informed decision-making [9].

For example, EHRs provide healthcare providers with access to a patient's complete medical history, ensuring that treatments and prescriptions are both appropriate and safe [10]. The adoption of IT has played a critical role in addressing the challenges of traditional healthcare systems [11]. The implementation of EHRs, for instance, has allowed providers to store and retrieve patient records electronically, reducing reliance on paper-based systems and improving both data accuracy and accessibility [12].

EHRs have further enhanced care coordination by enabling providers to share and access critical information, such as medical history, test results, and treatment plans [13]. This advancement has improved the continuity of care while also reducing the likelihood of redundant testing or conflicting prescriptions.

Emergence of Smart Healthcare Systems

The convergence of big data analytics, artificial intelligence (AI), and Internet of Things (IoT) technologies has driven the development of smart healthcare systems [14]. These systems utilize real-time data, predictive analytics, and interconnected devices to enable continuous monitoring, early diagnosis, and personalized treatment plans.

Smart healthcare systems aim to address gaps in traditional healthcare by creating a more connected and responsive environment. They offer solutions to improve care delivery while leveraging technological advancements for better patient outcomes. Smart healthcare systems facilitate seamless communication between different healthcare providers and systems, ensuring patient information is consistently up-to-date and accessible [15]. These systems mark a significant advancement in healthcare delivery by integrating cutting-edge technologies like IoT, AI, and big data analytics [16].

Designed to address the limitations of traditional healthcare models, these systems provide a more proactive, personalized, and efficient approach to patient care. A critical component of smart healthcare systems is the integration of IoT devices, such as wearable sensors and advanced medical equipment. These devices collect and transmit real-time patient data, enabling continuous monitoring and the early detection of potential health concerns [17].

By applying AI and machine learning algorithms, healthcare providers can analyse this data to identify patterns, predict health outcomes, and implement timely interventions and personalized treatment plans. Additionally, smart healthcare systems utilize big data analytics to process vast amounts of healthcare data from sources like EHRs, IoT devices, AI, and medical research [18]. This analysis uncovers valuable insights into disease trends, patient populations, and healthcare practices, supporting evidence-based decision-making and policy development.

KEY COMPONENTS OF SMART HEALTHCARE SYSTEMS

IoT in Healthcare

IoT devices in healthcare, such as wearable sensors and smart medical equipment as shown in (Fig. **2**), collect and transmit patient data in real time. This continuous stream of information allows healthcare providers to remotely monitor patients and respond promptly to any changes in their condition [19]. Wearable devices, including fitness trackers and smartwatches, track vital signs such as heart rate, blood pressure, and more, enhancing the efficiency of patient care.

The use of IoT devices in healthcare has revolutionized patient monitoring and data collection. Wearable devices, such as fitness trackers and smartwatches, continuously monitor vital signs like heart rate, blood pressure, and physical activity. This real-time data provides healthcare providers with valuable insights into a patient's health, enabling more proactive and personalized care.

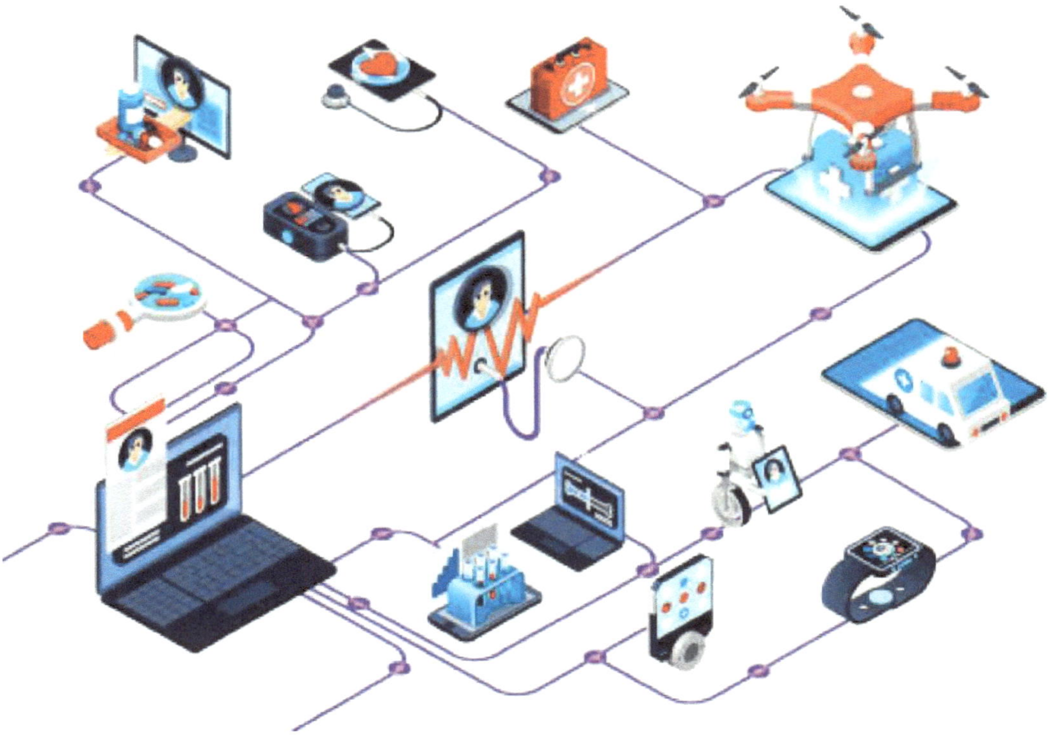

Fig. (2). IoT devices in healthcare.

For example, a patient with a chronic condition like diabetes or heart disease can utilize a continuous glucose monitor or cardiac monitor, which transmits data directly to their healthcare provider [20]. Providers can analyze this information to identify patterns, adjust treatment plans, and intervene promptly when necessary [21].

Electronic Health Records (EHRs)

They serve as the central hub for patient data, facilitating seamless information exchange among various healthcare providers [22]. By streamlining clinical workflows and administrative tasks, EHRs enhance efficiency and reduce errors. The ability to electronically access patient records has transformed healthcare delivery, simplified the tracking of patient progress and improved communication between specialists and other care providers. In smart healthcare systems, EHRs play a pivotal role as the central repository for patient information. They support the integration and sharing of data from multiple sources, such as IoT devices, diagnostic tests, and medical imaging [23]. This interoperability enables coordinated care and informed decision-making by allowing healthcare providers across specialties and locations to access a patient's complete medical history.

Furthermore, EHRs support clinical workflows and administrative tasks by minimizing errors and enhancing efficiency. For instance, electronic prescribing within EHR systems helps reduce medication errors by automatically checking for drug interactions and allergies. Additionally, EHRs streamline billing and coding processes, reducing administrative burdens and lowering associated costs.

Telemedicine and Remote Monitoring

Telemedicine platforms enable patients to have virtual consultations, reducing the need for in-person visits [24]. Remote monitoring tools, such as smartwatches and glucose monitors, help individuals manage chronic conditions and monitor their health from a distance. This has become particularly significant during the COVID-19 pandemic, allowing patients to receive care while minimizing the risk of virus exposure. Additionally, telemedicine increases healthcare accessibility for individuals in rural or underserved areas who may face challenges traveling to healthcare facilities.

Telemedicine and remote monitoring are integral components of smart healthcare systems, allowing healthcare providers to deliver and monitor care from afar. Virtual consultations offer patients the convenience of receiving medical advice and treatment from their homes or remote locations, reducing the reliance on in-person visits and improving healthcare access in more isolated regions [25].

Remote monitoring technologies, including wearable devices and home health systems, enable healthcare providers to track vital signs, medication adherence, and other critical health data in real time. This continuous stream of information supports predictive analytics, helping identify potential health risks and enabling early interventions to reduce complications and hospitalizations [26].

AI and ML algorithms

They process large volumes of healthcare data to identify patterns and predict outcomes. These advanced technologies enhance diagnostic accuracy, treatment planning, and personalized medicine. AI can analyse complex medical data, such as imaging scans and genomic information, to assist healthcare providers in making precise diagnoses and developing effective treatment strategies.

Machine learning models also help predict patient outcomes, such as the likelihood of readmission or disease progression, enabling timely interventions and proactive care.

AI and machine learning serve as foundational technologies powering the functionality of intelligent healthcare systems. These sophisticated algorithms process and analyse extensive healthcare data, such as patient records, medical imaging, and genomic information, to uncover patterns and forecast outcomes effectively. In diagnostics, AI can help healthcare providers analyse complex medical data, such as radiology images and pathology reports, with improved accuracy and efficiency. By training machine learning models on large datasets, these algorithms can detect subtle patterns and anomalies that might be challenging for human experts to identify, resulting in more precise diagnoses.

AI and machine learning are pivotal in advancing treatment planning and personalized medicine. By analysing a patient's medical history, genetic information, and other pertinent data, these technologies can predict potential health outcomes and recommend tailored treatment strategies. For example, machine learning models can forecast disease progression or determine a patient's likely response to specific therapies, enabling healthcare professionals to optimize treatment plans to meet individual needs effectively.

Big Data and Analytics

Big data analytics process extensive datasets to generate insights into patient populations, disease trends, and healthcare delivery. These insights shape evidence-based practices and policy-making. For instance, analysing data from EHRs can uncover patterns in patient outcomes and highlight best practices for treatment. Additionally, big data plays a vital role in public health, aiding in tracking disease outbreaks and assessing the effectiveness of public health interventions. Fig. (**3**) illustrates the Big Data Analytics Workflow in Healthcare.

Big data and analytics are integral to smart healthcare systems, enabling healthcare organizations to extract valuable insights from large volumes of data generated by sources such as EHRs, IoT devices, and medical research. Through advanced analytical methods, healthcare providers can uncover patterns and trends that may not be easily detected using traditional approaches.

A key application of big data analytics in healthcare is managing population health. By analysing data from EHRs, insurance claims, and other sources, organizations can identify high-risk patient groups and develop targeted interventions to enhance health outcomes and reduce costs. For example, predictive models can pinpoint patients at risk for chronic conditions like diabetes or heart disease, enabling early prevention strategies and personalized care plans.

Fig. (3). Big data analytics workflow in healthcare.

Additionally, big data analytics plays a crucial role in public health initiatives, such as tracking disease outbreaks, evaluating the effectiveness of public health interventions, and informing policy decisions. By leveraging data from sources like surveillance systems, social media, and environmental sensors, public health authorities can address challenges and guide their actions effectively.

PREDICTIVE MEDICINE: TRANSFORMING PATIENT CARE

Definition and Principles

Predictive medicine involves using data and statistical models to predict health outcomes and guide clinical decisions. It aims to anticipate health issues before they manifest, allowing for early intervention and prevention. Predictive medicine is founded on the principle that early detection and intervention can prevent complications and improve patient outcomes. By leveraging data from sources such as electronic health records (EHRs), wearable devices, and genetic profiles, predictive models identify individuals at risk for certain conditions and recommend proactive measures to mitigate potential health issues.

Predictive medicine represents a groundbreaking shift in medical care, utilizing data and advanced statistical models to forecast potential health concerns before they arise. This proactive strategy is rooted in the belief that early detection and timely intervention can significantly reduce complications and enhance patient outcomes.

Central to predictive medicine are sophisticated analytical methods, including machine learning and predictive modelling. These tools process data from diverse sources, such as electronic health records (EHRs), wearable technologies, genetic profiles, and environmental influences. By uncovering patterns and relationships within these data sets, predictive models can assess an individual's likelihood of developing specific conditions or facing adverse health events, paving the way for more targeted and effective preventive care.

Predictive Analytics in Healthcare

Predictive analytics combines both historical and real-time data to forecast future health outcomes, playing a critical role in identifying at-risk patients and creating targeted interventions. For instance, predictive models can assess data from diabetic patients to identify those at increased risk for complications such as diabetic retinopathy or kidney disease. Healthcare providers can then implement proactive measures, like regular screenings and lifestyle changes, to prevent these complications. This approach is vital in predictive medicine, as it enables healthcare providers to anticipate future health concerns by utilizing a wide range of data sources, allowing for more personalized and timely interventions.

One major use of predictive analytics in healthcare is in recognizing patients who may develop complications due to chronic conditions. For instance, predictive models can examine medical histories, genetic data, and lifestyle factors of diabetic patients to identify those more likely to experience complications such as diabetic retinopathy or kidney disease. With this insight, healthcare providers can take preventive actions, including regular checkups, adjusting medications, and encouraging lifestyle changes, all of which contribute to better health outcomes.

Moreover, predictive analytics can predict the probability of hospital readmissions. By examining data from EHRs, insurance claims, and additional sources, predictive models can pinpoint patients who are at a higher risk of being readmitted, based on factors such as existing health conditions, socioeconomic status, and previous healthcare use. This allows healthcare organizations to adopt strategies that reduce preventable readmissions and enhance overall care coordination.

Machine Learning Models for Predictive Medicine

Machine learning models, such as decision trees, neural networks, and support vector machines, play an essential role in predicting health outcomes by analysing intricate datasets. These models are designed to deliver accurate predictions based on patterns in the data. For example, a neural network can process imaging data to detect early signs of cancer, while a decision tree model can predict the likelihood of hospital readmissions using patient demographics and medical history.

At the core of predictive medicine, machine learning models empower healthcare professionals to make data-driven decisions by analysing complex information to predict various health outcomes. These models utilize advanced algorithms and statistical techniques to uncover patterns and relationships within data, thereby supporting early detection and improving clinical decision-making.

Different machine learning models are tailored for specific applications in predictive medicine. For instance, neural networks excel in analysing intricate data, such as medical images or genetic information. They are trained to detect subtle patterns and anomalies that may indicate the presence of specific diseases or conditions, enabling earlier diagnoses.

Decision trees are especially effective in predicting categorical outcomes, such as the likelihood of hospital readmission or the chances of developing a specific condition. These models generate decision-making rules based on factors like patient characteristics and medical history, assisting clinicians in making informed health risk predictions.

Support vector machines (SVMs) are another valuable tool in predictive medicine, especially for analysing high-dimensional data. SVMs can be used in tasks like disease classification, risk assessment, and predicting treatment responses.

A significant benefit of machine learning models in predictive medicine is their capacity to evolve and improve over time. With the continuous influx of patient data, these models adapt and refine their algorithms, enhancing the accuracy and reliability of their predictions with each update. This dynamic learning process allows for increasingly precise and effective healthcare interventions.

Case Studies: Predictive Medicine in Action

The practical applications of predictive medicine highlight its transformative potential. For instance, machine learning algorithms have been employed to anticipate hospital readmissions, enabling healthcare providers to take proactive steps to mitigate risks. These models analyze data from electronic health records (EHRs), incorporating patient demographics, medical history, and lab results to assign risk scores to patients. Providers then use these scores to create tailored care plans that include follow-up appointments, medication adjustments, and lifestyle recommendations. Such interventions have led to a significant decrease in readmission rates, showcasing the tangible benefits of predictive medicine.

Predictive medicine is far from a theoretical concept—it is being actively implemented in healthcare environments, delivering measurable improvements in patient outcomes and reducing costs. Numerous real-world case studies underscore its effectiveness and practicality in addressing complex healthcare challenges. Fig. (**4**) illustrates a typical Predictive Analytics Workflow, demonstrating how data is processed and utilized to generate actionable insights for patient care.

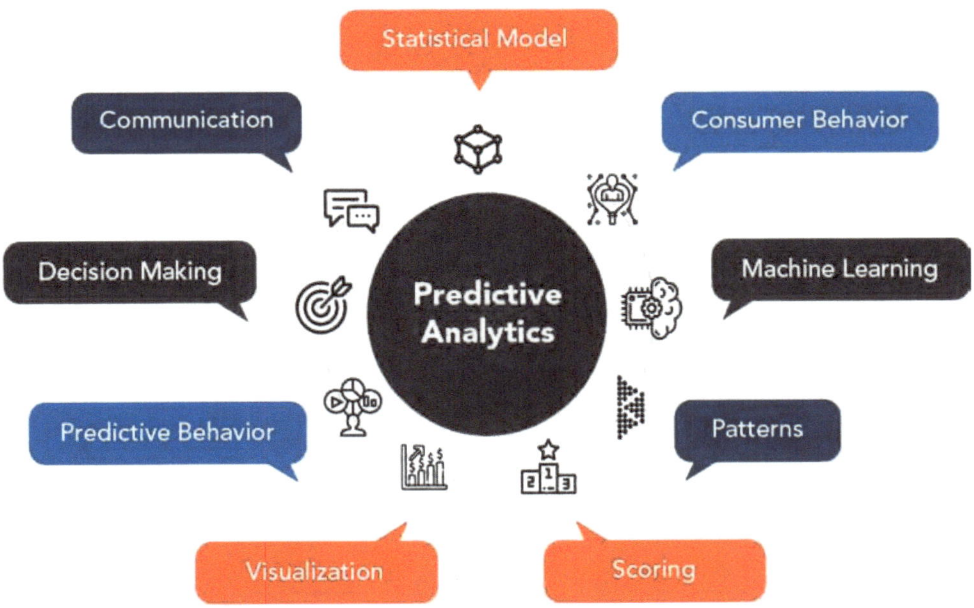

Fig. (4). Predictive analytics workflow.

A prominent example of predictive models in action is their application in forecasting hospital readmissions. Avoidable readmissions not only pose challenges for patients but also escalate healthcare costs significantly. In a study conducted within a leading healthcare system, researchers designed a machine learning model to identify patients at elevated risk of readmission due to heart failure. The model utilized data from electronic health records (EHRs), including patient demographics, medical history, laboratory results, and other pertinent variables, to calculate individualized risk scores for each patient. This approach allowed for targeted interventions, demonstrating the potential of predictive analytics to improve outcomes and reduce costs.

Armed with these risk scores, healthcare providers were able to develop individualized care plans for high-risk patients, including follow-up appointments, medication adjustments, and lifestyle interventions. The use of this predictive approach led to a notable decrease in heart failure readmission rates, highlighting the potential of predictive medicine to enhance patient outcomes and lower healthcare costs.

Another notable case study highlights the application of predictive models for the early identification of cancer. Researchers at a leading cancer centre developed a model. The algorithm was trained on an extensive dataset of mammograms, including images from both patients with and without breast cancer.

In a clinical trial, the predictive model showed greater accuracy in detecting breast cancer compared to traditional diagnostic methods. This ability to detect earlier has the potential to greatly enhance patient outcomes by facilitating earlier intervention and treatment, ultimately saving lives.

These case studies demonstrate the transformative impact of predictive medicine, showcasing the potential for better patient outcomes and cost savings within the healthcare system. By utilizing advanced analytical techniques and vast data sets, predictive models can identify at-risk patients, enable early interventions, and inform personalized treatment strategies, ultimately leading to improved health outcomes and more efficient healthcare delivery.

BENEFITS OF SMART HEALTHCARE SYSTEMS AND PREDICTIVE MEDICINE

Improved Patient Outcomes

Smart healthcare systems significantly enhance patient outcomes by facilitating timely interventions and enabling the development of personalized treatment plans. Predictive models play a critical role in identifying health risks before they escalate. For instance, a predictive model could detect a patient's heightened risk of a heart attack based on factors such as medical history and lifestyle choices. With this insight, healthcare providers can take pre-emptive actions, such as adjusting medications or suggesting lifestyle modifications, to mitigate the risk.

The integration of smart healthcare systems with predictive medicine offers a wealth of benefits, transforming patient care through real-time data analysis, advanced analytics, and individualized treatment strategies. These innovative methods support proactive care management, ensuring interventions are both timely and effective.

One of the standout advantages of this approach is the early identification of potential health concerns before they become critical. By analysing comprehensive data, including medical history, lifestyle habits, genetic profiles, and environmental influences, predictive models can assess an individual's risk for developing specific conditions or facing adverse health outcomes. For example, a predictive model might flag a patient with a high likelihood of a heart attack by evaluating factors such as their family history, diet, physical activity, and smoking habits. Armed with this information, healthcare providers can implement preventive strategies tailored to the patient's unique circumstances, significantly improving health outcomes and reducing the likelihood of severe events.

These interventions may include medication adjustments, lifestyle modifications (*e.g.*, diet and exercise recommendations), and regular monitoring to mitigate the identified risks and prevent the onset of severe complications.

Enhanced Efficiency and Reduced Costs

Automating routine tasks and streamlining workflows enhance operational efficiency. Predictive analytics help reduce unnecessary hospitalizations and procedures, thereby lowering healthcare costs. For example, a hospital may utilize predictive models to identify patients at risk for surgical complications. By implementing targeted interventions, such as preoperative assessments and postoperative monitoring, the hospital can mitigate the risk of complications and shorten hospital stays, resulting in substantial cost savings.

Smart healthcare systems and predictive medicine have the potential to improve efficiency and reduce costs within the healthcare system. By automating routine tasks, streamlining workflows, and minimizing unnecessary hospitalizations and procedures, these innovative methods can lead to substantial operational improvements and cost reductions.

One area where efficiency can be greatly enhanced is by automating repetitive tasks and processes. For instance, natural language processing (NLP) algorithms can automatically transcribe and process clinical notes, reducing the administrative burden on healthcare providers and minimizing the risk of errors. Additionally, machine learning models can be utilized to optimize scheduling and resource allocation, ensuring healthcare facilities operate at maximum efficiency.

Personalized Patient Care

Smart healthcare systems enable personalized care by customizing treatment plans to meet individual patient needs. This tailored approach enhances adherence to treatment and leads to better health outcomes. For instance, a personalized treatment plan for a diabetic patient may include regular blood sugar monitoring, medication adjustments, and lifestyle changes, such as diet and exercise. By adapting the treatment plan to each patient's specific needs, healthcare providers can improve blood sugar control and reduce the risk of complications.

Personalized patient care is a key element of smart healthcare systems and predictive medicine. By utilizing advanced technologies and data-driven insights, these methods empower healthcare providers to customize treatment plans based on the unique needs and characteristics of each patient.

In chronic disease management, personalized care can greatly improve patient outcomes and adherence to treatment plans. For example, a diabetic patient may receive a customized treatment plan that includes regular blood sugar monitoring, medication adjustments based on individual responses and genetics, as well as personalized lifestyle interventions, such as diet and exercise recommendations tailored to their preferences and physical abilities.

Personalized care goes beyond managing chronic conditions, encompassing a wide range of healthcare areas, including preventive care, diagnostics, and surgical procedures. By taking into account a patient’s medical history, genetic profile, and other pertinent details, healthcare providers can make well-informed decisions to determine the most suitable course of action. This approach not only reduces risks but also enhances the likelihood of achieving successful outcomes tailored to the individual's unique needs.

Early Disease Detection and Prevention

Predictive medicine facilitates the early identification of diseases, enabling timely preventive measures and prompt treatment. This proactive approach helps alleviate the burden of chronic illnesses. For example, predictive models can analyse a patient’s medical history and lab results to determine their risk of developing chronic kidney disease (CKD). Armed with this information, healthcare providers can implement preventive strategies, such as routine screenings and tailored medication adjustments, to slow the progression of CKD and reduce the likelihood of complications.

Early detection and prevention stand out as key advantages of predictive medicine, providing opportunities to reduce the burden of chronic diseases and improve overall public health. Through the use of predictive models and advanced analytics, healthcare providers can pinpoint individuals who are at risk of developing certain conditions, often before any symptoms arise, enabling timely interventions and proactive care.

A prominent example of predictive medicine's impact is the early detection of chronic kidney disease (CKD). Predictive models analyze data such as medical histories, lab results, and other relevant information to identify individuals at elevated risk for CKD. This is particularly critical since CKD often advances silently, showing no symptoms in its early stages. By identifying at-risk patients early, healthcare providers can take preventive steps, including lifestyle changes, medication adjustments, and routine monitoring, to slow disease progression and minimize the risk of complications like kidney failure or cardiovascular issues.

Similarly, predictive medicine has proven invaluable in the early detection of various cancers. Machine learning algorithms, trained on extensive datasets of medical images, biomarkers, and patient histories, can identify subtle patterns and anomalies that might signal early-stage cancer. Detecting cancer at these early stages dramatically improves the chances of successful treatment and better patient outcomes, as interventions can be initiated well before the disease progresses to more advanced and challenging stages. This proactive approach underscores the transformative potential of predictive medicine in improving healthcare delivery and patient well-being.

Improved Patient Engagement and Satisfaction

Patients play an active role in managing their health through remote monitoring and telemedicine, leading to greater satisfaction and adherence to treatment plans. For instance, individuals with chronic conditions can use wearable devices to track vital signs and share the data with their healthcare providers. This real-time exchange enables providers to make timely adjustments to treatment plans and offer personalized feedback and support, enhancing patient engagement and overall satisfaction.

Smart healthcare systems and predictive medicine bring significant benefits by fostering patient engagement and improving healthcare experiences. Through technologies such as remote monitoring and telemedicine, patients are empowered to take a proactive role in their health management, which leads to better compliance with treatment plans and a more satisfying care experience.

Remote monitoring tools, including wearable devices and home health systems, allow patients to record and share health data such as vital signs, medication adherence, and daily activity levels with their providers in real time. This continuous flow of information enables providers to monitor conditions remotely, make informed treatment decisions, and intervene when necessary, all while reducing the need for frequent in-person visits.

Telemedicine further enhances engagement by offering virtual consultations and follow-ups, particularly benefiting those in remote or underserved areas. The convenience of accessing high-quality care without the challenges of travel or long waiting times improves patient satisfaction and encourages adherence to medical advice and follow-up care.

Additionally, smart healthcare systems and predictive medicine promote active patient participation by delivering personalized insights, treatment suggestions, and educational resources. This collaborative approach strengthens the relationship between patients and providers, giving patients a deeper understanding of their conditions and treatments. With greater knowledge and involvement, patients feel more confident and satisfied with their care. Fig. (**5**) illustrates the comprehensive benefits of predictive medicine and its role in transforming patient care.

CHALLENGES AND CONSIDERATIONS

Data Privacy and Security

The widespread use of IoT devices and EHRs raises significant concerns about data privacy and security. Protecting patient data from breaches and unauthorized access is crucial. Healthcare providers must adopt robust cybersecurity measures, such as encryption and access controls, to safeguard patient data from unauthorized access and breaches. They must also adhere to regulations like the Health Insurance Portability and Accountability Act (HIPAA) in the United States, which sets standards for protecting patient data.

Data privacy and security are key challenges that must be addressed for the successful implementation of smart healthcare systems and predictive medicine. The widespread adoption of IoT devices, EHRs, and the exchange of sensitive patient data create potential vulnerabilities that could compromise data privacy and expose patients to risks such as identity theft, discrimination, or unauthorized access to their personal health information.

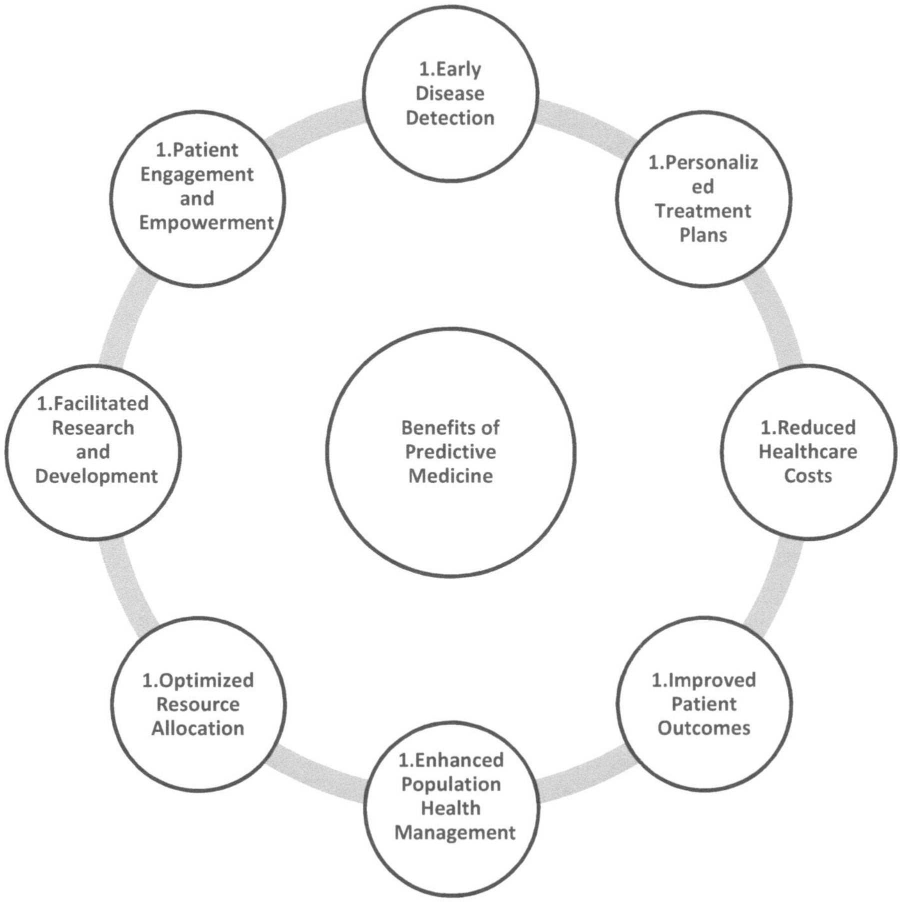

Fig. (5). Benefits of predictive medicine.

Healthcare organizations must implement robust cybersecurity measures to protect patient data from breaches and unauthorized access. This includes using advanced encryption techniques, access controls, and strict authentication protocols to secure data both at rest and during transmission. Regular security audits, employee training, and effective incident response plans are also critical components of a comprehensive cybersecurity strategy.

Compliance with data privacy regulations, such as the Health Insurance Portability and Accountability Act (HIPAA) in the United States, is another essential consideration. These regulations establish standards for safeguarding the privacy and security of patient health information. Healthcare organizations must ensure their practices and systems align with these standards. Non-compliance can result in significant financial penalties and harm to the organization's reputation.

Interoperability Issues

Achieving seamless data exchange among various healthcare systems and devices remains a persistent challenge, primarily due to the lack of standardization in data formats and communication protocols. For example, different electronic health record (EHR) systems often utilize proprietary formats, making it difficult to share patient data across platforms efficiently.

Interoperability—the ability of diverse healthcare systems and devices to exchange and interpret data seamlessly—is critical to the success of smart healthcare systems and predictive medicine. However, the healthcare industry operates within a complex landscape of heterogeneous systems, devices, and data structures, creating significant barriers to the free flow of information and often resulting in data silos.

A key obstacle to interoperability is the absence of standardized data formats and communication protocols across various EHRs, medical devices, and healthcare technologies. Each system's unique data format can impede the integration and sharing of information, leading to fragmented patient records, duplicated efforts, and delays or errors in clinical decision-making.

To overcome this challenge, the healthcare sector has increasingly adopted standardized frameworks such as the FHIR standard. FHIR outlines specifications for structuring and exchanging healthcare data, enabling systems and applications to interoperate more effectively.

In addition to FHIR, the implementation of APIs and cloud-based platforms plays a vital role in facilitating interoperability. APIs offer a standardized method for applications to communicate and share data, bridging gaps between disparate systems. Meanwhile, cloud-based platforms provide centralized storage and access to healthcare data, ensuring that information from multiple sources is consolidated and accessible in real time.

By addressing these interoperability challenges, healthcare providers can unlock the full potential of smart healthcare systems and predictive medicine, enhancing data sharing, improving patient care, and driving greater efficiency across the industry.

Ethical Consideration in AI

The integration of AI and predictive analytics in healthcare introduces several ethical challenges that require careful deliberation and action. These challenges primarily concern patient consent, data ownership, algorithmic bias, and the potential for unintended consequences, all of which must be addressed to ensure ethical and equitable healthcare delivery.

A cornerstone of ethical healthcare is obtaining informed consent, which also applies to the use of personal data for AI and predictive analytics. Healthcare organizations must ensure that patients are fully aware of how their data will be utilized, including its role in predictive models and machine learning systems. Providing clear, accessible information about data use and offering patients the ability to opt out respects their autonomy and builds trust in the use of advanced technologies.

The question of data ownership is central to ethical discussions, given the sensitive nature of healthcare information. Establishing clear policies on who owns, controls, and accesses patient data is critical to protecting privacy and autonomy. These policies should strike a balance between safeguarding personal information and enabling its responsible use in research and innovation, fostering advancements in healthcare while upholding ethical principles.

One of the most significant ethical concerns in AI and predictive analytics is algorithmic bias. Machine learning models are often trained on historical data that may reflect existing societal inequalities, such as disparities in access to healthcare services. If these biases go unaddressed, predictive models risk perpetuating or even exacerbating inequities, leading to unfair treatment or inaccurate predictions for underrepresented populations.

To combat this, healthcare organizations must:

- Ensure diversity and representativeness in training datasets.
- Conduct rigorous testing and auditing of algorithms to identify and mitigate biases.
- Develop strategies for ongoing monitoring to ensure equitable outcomes across all patient groups.

Transparency in the development and deployment of AI and predictive analytics is essential to maintaining ethical integrity. Patients and healthcare providers must understand how predictive models work, what data they rely on, and how decisions are made. Additionally, accountability mechanisms should be established to address errors, unintended consequences, or harm resulting from these technologies.

By addressing these ethical considerations, healthcare organizations can ensure that the integration of AI and predictive analytics not only enhances patient care but also upholds principles of fairness, privacy, and transparency. This commitment is critical to building trust and ensuring the responsible use of advanced technologies in the evolving landscape of healthcare.

Technical Challenges

Implementing and maintaining smart healthcare systems require significant technical expertise and resources. Ensuring system reliability, scalability, and integration with existing infrastructure are critical challenges. For example, healthcare providers must invest in robust IT infrastructure, such as servers and network equipment, to support smart healthcare systems. Ensuring system reliability and scalability is essential to accommodate the growing volume of healthcare data and provide timely and accurate predictions.

The implementation and maintenance of smart healthcare systems and predictive medicine pose significant technical challenges that require substantial expertise and resources. These challenges span various aspects, including system reliability, scalability, and integration with existing infrastructure.

Ensuring the reliability of smart healthcare systems and predictive models is of paramount importance, particularly since these technologies are often deployed in mission-critical applications where accuracy and availability directly impact patient safety. Any system downtime, malfunction, or error can lead to delayed or incorrect diagnoses, posing serious risks to patient outcomes.

To mitigate these risks, healthcare organizations should:

1. **Invest in Robust IT Infrastructure**: Build resilient systems with redundancy to ensure uninterrupted operation, even in the event of hardware or software failures.
2. **Implement Rigorous Testing Protocols**: Conduct extensive testing of predictive models and system functionalities to identify potential vulnerabilities and address them proactively.
3. **Adopt Real-Time Monitoring**: Use advanced monitoring tools to detect and resolve issues before they escalate into system failures.
4. **Develop Contingency Plans**: Establish comprehensive backup strategies and disaster recovery plans to safeguard against data loss and minimize the impact of system downtime.

As the volume of healthcare data grows exponentially, driven by advancements in IoT devices, electronic health records (EHRs), and other digital health technologies, scalability emerges as a critical technical challenge. Smart healthcare systems must be capable of efficiently handling this surge in data while maintaining high performance and responsiveness.

To achieve scalability, healthcare organizations should:

1. **Leverage Advanced Computing Resources**: Invest in high-performance computing technologies capable of processing vast amounts of data in real time.
2. **Utilize Distributed Data Processing Frameworks**: Implement frameworks like Apache Hadoop or Apache Spark to enable efficient parallel processing of large datasets.
3. **Adopt Scalable Storage Solutions**: Use cloud-based or hybrid storage systems that can expand dynamically to accommodate growing data volumes.
4. **Optimize Data Workflows**: Design data collection, processing, and analysis pipelines that prioritize efficiency and minimize latency.
5. **Incorporate Artificial Intelligence**: Use AI algorithms optimized for large-scale data handling to identify patterns and generate insights without compromising system performance.

Integrating smart healthcare systems with existing infrastructure and legacy systems can be a challenging process. Many healthcare organizations have made significant investments in traditional IT systems and workflows, and ensuring that new technologies work seamlessly with these established systems can be difficult. Thorough planning, thoughtful system architecture design, and effective change management strategies are crucial to ensure a smooth transition and minimize disruptions to daily operations.

Accurate and consistent data is essential for smart healthcare systems and predictive models to provide reliable insights. However, healthcare data is often complex and prone to errors due to manual entry and inconsistent sources. To address these challenges:

1. **Implement Data Governance**: Establish policies to standardize data collection, ensure accountability, and maintain consistency.
2. **Automate Validation**: Use tools and algorithms to detect errors and anomalies in real time.
3. **Data Cleansing**: Regularly audit and correct errors, redundancies, and inconsistencies.
4. **Improve Integration**: Utilize standardized protocols like FHIR to harmonize data from various systems.
5. **Train Staff**: Educate teams on accurate data entry and best practices

Regulatory and Legal Issues

Regulatory frameworks and legal considerations for smart healthcare systems and predictive medicine are still evolving. Compliance with regulations and addressing liability issues are essential for widespread adoption. Healthcare providers must address complex regulatory requirements, including obtaining approval from regulatory bodies for implementing AI and predictive analytics. Additionally, liability concerns must be resolved, such as identifying accountability for errors or adverse outcomes caused by predictive models. Addressing these issues is crucial for fostering trust and enabling wider adoption of these technologies.

The swift progression of smart healthcare systems and predictive medicine has outstripped the establishment of comprehensive regulatory frameworks and legal guidelines. As these cutting-edge technologies advance, it is imperative for healthcare organizations and policymakers to tackle emerging regulatory and legal challenges to ensure their adoption is safe, ethical, and widely accepted.

One of the primary regulatory challenges is the lack of clear guidelines and standards for the use of AI and predictive analytics in healthcare. While regulatory bodies like the FDA have begun to develop frameworks for evaluating and approving AI-based medical devices, the landscape remains complex and fragmented. Healthcare providers must navigate a patchwork of regulations and obtain necessary approvals from multiple regulatory bodies.

The adoption of smart healthcare systems and predictive medicine introduces several legal and ethical challenges that must be addressed to ensure their safe and effective use.

Determining accountability for errors or adverse outcomes from AI-based predictive models is a significant legal challenge. Questions arise about whether responsibility lies with the healthcare provider, the technology vendor, or the algorithm itself. To address these concerns:

- Establish clear liability frameworks to delineate responsibilities among stakeholders.
- Protect both healthcare providers and patients by implementing safeguards for fair accountability.
- Encourage innovation by creating balanced legal standards that reduce ambiguity.

Smart healthcare systems rely on the collection and sharing of sensitive patient data, making compliance with regulations like HIPAA and GDPR crucial. To mitigate risks:

- Implement strong data governance practices and advanced security measures to prevent breaches.

- Ensure transparency about data usage to build patient trust.
- Establish protocols for secure storage, transmission, and access to sensitive health information.

Legal issues related to ethical considerations, such as patient autonomy, informed consent, and algorithmic bias, require thoughtful policies:

- Patients must retain control over their personal data, including the ability to opt out of data-driven practices.
- Develop guidelines to ensure predictive models and algorithms are free of bias by training them on diverse datasets.
- Uphold ethical standards by making the use of AI and predictive analytics transparent and fully consent-driven.

FUTURE DIRECTIONS

Advances in Technology

Quantum computing has the potential to revolutionize healthcare by processing vast datasets in real-time, enabling faster and more accurate predictions. Its unmatched computational power can drive advancements in genomics, drug discovery, and disease modelling, paving the way for breakthroughs in personalized medicine and precision therapies.

5G's high-speed, low-latency connectivity is set to transform telemedicine and remote patient monitoring. It facilitates real-time transmission of high-definition medical images, seamless virtual consultations, and instant data sharing from wearable devices, significantly enhancing the efficiency and effectiveness of remote care.

Integration with Genomic Data

Integrating genomic data into predictive models is a transformative step in healthcare, enabling more precise and personalized treatments. Precision medicine leverages an individual's genetic profile alongside clinical and lifestyle data to create tailored treatment plans, enhancing outcomes and reducing costs.

By analysing genetic information, healthcare providers can identify mutations or variants that increase the risk of specific conditions, such as cancers or neurological disorders. This knowledge allows for targeted screenings and preventive measures, enabling early detection and improving treatment success rates.

Genomic data also informs the selection of therapies that align with a patient's unique genetic makeup. This is particularly impactful in oncology, where genetic mutations can influence responses to cancer treatments. Personalized therapies maximize treatment efficacy while minimizing side effects, ensuring more effective and patient-centred care.

Integrating genomic insights with predictive analytics marks a new era in precision medicine, promising better outcomes through highly customized interventions.

Expansion of Telemedicine

Future advancements in telemedicine technology will enhance remote consultations, making healthcare more accessible. For instance, telemedicine platforms can integrate AI and machine learning algorithms to analyse patient data and offer real-time feedback and recommendations. Additionally, advancements in virtual reality (VR) and augmented reality (AR) can improve telemedicine consultations by creating immersive and interactive experiences.

The expansion of telemedicine is reshaping healthcare, providing patients with greater access to high-quality care regardless of location or mobility challenges. The COVID-19 pandemic significantly accelerated telemedicine adoption, with virtual consultations becoming a safe and effective alternative to in-person visits.

Future Advancements in Telemedicine

- **AI Integration**: Incorporating AI and machine learning into telemedicine platforms will enable real-time data analysis, offering healthcare providers valuable insights and recommendations during virtual consultations.
- **Virtual and Augmented Reality**:
  - **VR**: Creates immersive environments that simulate in-person interactions, fostering more natural and engaging patient-provider communication.
  - **AR**: Enhances consultations by overlaying digital information onto the real-world environment, helping providers explain medical conditions and treatment options with greater clarity.

These technological advancements will enhance the quality of remote consultations while broadening the scope of services that telemedicine can provide, making healthcare more versatile and accessible. For example, remote surgical consultations, where surgeons can guide procedures using AR or VR, could become a reality, enabling access to specialized medical expertise in areas where it may not be readily available.

Enhanced Data Analytics

The improved quantity/volume of medical data will drive the requirement for more sophisticated data analytics tools. Advanced analytics can provide deeper insights into patient populations and disease trends. For example, predictive models can evaluate social determinants of health, such as income and education, to identify disparities in healthcare access and outcomes. Additionally, advanced analytics can aid in population health management by pinpointing high-risk individuals and enabling the development of tailored interventions.

As the volume of healthcare data increasing exponentially, fuelled by the rapid expansion of IoT devices, EHRs, and other data sources, the need for upgraded data cleaning devices and techniques becomes increasingly crucial. These advanced analytics capabilities will allow healthcare organizations to achieve minute details, identify complex patterns, and predict correct results, improving patient diagnosis and medical treatment.

One area where advanced analytics will serve a crucial role is in understanding the influence of SDoH on access to healthcare and health outcomes. Factors like income, education, housing, and transportation play a crucial role in shaping an individual's overall health and their ability to access necessary medical care. By leveraging advanced analytics and predictive modelling techniques, healthcare organizations can analyse these socioeconomic and environmental factors, identify disparities in healthcare access and outcomes among different populations, and develop targeted interventions to address these inequalities.

Advanced analytics will play a key role in population health management efforts. By examining extensive datasets from sources like electronic health records (EHRs), claims data, and public health registries, healthcare organizations can pinpoint high-risk patient groups and develop customized strategies to better prevent or manage chronic illnesses. Predictive models can categorize populations by risk levels, allowing healthcare providers to optimize resource allocation and deliver targeted, preventive care to those with the greatest need.

Advanced analytics can reveal critical insights into disease trends and patterns, aiding in the development of public health policies and the optimization of resource allocation strategies. By evaluating data from surveillance systems, clinical studies, and research databases, public health officials can track and manage the spread of infectious diseases effectively.

Focus on Preventive Care

Predictive medicine shifts the focus from reactive to preventive care, aiming to stop diseases before they arise and reduce the strain on healthcare systems. For instance, predictive models can pinpoint patients at risk for chronic conditions like diabetes and hypertension, enabling the recommendation of preventive strategies such as lifestyle changes and regular screenings. This approach not only enhances patient outcomes but also lowers healthcare costs by addressing diseases before they progress to more severe stages.

The adoption of smart healthcare systems and predictive medicine is driving a significant shift toward preventive care, moving away from the traditional reactive approach. This focus on disease prevention has the ability to enhance patient outcomes, ease the strain on healthcare systems, and reduce overall healthcare costs.

Predictive models and advanced analytics are essential in identifying individuals at risk of developing chronic conditions like diabetes, hypertension, and cardiovascular diseases. By evaluating various factors, including medical history, lifestyle choices, genetic predispositions, and environmental exposures, these models can classify populations based on risk levels and recommend tailored preventive measures.

For high-risk individuals, healthcare providers can implement targeted strategies such as lifestyle modifications (*e.g.*, dietary changes, exercise routines), regular screenings, and preventive medications or treatments. These proactive measures can help delay or even prevent the onset of chronic conditions, reducing the need for more intensive and expensive treatments in the future.

Preventive care strategies will play a crucial role in tackling public health challenges, such as infectious disease outbreaks and epidemics. By utilizing predictive models and real-time data from surveillance systems, public health officials can identify potential hotspots and high-risk populations. This enables targeted interventions, such as vaccination drives or community education programs, to control and reduce the spread of diseases. Fig. (**6**) highlights the future direction of predictive medicine.

Future Directions in Predictive Medicine

- **Integration of AI and Machine Learning**
- **Multi-omics Data Integration**
- **Real-time Data Analysis**
- **Wearable and IoT Devices**
- **Precision Prevention**
- **Population Health Management**
- **Interoperability and Data Sharing**
- **Ethical and Regulatory Considerations**
- **Patient-Centered Approaches**
- **Translation to Clinical Practice**

Fig. (6). Future directions in predictive medicine.

Healthcare systems will need to transition from a fee-for-service model, which rewards reactive treatments, to a value-based care model that incentivizes preventive care and promotes population health management. This shift will require changes in healthcare policies, reimbursement structures, and care delivery models to encourage and support proactive, preventive approaches.

CONCLUSION

Smart healthcare systems and predictive medicine are set to revolutionize the future of healthcare. By leveraging advanced technologies, these systems offer the potential to enhance patient outcomes, boost efficiency, and reduce costs. However, addressing challenges related to data privacy, interoperability, and ethical considerations is essential for successful implementation. As technology advances, the integration of smart healthcare systems and predictive medicine will play a pivotal role in transforming the healthcare landscape.

REFERENCES

[1] E.J. Topol, "High-performance medicine: the convergence of human and artificial intelligence", *Nat. Med.,* vol. 25, no. 1, pp. 44-56, 2019. [http://dx.doi.org/10.1038/s41591-018-0300-7] [PMID: 30617339]

[2] C.H. Lee, and H.J. Yoon, "Medical big data: promise and challenges", *Kidney Res. Clin. Pract.,* vol. 36, no. 1, pp. 3-11, 2017. [http://dx.doi.org/10.23876/j.krcp.2017.36.1.3] [PMID: 28392994]

[3] E.W. Johnson, "Machine learning and decision support in critical care", *Proc. IEEE,* vol. 108, no. 3, pp. 264-282, 2020.

[4] W. Raghupathi, and V. Raghupathi, "Big data analytics in healthcare: promise and potential", *Health Inf. Sci. Syst.,* vol. 2, no. 1, p. 3, 2014. [http://dx.doi.org/10.1186/2047-2501-2-3] [PMID: 25825667]

[5] A. A. Pise, *Enabling artificial intelligence of things (AIoT) healthcare architectures and security issues.,* 2022. [http://dx.doi.org/10.1155/2022/8421434]

[6] D.W. Bates, S. Saria, L. Ohno-Machado, A. Shah, and G. Escobar, "Big data in health care: using analytics to identify and manage high-risk and high-cost patients", *Health Aff. (Millwood),* vol. 33, no. 7, pp. 1123-1131, 2014. [http://dx.doi.org/10.1377/hlthaff.2014.0041] [PMID: 25006137]

[7] J. He, S.L. Baxter, J. Xu, J. Xu, X. Zhou, and K. Zhang, "The practical implementation of artificial intelligence technologies in medicine", *Nat. Med.,* vol. 25, no. 1, pp. 30-36, 2019. [http://dx.doi.org/10.1038/s41591-018-0307-0] [PMID: 30617336]

[8] Z. Obermeyer, and E.J. Emanuel, "Predicting the future—big data, machine learning, and clinical medicine", *N. Engl. J. Med.,* vol. 375, no. 13, pp. 1216-1219, 2016. [http://dx.doi.org/10.1056/NEJMp1606181] [PMID: 27682033]

[9] T. Panch, H. Mattie, and L.A. Celi, "The “inconvenient truth” about AI in healthcare", *NPJ Digit. Med.,* vol. 2, no. 1, p. 77, 2019. [http://dx.doi.org/10.1038/s41746-019-0155-4] [PMID: 31453372]

[10] E.H. Shortliffe, and M.J. Sepúlveda, "Clinical decision support in the era of artificial intelligence", *JAMA,* vol. 320, no. 21, pp. 2199-2200, 2018. [http://dx.doi.org/10.1001/jama.2018.17163] [PMID: 30398550]

[11] A. Rajkomar, J. Dean, and I. Kohane, "Machine learning in medicine", *N. Engl. J. Med.,* vol. 380, no. 14, pp. 1347-1358, 2019. [http://dx.doi.org/10.1056/NEJMra1814259] [PMID: 30943338]

[12] A. A. Pise, *Internet of Things-System.,* 2023.

[13] P. Hamet, and J. Tremblay, "Artificial intelligence in medicine", *Metabolism,* vol. 69, pp. S36-S40, 2017. [http://dx.doi.org/10.1016/j.metabol.2017.01.011] [PMID: 28126242]

[14] I. Sim, "Mobile devices and health", *N. Engl. J. Med.,* vol. 381, no. 10, pp. 956-968, 2019. [http://dx.doi.org/10.1056/NEJMra1806949] [PMID: 31483966]

[15] A. Coravos, S. Khozin, and K.D. Mandl, "Developing and adopting safe and effective digital biomarkers to improve patient outcomes", *NPJ Digit. Med.,* vol. 2, no. 1, p. 14, 2019. [http://dx.doi.org/10.1038/s41746-019-0090-4] [PMID: 30868107]

[16] A.L. Fogel, and J.C. Kvedar, "Artificial intelligence powers digital medicine", *NPJ Digit. Med.,* vol. 1, no. 1, p. 5, 2018. [http://dx.doi.org/10.1038/s41746-017-0012-2] [PMID: 31304291]

[17] A.L. Beam, and I.S. Kohane, "Big data and machine learning in health care", *JAMA,* vol. 319, no. 13, pp. 1317-1318, 2018. [http://dx.doi.org/10.1001/jama.2017.18391] [PMID: 29532063]

[18] A. A. Pise, *Adoption of blockchain technology for privacy and security in Industry 4.0,* 2022.

[19] D. Singh, Manju, and U. Gupta, "Smart healthcare: A breakthrough in the growth of technologies", In: *In Artificial intelligence-based healthcare systems* Cham: Springer Nature Switzerland, 2023, pp. 73-85. [http://dx.doi.org/10.1007/978-3-031-41925-6_5]

[20] M. Hartmann, U.S. Hashmi, and A. Imran, "Edge computing in smart health care systems: Review, challenges, and research directions", *Transactions on Emerging Telecommunications Technologies,* vol. 33, no. 3, p. e3710, 2022. [http://dx.doi.org/10.1002/ett.3710]

[21] J.B. Awotunde, S.O. Folorunso, S.A. Ajagbe, J. Garg, and G.J. Ajamu, "AiIoMT: IoMT-based system-enabled artificial intelligence for enhanced smart healthcare systems", *Machine learning for critical Internet of Medical Things: Applications and use cases,* pp. 229-254, 2022. [http://dx.doi.org/10.1007/978-3-030-80928-7_10]

[22] S. Mohapatra, S. Satpathy, and S. N. Mohanty, "A comparative knowledge base development for cancerous cell detection based on deep learning and fuzzy computer vision approach", *Multimed Tools Appl,* vol. 81, no. 17, pp. 24799-24814, 2022. [http://dx.doi.org/10.1007/s11042-022-12824-0]

[23] Y. Yang, K. Siau, W. Xie, and Y. Sun, "Smart health: Intelligent healthcare systems in the metaverse, artificial intelligence, and data science era", *J Organ End User Comput,* vol. 34, no. 1, pp. 1-14, 2022. [http://dx.doi.org/10.4018/JOEUC.308814]

[24] M. Akter, N. Moustafa, T. Lynar, and I. Razzak, "Edge intelligence: Federated learning-based privacy protection framework for smart healthcare systems", *IEEE J Biomed Health Inform,* vol. 26, no. 12, pp. 5805-5816, 2022. [http://dx.doi.org/10.1109/JBHI.2022.3192648]

[25] R. Dwivedi, D. Mehrotra, and S. Chandra, "Potential of internet of medical things (IoMT) applications in building a smart healthcare system: A systematic review", *J Oral Biol Craniofac Res,* vol. 12, no. 2, pp. 302-318, 2022. [http://dx.doi.org/10.1016/j.jobcr.2021.11.010]

[26] S. Rani, M. Chauhan, A. Kataria, and A. Khang, "IoT equipped intelligent distributed framework for smart healthcare systems", In: *Towards the integration of IoT, cloud and big data: Services, applications and standards* Springer Nature Singapore: Singapore, 2023, pp. 97-114. [http://dx.doi.org/10.1007/978-981-99-6034-7_6]

CHAPTER 4

An Empirical Analysis of Machine Learning Models Used for Dental Image Processing from a Statistical Perspective

Mohit K. Popat[1,*] and **Dinesh N. Chaudhari**[1]

[1] *Department of Computer Science and Engineering J.D.I.E.T, Yavatmal, India*

Abstract: Due to the advent of machine learning models, dental image processing has made significant strides in recent years. The accuracy, precision, recall, and scalability of dental image analysis may be improved with the help of these models. Nonetheless, a statistically exhaustive evaluation and comparison of these models is lacking. This study aims to fill this void by comprehensively reviewing existing machine learning models used for dental image processing and analyzing their performance based on key metrics, such as precision, accuracy, recall, delay, and scalability. In addition, we propose a novel Dental Rank Metric (DRM) that combines these parameters to determine the most effective models for dental image processing scenarios. This study is necessary due to the growing significance of dental image analysis in clinical practice. Dental images are crucial in the diagnosis, treatment planning, and progression monitoring of various dental conditions. Accurate and efficient analysis of these images is essential for enhancing patient care, decreasing human error, and improving the overall efficiency of dental procedures.

In this paper, various machine learning models used for dental image processing are surveyed and evaluated. We investigate the advantages and disadvantages of multiple models, shedding light on their suitability for specific dental imaging tasks. The results of this study will aid dentists and researchers in selecting the most appropriate machine-learning models for dental image analysis. By identifying optimal models *via* the DRM, the efficiency and precision of dental image processing can be significantly enhanced, ultimately benefiting patients and advancing the dental healthcare process.

Keywords: Deep learning, DRM, Feature, Neural networks, Scenarios, X-rays CBCT.

INTRODUCTION

Dental image analysis is an indispensable component of modern dentistry, pivotal in ensuring precise diagnoses, meticulous treatment planning, and vigilant

* **Corresponding author Mohit K. Popat:** Department of Computer Science and Engineering J.D.I.E.T, Yavatmal, India; E-mail: mohit.popat872@gmail.com

Suneeta Satpathy, Sachi Nandan Mohanty & Subhendu Kumar Pani (Eds.)

monitoring of various dental conditions. The integration of machine learning techniques, coupled with the advancements in image processing, has sparked a burgeoning interest in leveraging these methodologies to enhance the efficacy and accuracy of dental image analysis.

The proliferation of machine learning models presents a promising avenue for automating repetitive tasks, assisting clinicians in making well-informed decisions, and mitigating the inherent subjectivity in manual interpretations. However, despite the notable surge in the development and application of machine learning models in dental image processing, a comprehensive and statistically exhaustive evaluation and comparison of these models remains conspicuous by their absence. Given the nuanced performance variations influenced by myriad factors, clinicians and researchers encounter challenges in discerning the most suitable models for their specific dental imaging tasks.

Key considerations such as precision (the capability to accurately identify positive instances), accuracy (the overall correctness of predictions), recall (the ability to capture all positive instances), delay (the temporal requirement for analysis), and scalability (the adeptness in handling larger datasets) underscore the complexity in model selection. Hence, the fundamental objective of this paper is to bridge this gap by undertaking an exhaustive review of extant machine learning models employed in dental image processing and meticulously analyzing their performance utilizing pivotal statistical metrics.

By formulating a robust Dental Rank Metric (DRM), we endeavor to illuminate the strengths and weaknesses inherent in diverse models, aiming to delineate the optimal models for dental image analysis. This endeavor harbors the potential to revolutionize dental image analysis, thereby augmenting patient care and outcomes. By pinpointing models characterized by superior precision, accuracy, recall, minimal delay, and scalability, dental practitioners can harness automated systems to render more precise and expeditious diagnoses.

Furthermore, researchers stand to glean more profound insights into the performance nuances of various models, thus catalyzing advancements in the field. The subsequent sections of this paper will expound upon the imperative need for this study, elucidate the manifold applications of machine learning in dental image analysis, scrutinize pertinent related works, delineate the methodological framework employed for evaluating the models, present the ensuing results and analysis, introduce the Dental Rank Metric (DRM), and culminate with prognostications for future research directions in this domain. Selected models pertinent to this discourse will be succinctly examined in this section of the narratives.

IN-DEPTH REVIEW OF EXISTING MODELS USED FOR DENTAL IMAGE ANALYSIS

Numerous models have been proposed for dental image analysis, addressing various dental applications. These models leverage advanced techniques, including deep learning, to enhance dental diagnosis and treatment. For instance, Dental-YOLO aids in precise alveolar bone and mandibular canal identification using CBCT images [1]. A novel network for dental plaque segmentation improves early detection without the need for coloring substances [2]. Spectral imaging offers non-ionizing, affordable imaging with enhanced contrast for various dental features [3]. Deep convolutional neural networks assist in person identification through panoramic dental X-rays [4]. Semantic decomposition networks facilitate dental plaque segmentation [5]. Enhanced truncated-correlation photothermal coherence tomography enables early caries detection [6]. Efficient Convolutional Neural Networks are used for tooth numbering and dental condition detection [7]. Computer vision and machine learning are employed for dental impression tray selection [8]. Deep learning architecture enhances craniomaxillofacial landmark identification [9]. Near-infrared dental fluorescence imaging aids in dental disease identification [10].

Furthermore, researchers have developed MI-DCNNE for automated dental caries diagnosis with impressive accuracy [11]. The DAIS system effectively restores dental inlay surfaces [12]. A novel algorithm combining oral-anatomical information and a data-driven Dense U-Net enhances lesion identification in dental cone beam CT images [13]. Dual-attention domain-adaptive segmentation networks improve medical image segmentation [14]. The Category-Guide Attention U-Net aids in orthodontic therapy by detecting dental defects [15]. Automated dental classification of retained roots, endodontically treated teeth, and implants are achieved with contemporary CNNs [16]. A semantic graph-based technique offers precise tooth segmentation in clinical dentistry CBCT images [17]. Tooth segmentation from intra-oral scanning images is improved using the two-stream graph convolutional network [18]. Finally, ceramic X-ray tubes show promise for dental imaging due to their smaller size [19], and deep learning models automate dental forensic investigations, including sex assessment, age prediction, and tooth type determination [20].

In the study referenced by another study [21], the primary objective is enhancing dental diagnostic efficiency through integrating image analysis techniques and artificial intelligence (AI). The proposed methodology involves several preprocessing steps, such as histogram equalization and flat-field correction, which are aimed at refining pixel assignment and improving the resolution of bone structural features. These enhancements facilitate more accurate identification of teeth. Convolutional neural networks (CNNs) such as Google Net and Squeeze Net are employed to analyze lesion areas in clinical panoramas, achieving notable accuracy in evaluating dental restorations and identifying missing teeth.

Another study [22] delves into periodontitis, presenting an image processing-based diagnostic tool utilizing dental panoramic radiography (DPR) alongside landmark identification for differentiating periodontitis stages. DPR accuracy is improved by employing preprocessing techniques like Histogram Equalization and Median filter, while a novel pixel-based classification approach achieves high accuracy rates.

Moreover, another study [23] introduces a methodology for transforming web vector images into a searchable knowledge graph through incremental clustering based on similarity criteria such as form, area, color, and location. This method adheres to web standards, enabling image representation and search using Semantic Web technologies.

Additionally, a study [24] explores grain development in 3Y-TZP material utilized in dental restorations, providing insights into optimal experimental techniques. Challenges in dental imaging stemming from uneven brightness and low contrast by proposing an adaptive image enhancement technique were addressed [25].

Furthermore, several studies propose advancements in dental treatment planning and diagnosis. These include techniques for dental implant recognition [26], improving implant placement accuracy in augmented reality surgery [27], automating the location of the inferior alveolar nerve canal (IAC) during implant surgery [28], and automating 2D or 3D image registration for treatment planning [29].

Moreover, there are efforts towards identifying dental materials in X-ray images [30], automated clinical quality assessment in dental radiography [31], and diagnosing dental caries using deep learning models [32]. Segmentation methods for panoramic dental radiographs [33], utilization of fractal dimension in oral radiographs for osteoporosis screening [34], and evaluating photon-counting detector (PCD) CT for dental implants [35] are also explored.

Studies also assess the picture quality of dental CBCT systems [36], propose tooth segmentation techniques [37, 38], investigate artifact removal in CBCT images [39], and explore the use of 3D models in dental implant planning [40]. A pipeline called Image to Dental Point Cloud (I2DP) has also been introduced to enhance CBCT-based dental 3D models [41].

In another study [42], a method utilizing CNNs is developed for identifying common lesions in digital dental X-ray images. Traditional diagnostic methods are prone to errors and consume a significant amount of energy. The proposed method, employing CNNs like VGG-16 and DenseNet-121, achieves remarkable accuracy in classifying dental lesions.

Furthermore, a computerized method is presented for identifying tooth regions in panoramic dental X-ray images, enhancing CAD systems' accuracy in dental analysis [43]. Similarly, an automated process for grading dental decay in panoramic photos proposes the improvement of clinical diagnosis efficiency [44].

Moreover, a deep learning framework for identifying dental treatment modalities and accurate dentition numbering introduces aiding in treatment planning [45]. Bone remodeling around dental implants is analyzed in a study [46], emphasizing its importance in successful implant procedures.

Additionally, a study [47] explores person identification using dental X-ray images, introducing a novel attention-based network for precise human identification. Deep learning methods for dental caries identification [48], self-training-based strategies for caries segmentation, and computerized methods for tooth identification in panoramic radiographs are also discussed for different use cases [49 - 54].

These studies showcase various methodologies and advancements in dental image analysis, catering to multiple aspects of diagnosis, treatment planning, and forensic applications. An iterative statistical analysis of these models will be discussed in the subsequent section of this text.

STATISTICAL ANALYSIS

In this section, we compare the reviewed models in terms of Precision (P), Accuracy (A), Recall (R), Delay (D), and Scalability (S) Metrics. While the values of Precision, Accuracy, and Recall were directly referred from the reviewed texts, the values for delay & scalability were converted into Fuzzy Ranges of Low (L), Medium (M), High (H), and Very High (VH) based on their performance under real-time scenarios. Based on this strategy, the statistical performance of these models can be observed in Table **1** as follows:

Table 1. Statistical analysis of the reviewed models.

Method	P	A	R	D	S
PointNet	85%	90%	80%	L	H
PointNet++	88%	92%	85%	L	H

(Table 1) cont.....

Method	P	A	R	D	S
PointCNN	87%	91%	82%	M	M
RS-CNN	92%	94%	89%	L	H
3D ContextNet	90%	93%	87%	H	L
SO-Net	89%	92%	86%	L	L
3DmFV-Net	91%	93%	88%	M	L
ShellNet	92%	94%	89%	M	H
A-CNN	91%	93%	88%	H	H
PU-GCN	93%	95%	90%	H	H
ToothNet	92%	94%	89%	H	VH
DeepTeeth	90%	93%	87%	H	H
DentalCaps	91%	94%	88%	L	H
Teeth-Capsule	89%	92%	86%	M	H
Tooth-Net	91%	93%	88%	H	M
ToothNet++	92%	94%	89%	H	H
PanoramicU-Net	90%	93%	87%	H	H
DENTIST	92%	94%	89%	M	H
ToothUnet	90%	93%	87%	L	VH
DeepTooth	91%	93%	88%	H	H
ToothNetwith graph	92%	94%	89%	H	H
DentCaps	91%	93%	88%	M	H
TNetV2	90%	93%	87%	L	H
Endo-Teeth	89%	92%	86%	VH	H
DentistryGAN	91%	94%	88%	VH	VH
ToothCapsule	92%	94%	89%	H	M
Carie-Net	90%	93%	87%	H	H
ToothNet with landmark points	91%	94%	88%	H	M
DeepToothNet	92%	94%	89%	H	H
ToothCaps-Net	90%	93%	87%	VH	H
Denture-capsule	91%	93%	88%	H	M
Ortho-Capsule	92%	94%	89%	VH	H
SegToothNet	90%	93%	87%	H	VH
Teeth-CAD-Net	92%	94%	89%	H	H
RNN for Tooth detection	90%	93%	87%	VH	H
Dental-CT-Net	91%	94%	88%	M	H

(Table 1) cont.....

Method	P	A	R	D	S
Capsule with AutoSeg	92%	94%	89%	L	VH
AutoToothNet	90%	93%	87%	M	H
PointCapsNet	91%	94%	88%	H	M
ToothGAN	92%	94%	89%	H	H
Imageto Dental point cloud (I2DP)	90%	93%	87%	H	H
CNN for dental X-ray classification	91%	94%	88%	H	M
Tooth region detection in X-ray	92%	94%	89%	VH	H
Automatic caries grading	90%	93%	87%	H	M
DENTECT	92%	94%	89%	VH	M
Bone Remodeling rate analysis	91%	94%	88%	H	H
Human Identification from X-ray	90%	93%	87%	H	H
Dental Caries Detection in Panoramic images	92%	94%	89%	H	VH
Dental Caries Detection and Segmentation	91%	94%	88%	H	H
Computerized tooth classification	92%	94%	89%	H	H

As per Table **1**, PU-GCN, RS-CNN, ShellNet, ToothNet, ToothNet++, DENTIST, ToothNet with graph, ToothCapsule, and DeepToothNet have higher precision levels, thus can be used for precise evaluation of dental images. PU-GCN, RS-CNN, ShellNet, ToothNet, DentalCaps, ToothNet++, DENTIST, ToothNet with graph, and DentistryGAN have higher accuracy and thus can be used for high-performance scenarios. It was also observed that PU-GCN, RS-CNN, ShellNet, ToothNet, ToothNet++, DENTIST, ToothNet with graph, and ToothCapsule have better performance when compared with existing methods.

Regarding the delay, PointNet++, RS-CNN, SO-Net, DentalCaps, ToothUnet, TNetV2, and ToothCapsule with AutoSeg outperform others and thus can be used for high-speed scenarios. Meanwhile, the scalability of ToothNet, SegToothNet, Dental Caries Detection in Panoramic images, ToothUnet, ToothCapsule with AutoSeg, and DentistryGAN is found to be better; they can be used for highly scalable dental image processing applications. Based on these metrics, an augmented Dental Rank Metric (DRM) was evaluated *via* equation 1:

$$DRM = \frac{P + A + R}{300} + \frac{1}{D} + \frac{S}{4} \tag{1}$$

As per this evaluation, Tooth Capsule with AutoSeg, ToothUnet, RS-CNN, DentalCaps, TNetV2, PointNet++, and PointNet have superior performance, thus can be used for high precision, high accuracy, high recall, low delay, and high scalability scenarios. Researchers can also extend these models to enhance their existing deployments for efficient dental image processing under clinical use cases.

CONCLUSION & FUTURE SCOPE

This study conducted an empirical statistical analysis of dental image processing machine learning models. We considered precision, accuracy, recall, latency, and scalability levels to evaluate and compare these models. Analysis revealed machine learning models that outperformed others in specific assessment criteria. Models like PU-GCN, RS-CNN, ShellNet, ToothNet, ToothNet++, DENTIST, ToothNet with graph, ToothCapsule, and DeepToothNet improved dental image interpretation precision. High-performance models like PU-GCN, RS-CNN, ShellNet, ToothNet, DentalCaps, ToothNet++, DENTIST, ToothNet with graph, and DentistryGAN had improved accuracy. For high-consistency use cases, models with higher recall, such as PU-GCN, RS-CNN, ShellNet, ToothNet, ToothNet++, DENTIST, ToothNet with graph, and ToothCapsule, are suitable. In high-speed scenarios, PointNet++, RS-CNN, SO-Net, DentalCaps, ToothUnet, TNetV2, and ToothCapsule with AutoSeg performed best in terms of delay. Models like ToothNet, SegToothNet, Dental Caries Detection in Panoramic Pictures, ToothUnet, ToothCapsule with AutoSeg, and DentistryGAN were more scalable for dental image processing.

We proposed the Dental Rank Metric (DRM) equation (1) to standardize model ranking across assessment criteria sets. This metric revealed that ToothCapsule with AutoSeg, ToothUnet, RS-CNN, DentalCaps, TNetV2, PointNet++, and PointNet performed best and were suitable for high precision, accuracy, recall, low delay, and scalability scenarios.

FUTURE VISION & SCOPES

This study lays the groundwork for dental image processing breakthroughs using machine learning models. Many paths are worth exploring and developing, including adding dental photos from other populations and diseases to the evaluation dataset, which could generalize the findings. Applying transfer learning approaches to dental image processing may improve current models using medical imaging domain knowledge sets. Interpreting and explaining machine learning model decisions can increase their adoption in healthcare. Interpretable models can help doctors and AI systems collaborate. Clinical trials and pilot studies of the superior models in dentistry offices will reveal their efficacy and drawbacks. Like any healthcare AI application, these models' development and implementation must consider patient privacy, bias, and fairness. To accept and use the selected high-performance models, dental professionals must smoothly integrate the selected high-performance models into current clinical processes. We may find more precise and effective solutions by researching and developing new dental image processing model architectures.

In conclusion, this empirical study and the Dental Rank Metric (DRM) provide important information about the performance of dental image processing machine learning models. Dental imaging experts, scientists, and software developers can use the findings to choose the best models. By improving these models and covering future research areas for real-time scenarios, we can enhance dental image analysis' effectiveness, accuracy, and impact, helping patients and advancing dental healthcare.

REFERENCES

[1] M. Widiasri, A.Z. Arifin, N. Suciati, C. Fatichah, E.R. Astuti, R. Indraswari, R.H. Putra, and C. Za'in, "Dental-YOLO: alveolar bone and mandibular canal detection on cone beam computed tomography images for dental implant planning", *IEEE Access,* vol. 10, pp. 101483-101494, 2022. [http://dx.doi.org/10.1109/ACCESS.2022.3208350]

[2] S. Li, Y. Guo, Z. Pang, W. Song, A. Hao, B. Xia, and H. Qin, "Automatic dental plaque segmentation based on local-to-global features fused self-attention network", *IEEE J. Biomed. Health Inform.,* vol. 26, no. 5, pp. 2240-2251, 2022. [http://dx.doi.org/10.1109/JBHI.2022.3141773] [PMID: 35015655]

[3] J. Hyttinen, P. Fält, H. Jäsberg, A. Kullaa, and M. Hauta-Kasari, "Computational filters for dental and oral lesion visualization in spectral images", *IEEE Access,* vol. 9, pp. 145148-145160, 2021. [http://dx.doi.org/10.1109/ACCESS.2021.3121815]

[4] Y. Lai, F. Fan, Q. Wu, W. Ke, P. Liao, Z. Deng, H. Chen, and Y. Zhang, "LCANet: Learnable connected attention network for human identification using dental images", *IEEE Trans. Med. Imaging,* vol. 40, no. 3, pp. 905-915, 2021. [http://dx.doi.org/10.1109/TMI.2020.3041452] [PMID: 33259294]

[5] J. Shi, B. Sun, X. Ye, Z. Wang, X. Luo, J. Liu, H. Gao, and H. Li, "Semantic decomposition network with contrastive and structural constraints for dental plaque segmentation", *IEEE Trans. Med. Imaging,* vol. 42, no. 4, pp. 935-946, 2023. [http://dx.doi.org/10.1109/TMI.2022.3221529] [PMID: 36367911]

[6] D. Thapa, R. Welch, R.P. Dabas, M. Salimi, P. Tavakolian, K. Sivagurunathan, K. Ngai, B. Huang, Y. Finer, S. Abrams, A. Mandelis, and N. Tabatabaei, "Comparison of long-wave and mid-wave infrared imaging modalities for photothermal coherence tomography of human teeth", *IEEE Trans. Biomed. Eng.,* vol. 69, no. 9, pp. 2755-2766, 2022. [http://dx.doi.org/10.1109/TBME.2022.3153209] [PMID: 35196221]

[7] S.Y. Lin, and H.Y. Chang, "Tooth numbering and condition recognition on dental panoramic radiograph images using CNNs", *IEEE Access,* vol. 9, pp. 166008-166026, 2021. [http://dx.doi.org/10.1109/ACCESS.2021.3136026]

[8] M.A. Hasan, N.A. Abdullah, M.M. Rahman, M.Y.I.B. Idris, and O.F. Tawfiq, "Dental impression tray selection from maxillary arch images using multi-feature fusion and ensemble classifier", *IEEE Access,* vol. 9, pp. 30573-30586, 2021. [http://dx.doi.org/10.1109/ACCESS.2021.3059785]

[9] Y. Lang, C. Lian, D. Xiao, H. Deng, K.H. Thung, P. Yuan, J. Gateno, T. Kuang, D.M. Alfi, L. Wang, D. Shen, J.J. Xia, and P.T. Yap, "Localization of craniomaxillofacial landmarks on CBCT images using 3D mask R-CNN and local dependency learning", *IEEE Trans. Med. Imaging,* vol. 41, no. 10,

pp. 2856-2866, 2022.
[http://dx.doi.org/10.1109/TMI.2022.3174513] [PMID: 35544487]

[10] Z. Li, Z. Li, Y.V. Holamoge, W. Zaid, M.L. Osborn, S. Bhatta, A. Jeyaseelan, Y. Li, S. Yao, and J. Xu, "Mouthwash to deliver indocyanine green for near infrared dental fluorescence imaging", *IEEE J. Sel. Top. Quantum Electron.,* vol. 27, no. 4, pp. 1-8, 2021.
[http://dx.doi.org/10.1109/JSTQE.2020.2995659]

[11] A. Imak, A. Celebi, K. Siddique, M. Turkoglu, A. Sengur, and I. Salam, "Dental caries detection using score-based multi-input deep convolutional neural network", *IEEE Access,* vol. 10, pp. 18320-18329, 2022.
[http://dx.doi.org/10.1109/ACCESS.2022.3150358]

[12] S. Tian, M. Wang, F. Yuan, N. Dai, Y. Sun, W. Xie, and J. Qin, "Efficient computer-aided design of dental inlay restoration: A deep adversarial framework", *IEEE Trans. Med. Imaging,* vol. 40, no. 9, pp. 2415-2427, 2021.
[http://dx.doi.org/10.1109/TMI.2021.3077334] [PMID: 33945473]

[13] Z. Zheng, H. Yan, F.C. Setzer, K.J. Shi, M. Mupparapu, and J. Li, "Anatomically constrained deep learning for automating dental CBCT segmentation and lesion detection", *IEEE Trans. Autom. Sci. Eng.,* vol. 18, no. 2, pp. 603-614, 2021.
[http://dx.doi.org/10.1109/TASE.2020.3025871]

[14] X. Chen, T. Kuang, H. Deng, S.H. Fung, J. Gateno, J.J. Xia, and P.T. Yap, "Dual adversarial attention mechanism for unsupervised domain adaptive medical image segmentation", *IEEE Trans. Med. Imaging,* vol. 41, no. 11, pp. 3445-3453, 2022.
[http://dx.doi.org/10.1109/TMI.2022.3186698] [PMID: 35759585]

[15] X. Wang, "CGA-UNet: Category-guide attention U-net for dental abnormality detection and segmentation from dental-maxillofacial images", *IEEE Transactions on Instrumentation and Measurement,* vol. 72, pp. 1-11, 2023.
[http://dx.doi.org/10.1109/TIM.2023.3288256]

[16] S. -L. Chen, "Automated detection system based on convolution neural networks for retained root, endodontic treated teeth, and implant recognition on dental panoramic images", *IEEE Sensors Journal,* vol. 22, no. 23, pp. 23293-23306, 20222022.
[http://dx.doi.org/10.1109/JSEN.2022.3211981]

[17] P. Li, Y. Liu, Z. Cui, F. Yang, Y. Zhao, C. Lian, and C. Gao, "Semantic graph attention with explicit anatomical association modeling for tooth segmentation from CBCT images", *IEEE Trans. Med. Imaging,* vol. 41, no. 11, pp. 3116-3127, 2022.
[http://dx.doi.org/10.1109/TMI.2022.3179128] [PMID: 35635829]

[18] Y. Zhao, L. Zhang, Y. Liu, D. Meng, Z. Cui, C. Gao, X. Gao, C. Lian, and D. Shen, "Two-stream graph convolutional network for intra-Oral scanner image segmentation", *IEEE Trans. Med. Imaging,* vol. 41, no. 4, pp. 826-835, 2022.
[http://dx.doi.org/10.1109/TMI.2021.3124217] [PMID: 34714743]

[19] A.P. Gupta, J. Choi, M. Mativenga, K. Park, J. Jang, S.J. Yeo, J. Jung, M.S. Chae, B-N. Lee, J. Jang, H-L. Zhu, H-S. Mok, J.S. Ahn, and J. Ryu, "Compact X-ray tube with ceramic vacuum seal for portable and robust dental imaging", *IEEE Trans. Electron Dev.,* vol. 68, no. 9, pp. 4705-4710, 2021.
[http://dx.doi.org/10.1109/TED.2021.3097315]

[20] D. Milošević, M. Vodanović, I. Galić, and M. Subasic, "A Comprehensive exploration of neural networks for forensic analysis of adult single tooth X-ray images", *IEEE Access,* vol. 10, pp. 70980-71002, 2022.
[http://dx.doi.org/10.1109/ACCESS.2022.3187959]

[21] S.L. Chen, T-Y. Chen, Y-C. Huang, C-A. Chen, H-S. Chou, Y-Y. Huang, W-C. Lin, T-C. Li, J-J. Yuan, P.A.R. Abu, and W-Y. Chiang, "Missing teeth and restoration detection using dental panoramic radiography based on transfer learning with CNNs", *IEEE Access,* vol. 10, pp. 118654-118664, 2022.

[http://dx.doi.org/10.1109/ACCESS.2022.3220335]

[22] M.S.A. Vigil, and V.S. Bharathi, "Classification of periodontitis stages in mandibular area from dental panoramic radiograph using Adaptive Center Line-Distance Based image processing approach", *J. Ambient Intell. Humaniz. Comput.,* vol. 14, no. 7, pp. 8859-8869, 2023. [http://dx.doi.org/10.1007/s12652-021-03634-7]

[23] K. Salameh, F.E. Akoum, and J. Tekli, "Unsupervised knowledge representation of panoramic dental X-ray images using SVG image-and-object clustering", *Multimedia Syst.,* vol. 29, no. 4, pp. 2293-2322, 2023. [http://dx.doi.org/10.1007/s00530-023-01099-6]

[24] R. Shahmiri, O.C. Standard, J.N. Hart, G. Bahmanrokh, Y. Yin, A. Samiee, N. Gharagozlu, and C.C. Sorrell, "Critical effects of thermal processing conditions on grain size and microstructure of dental Y-TZP during layering and glazing", *J. Mater. Sci.,* vol. 58, no. 9, pp. 3854-3878, 2023. [http://dx.doi.org/10.1007/s10853-023-08227-7]

[25] R. Khan, S. Akbar, A. Khan, M. Marwan, Z.H. Qaisar, A. Mehmood, F. Shahid, K. Munir, and Z. Zheng, "Dental image enhancement network for early diagnosis of oral dental disease", *Sci. Rep.,* vol. 13, no. 1, p. 5312, 2023. [http://dx.doi.org/10.1038/s41598-023-30548-5] [PMID: 37002256]

[26] A. Kohlakala, J. Coetzer, J. Bertels, and D. Vandermeulen, "Deep learning-based dental implant recognition using synthetic X-ray images", *Med. Biol. Eng. Comput.,* vol. 60, no. 10, pp. 2951-2968, 2022. [http://dx.doi.org/10.1007/s11517-022-02642-9] [PMID: 35978215]

[27] L. Shrestha, A. Alsadoon, P.W.C. Prasad, N. AlSallami, and S. Haddad, "Augmented reality for dental implant surgery: enhanced ICP", *J. Supercomput.,* vol. 77, no. 2, pp. 1152-1176, 2021. [http://dx.doi.org/10.1007/s11227-020-03322-x]

[28] U.M. Pandyan, B. Arumugam, U. Gurunathan, and S.H. Kopuli Ashkar Ali, "Automatic localization of inferior alveolar nerve canal in panoramic dental images", *Signal Image Video Process.,* vol. 16, no. 5, pp. 1389-1397, 2022. [http://dx.doi.org/10.1007/s11760-021-02091-1]

[29] C. Rannulu, S. Onal, and M. Omran, "A graphical user interface for automated 2- or 3-dimensional image registration in dental treatment recovery planning: the DentIR application", *Oral Radiol.,* vol. 37, no. 1, pp. 101-108, 2021. [http://dx.doi.org/10.1007/s11282-020-00431-4] [PMID: 32189132]

[30] A. Imak, A. Çelebi, M. Türkoğlu, and A. Şengür, "Dental material detection based on faster regional convolutional neural networks and shape features", *Neural Process. Lett.,* vol. 54, no. 3, pp. 2107-2126, 2022. [http://dx.doi.org/10.1007/s11063-021-10721-5]

[31] K.S. Sankaran, "An improved multipath residual CNN-based classification approach for periapical disease prediction and diagnosis in dental radiography", *Neural Comput. Appl.,* vol. 34, no. 22, pp. 20067-20082, 2022. [http://dx.doi.org/10.1007/s00521-022-07556-z]

[32] P. Singh, and P. Sehgal, "G.V Black dental caries classification and preparation technique using optimal CNN-LSTM classifier", *Multimedia Tools Appl.,* vol. 80, no. 4, pp. 5255-5272, 2021. [http://dx.doi.org/10.1007/s11042-020-09891-6]

[33] M.H. Bozkurt, and S. Karagol, *Statistical elimination based approach to jaw and tooth separation on panoramic radiographs for dental human identification.* Multimed Tools Appl, 2023. [http://dx.doi.org/10.1007/s11042-023-14746-x]

[34] R. Franciotti, M. Moharrami, A. Quaranta, M.E. Bizzoca, A. Piattelli, G. Aprile, and V. Perrotti, "Use of fractal analysis in dental images for osteoporosis detection: a systematic review and meta-analysis", *Osteoporos. Int.,* vol. 32, no. 6, pp. 1041-1052, 2021.

[http://dx.doi.org/10.1007/s00198-021-05852-3] [PMID: 33511446]

[35] T.S. Patzer, A.S. Kunz, H. Huflage, P. Gruschwitz, P. Pannenbecker, S. Afat, J. Herrmann, B. Petritsch, T.A. Bley, and J.P. Grunz, "Combining virtual monoenergetic imaging and iterative metal artifact reduction in first-generation photon-counting computed tomography of patients with dental implants", *Eur. Radiol.,* vol. 33, no. 11, pp. 7818-7829, 2023. [http://dx.doi.org/10.1007/s00330-023-09790-y] [PMID: 37284870]

[36] T.T.N. Tran, D.S.C. Jin, K.L. Shih, M-L. Hsu, and J-C. Chen, "An image quality comparison study between homemade and commercial dental cone-beam CT systems", *J. Med. Biol. Eng.,* vol. 41, no. 6, pp. 870-880, 2021. [http://dx.doi.org/10.1007/s40846-021-00663-7]

[37] Q. Chen, Y. Zhao, Y. Liu, Y. Sun, C. Yang, P. Li, L. Zhang, and C. Gao, "MSLPNet: multi-scale location perception network for dental panoramic X-ray image segmentation", *Neural Comput. Appl.,* vol. 33, no. 16, pp. 10277-10291, 2021. [http://dx.doi.org/10.1007/s00521-021-05790-5]

[38] Y. Nishitani, R. Nakayama, D. Hayashi, A. Hizukuri, and K. Murata, "Segmentation of teeth in panoramic dental X-ray images using U-Net with a loss function weighted on the tooth edge", *Radiological Phys. Technol.,* vol. 14, no. 1, pp. 64-69, 2021. [http://dx.doi.org/10.1007/s12194-020-00603-1] [PMID: 33398671]

[39] S. Anil, S. Vellappally, A.A. Al Kheraif, D.D. Divakar, W. Said, and A.S. Hassanein, "Retracted Article: Improving the reconstruction of dental occlusion using a reconstructed-based identical matrix point technique", *J. Ambient Intell. Humaniz. Comput.,* vol. 14, no. 3, pp. 1937-1950, 2023. [http://dx.doi.org/10.1007/s12652-021-03404-5]

[40] H. Sayin, B. Aksoy, and K. Özsoy, "Optimization of CBCT data with image processing methods and production with fused deposition modeling 3D printing", *Med. Biol. Eng. Comput.,* 2023. [http://dx.doi.org/10.1007/s11517-023-02889-w] [PMID: 37505414]

[41] Z. Yu, M. Li, J. Yang, Z. Chen, H. Zhang, W. Liu, F.K. Han, and J. Liu, "A Benchmark dual-modality dental imaging dataset and a novel cognitively inspired pipeline for high-resolution dental point cloud synthesis", *Cognit. Comput.,* vol. 15, no. 6, pp. 1922-1933, 2023. [http://dx.doi.org/10.1007/s12559-023-10161-4]

[42] F. Liu, L. Gao, J. Wan, Z.L. Lyu, Y.Y. Huang, C. Liu, and M. Han, "Recognition of digital dental X-ray images using a convolutional neural network", *J. Digit. Imaging,* vol. 36, no. 1, pp. 73-79, 2022. [http://dx.doi.org/10.1007/s10278-022-00694-9] [PMID: 36109403]

[43] Y. Mima, R. Nakayama, A. Hizukuri, and K. Murata, "Tooth detection for each tooth type by application of faster R-CNNs to divided analysis areas of dental panoramic X-ray images", *Radiological Phys. Technol.,* vol. 15, no. 2, pp. 170-176, 2022. [http://dx.doi.org/10.1007/s12194-022-00659-1] [PMID: 35507126]

[44] Q. Chen, J. Huang, H. Zhu, L. Lian, K. Wei, and X. Lai, "Automatic and visualized grading of dental caries using deep learning on panoramic radiographs", *Multimedia Tools Appl.,* vol. 82, no. 15, pp. 23709-23734, 2023. [http://dx.doi.org/10.1007/s11042-022-14089-z]

[45] A.E. Yüksel, S. Gültekin, E. Simsar, Ş.D. Özdemir, M. Gündoğar, S.B. Tokgöz, and İ.E. Hamamcı, "Dental enumeration and multiple treatment detection on panoramic X-rays using deep learning", *Sci. Rep.,* vol. 11, no. 1, p. 12342, 2021. [http://dx.doi.org/10.1038/s41598-021-90386-1] [PMID: 34117279]

[46] B. Trochim, M. Borowska, and J. Szarmach, "Analysis of X-rays in bone remodelling around Active and Replace dental implants", *Signal Image Video Process.,* vol. 16, no. 1, pp. 111-118, 2022. [http://dx.doi.org/10.1007/s11760-021-01971-w]

[47] Q. Wu, F. Fan, P. Liao, Y. Lai, W. Ke, W. Du, H. Chen, Z. Deng, and Y. Zhang, "Human identification with dental panoramic images based on deep learning", *Sens. Imaging,* vol. 22, no. 1, p.

4, 2021.
[http://dx.doi.org/10.1007/s11220-020-00326-y]

[48] A. Haghanifar, M.M. Majdabadi, S. Haghanifar, Y. Choi, and S.B. Ko, "PaXNet: Tooth segmentation and dental caries detection in panoramic X-ray using ensemble transfer learning and capsule classifier", *Multimedia Tools Appl.,* vol. 82, no. 18, pp. 27659-27679, 2023.
[http://dx.doi.org/10.1007/s11042-023-14435-9]

[49] S. Nandan Mohanty, S.K. Saxena, S. Satpathy, J.M. Chatterjee, Ed., *Applications of artificial intelligence in COVID-19. Medical virology: From pathogenesis to disease control.* Springer: Singapore.
[http://dx.doi.org/10.1007/978-981-15-7317-0]

[50] S. Satpathy, S. Nandan Mohanty, J.M. Chatterjee, and A. Swain, "Comprehensive claims of AI for healthcare applications-coherence towards COVID-19", In: *Applications of artificial intelligence in COVID-19. Medical virology: From pathogenesis to disease control.,* S. Nandan Mohanty, S.K. Saxena, S. Satpathy, J.M. Chatterjee, Eds., Springer: Singapore, 2021.
[http://dx.doi.org/10.1007/978-981-15-7317-0_1]

[51] S. Mohapatra, S. Satpathy, and S.N. Mohanty, "A comparative knowledge base development for cancerous cell detection based on deep learning and fuzzy computer vision approach", *Multimedia Tools Appl.,* vol. 81, no. 17, pp. 24799-24814, 2022.
[http://dx.doi.org/10.1007/s11042-022-12824-0]

[52] S. Satpathy, M. Mangla, N. Sharma, H. Deshmukh, and S. Mohanty, "Predicting mortality rate and associated risks in COVID-19 patients", *Spatial Information Research,* vol. 29, no. 4, pp. 455-464, 2021.
[http://dx.doi.org/10.1007/s41324-021-00379-5]

[53] A. Qayyum, A. Tahir, M.A. Butt, A. Luke, H.T. Abbas, J. Qadir, K. Arshad, K. Assaleh, M.A. Imran, and Q.H. Abbasi, "Dental caries detection using a semi-supervised learning approach", *Sci. Rep.,* vol. 13, no. 1, p. 749, 2023.
[http://dx.doi.org/10.1038/s41598-023-27808-9] [PMID: 36639724]

[54] C. Muramatsu, T. Morishita, R. Takahashi, T. Hayashi, W. Nishiyama, Y. Ariji, X. Zhou, T. Hara, A. Katsumata, E. Ariji, and H. Fujita, "Tooth detection and classification on panoramic radiographs for automatic dental chart filing: improved classification by multi-sized input data", *Oral Radiol.,* vol. 37, no. 1, pp. 13-19, 2021.
[http://dx.doi.org/10.1007/s11282-019-00418-w] [PMID: 31893343]

CHAPTER 5

Diabetes Prediction using Machine Learning

N. Raghava Rao[1], **Mantri Yakshitha**[1,*] and **Kulakarni Shreya**[1]

[1] *Department of Information Technology, Institute of Aeronautical Engineering, Hyderabad, India*

Abstract: Chronic diabetes might cause a worldwide health crisis. The International Diabetes Federation estimates that there are approximately 382 million people with diabetes worldwide and this number is projected to rise to 592 million by 2035. We are creating a website that uses machine learning to help with early detection. In response, this project is developing a website based on machine learning for early detection. It combines fresh data with preexisting information using a Support Vector Machine Algorithm to create a strong prediction framework. The service provides customers with customized food regimens based on their diabetes forecast, considering their unique health requirements. Furthermore, the system makes national and local recommendations for the best diabetes management physicians. The initiative broadens its effect by finding top diagnostic experts nationwide and creating an extensive healthcare support network. To better classify diabetes, we have suggested a prediction model incorporating regular parameters such as age, insulin, BMI, glucose, and so on, coupled with a few exogenous factors known to cause diabetes. Our website encourages proactive health management by correctly predicting diabetes using machine learning. Diabetes prediction algorithms have advanced significantly due to the individualized dietary suggestions and expedited access to highly skilled healthcare providers offered to users. This initiative, which combines technological innovation with functional healthcare solutions, represents a substantial leap in diabetes prediction approaches.

Keywords: Medical Diagnosis, Diabetes, Machine Learning, Support Vector Machine, Medical Computation.

INTRODUCTION

In the field of healthcare, overcoming the difficulties presented by long-term illnesses like diabetes calls for creative solutions that make use of state-of-the-art technology. Using a cutting-edge method called "Diabetes prediction using machine learning," this initiative aims to transform diabetes treatment completely. Diabetes is a common health issue that affects people all over the globe. It is defined by the body's impaired capacity to turn food into energy owing to inad-

* **Corresponding author Mantri Yakshitha:** Department of Information Technology, Institute Of Aeronautical Engineering, Hyderabad, India; E-mail: yakshithamantri@gmail.com

Suneeta Satpathy, Sachi Nandan Mohanty & Subhendu Kumar Pani (Eds.)

equate insulin synthesis. This research has built a complex web-based platform that uses machine learning algorithms, especially the potent Support Vector Machine (SVM), to forecast diabetes outcomes with fantastic precision, realizing the need for accurate prediction tools.

The system's core functionality is its capacity to provide individualized predictions by projecting newly acquired patient data onto an existing dataset. The SVM algorithms have better categorization capabilities than other machine learning models, offering a strong basis for precise diabetes diagnosis. However, the innovation goes beyond forecasting; this all-encompassing framework creates customized eating regimens according to specific diagnoses. Additionally, it makes it easier for people to get in contact with the best medical specialists by suggesting excellent physicians for diabetes treatment in Hyderabad, India, as well as around India. This innovative and comprehensive method of diabetes management combines machine learning, individualized health advice, and healthcare referrals.

Through the seamless integration of technology and healthcare, this effort seeks to improve the diabetes management process from diagnosis to completion. The use of machine learning guarantees accurate forecasts and enables the delivery of customized health advice *via* individualized meal plans and focused suggestions from medical professionals.

LITERATURE SURVEY

Through a selection of pertinent papers, this literature review offers a thorough investigation of diabetes prediction.

V. Anuja Kumari *et al.*'s work [1] uses the Pima Indian Diabetes dataset and Support Vector Machines (SVM) to create a classifier that predicts diabetes illness with the best possible cost and performance. MATLAB R2010a is used for the implementation, yielding a 78.2% accuracy. Achieving cost-effectiveness and enhanced forecasting capacities are the main goals. By tackling the problem of diabetes prediction, demonstrating the effectiveness of SVM, and highlighting the importance of maximizing both cost and performance in disease classification models, this study advances the field.

An experiment on the use of data mining techniques and methods for diabetes diagnosis research was carried out by K. Rajesh and V. Sangeetha [2] to improve diabetes diagnosis using trustworthy prediction methodologies. They focus on data mining, the process of examining data from several perspectives to extract insightful knowledge. The main goal is finding new patterns and interpreting them to get valuable user data. In the medical area, data mining is essential for diagnosis and therapy. This study investigates various data mining approaches and strategies for effectively categorizing diabetes data. The goal is to find appropriate categorization strategies and extract valuable patterns to improve medical knowledge of diabetes.

A research paper describing a secure and privacy-preserving IoT-edge Artificial Intelligence (AI)-blockchain system for diabetes prediction was suggested by Alain Hennebelle and Leila Ismail [3]. Using blockchain to provide coherent risk factor data across hospitals, the solution ensures user data confidentiality and privacy. A comparative study was carried out using a range of medical sensors and techniques, including real-world diabetes datasets using the random forest (RF) model, logistic regression (LR), and support vector machine (SVM). According to the results, the RF model performs 4.57% more accurately than LR and SVM, even though its execution time is 2.87 times longer. The effects of feature selection on performance vary depending on the dataset.

According to Thi Ngoc Trang Tran *et al.* (2021) [4], a thorough assessment of healthcare recommender system research may be presented using their approach: Previous key summaries give information on recommended scenarios and strategies according to our study. Nutrition, medicine, health status estimates, services, and medical expert advice are included in this plan. They use real-world examples to help students grasp recommendation algorithms. Iigo Orue Saiz *et al.* [5] searched for previous research on this topic and recommendation algorithms used in the central databases throughout the last 5 years. Based on the findings, the following deduction is made: Previous studies pay less attention to the data or sample description and more attention to the recommendation system, which is usually collaborative filtering; it is unclear whether indices are utilized to compute calories or nutrients. Working with open data or globally represented data is thus essential for viability, allowing the experience to be repeated by many groups or, at the very least, comparable.

METHODOLOGY

The online service “Diabetes Prediction” employs support vector machine learning, a cutting-edge machine learning, to transform the evaluation and treatment of diabetes. The algorithm predicts diabetes risk based on user-inputted health factors and outputs findings that are simple to interpret. Customized eating programs that fit specific age ranges and tastes are created if the forecast is accurate. This guarantees efficient control of blood sugar levels [6]. A separate recommendations page provides details about leading diabetic physicians and facilities in India, with a particular emphasis on Hyderabad, for continuing medical concerns. This facilitates access to qualified providers for specialized treatment and allows people to make educated healthcare choices. The site improves diabetes prevention and care holistically by combining an easy-to-use interface, SVM's strong classification, and tailored advice [7].

A machine learning-driven system architecture is used by the diabetes prediction model and healthcare support platform to enable early diabetes identification and proactive treatment [8, 9]. In Fig. (**1**), the model uses the Support Vector Machine (SVM) method to include external indications linked to diabetes risk and standard health parameters. The Django web framework drives the backend for effective data processing, while HTML, CSS, and JavaScript were used to create the user-friendly front end, which allows for smooth data entry. When users enter essential health factors, the SVM system predicts diabetes and provides accurate results quickly [10, 11]. If the outcome is favorable, the platform automatically creates age-appropriate customized meal regimens to help individuals control their blood sugar levels. A suggestions page points readers to the best hospitals and physicians in India, especially in Hyderabad, for chronic health problems. Software needs include Django as the backend framework, HTML, CSS, and JavaScript for the front end, and Git and Visual Studio Code as necessary development tools. Low hardware requirements provide accessibility for a large number of users.

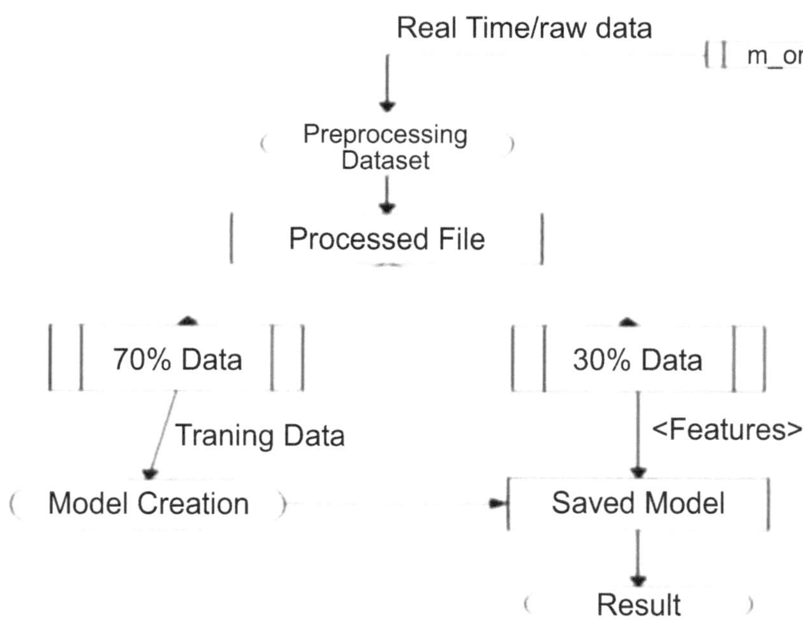

Fig. (1). System architecture.

The `pd.read_csv` function from the pandas package integrates the "diabetes.csv" dataset to acquire data for the diabetes prediction model. This makes it easier for real-world data to integrate into the machine-learning code seamlessly. Using this technique, the device gathers a wide range of data, including vital health indicators like blood pressure, age, skin thickness, insulin levels, glucose levels, body mass index, and diabetic function. The "diabetes.csv" dataset is an essential tool that gives the algorithm, as in Fig. (**2**), a strong base on which to train and test. The model effectively handles the structured data with the help of the pandas library, enabling precise predictions and tailored suggestions. This method guarantees the inclusion of pertinent and representative data for a more thorough and efficient diabetes risk assessment and improves the system's prediction skills.

```
In [4]: data=pd.read_csv(r"C:\Users\HP\Downloads\diabetes.csv")
        data
```

Out[4]

	Pregnancies	Glucose	BloodPressure	SkinThickness	Insulin	BMI	DiabetesPedigreeFunction	Age	Outcome
0	6	148	72	35	0	33.6	0.627	50	1
1	1	85	66	29	0	26.6	0.351	31	0
2	8	183	64	0	0	23.3	0.672	32	1
3	1	89	66	23	94	28.1	0.167	21	0
4	0	137	40	35	168	43.1	2.288	33	1
...	...	...	...	...	...	...	...	...	...
763	10	101	76	48	180	32.9	0.171	63	0
764	2	122	70	27	0	36.8	0.340	27	0
765	5	121	72	23	112	26.2	0.245	30	0

Fig. (2). Data collection.

The "diabetes.csv" dataset's missing values must be carefully handled throughout the data preparation step of the diabetes prediction procedure. Using imputation methods, the preprocessing (Fig. **3**) fills in the blanks for important health metrics like insulin, body mass index, and glucose levels. Detecting and managing missing values is made possible by the pandas package, which guarantees a complete and trustworthy dataset for machine learning. This thorough preprocessing guarantees reliable and impartial predictions and improves the model's accuracy. The diabetes prediction system may provide customers with more reliable and clinically relevant risk assessments by methodically correcting missing data.

The "Outcome" column, which usually represents the target variable showing the existence of diabetes, is excluded from the dataset 'X' to prepare it for the diabetes prediction study. This distinction is essential for training and assessing the prediction model. Fig. (**4**) shows an 80-20 split used to divide the dataset into training and testing sets, with 80% of the data allocated to the training set where the model is developed. The machine learning algorithm discovers patterns and correlations that improve its predictive power within this subset. The testing set, which consists of the remaining 20% of the data, is then used to assess how well the model generalizes. This method aids in evaluating the model's performance on unknown data and its capacity to provide precise predictions outside the training set. The 80/20 split promotes a strong and trustworthy diabetes prediction model by ensuring a fair trade-off between training and assessment.

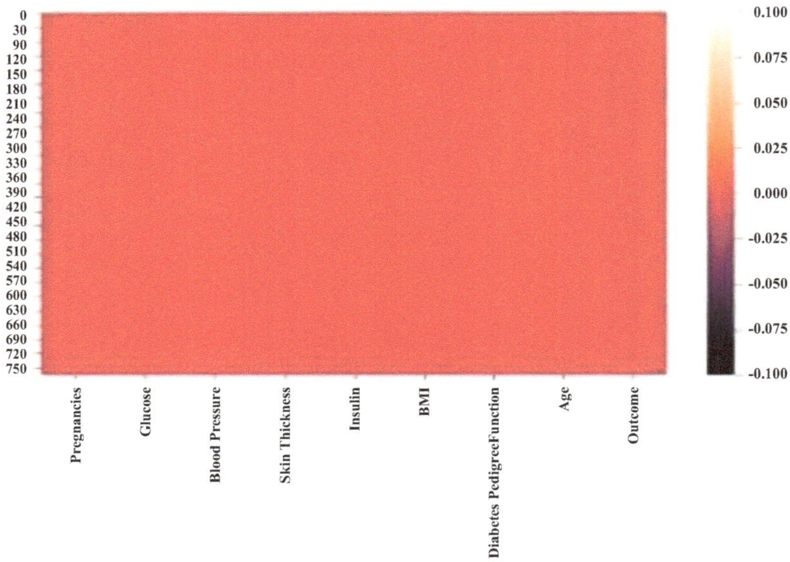

Fig. (3). Preprocessing.

Algorithms

SVM: The Support Vector Machine (SVM) is a popular supervised learning approach. Classification and regression are possible. Classification problems are the primary application of Machine Learning.

The SVM method (Fig. **5**) finds the optimal decision boundary or line to classify an n-dimensional space. This will make grouping new data points easier in the future. This best option limit is a hyperplane. SVM selects extreme vectors to form the hyperplane. Since only a few key vectors are used in this process, the method is known as a Support Vector Machine. The image below displays two groups divided by a chosen boundary or hyperplane:

	Pregnancies	Glucose	Blood Pressure	Skin Thickness	Insulin	BMI	Diabetes Pedigree Function	Age
326	1	122	64	32	156	35.1	0.692	30
142	2	108	52	26	63	32.5	0.318	22
732	2	174	88	37	120	44.5	0.646	24
581	6	109	60	27	0	25.0	0.206	27
212	7	179	99	31	0	34.2	0.164	60
...	...	...	...	...	...	...	...	...
703	2	129	0	0	0	38.5	0.304	41
180	6	87	80	0	0	23.2	0.084	32
330	8	118	72	19	0	23.1	1,476	46
549	4	189	110	31	0	28.5	0.680	37
474	4	114	64	0	0	28.9	0.126	24

614 rows x 8 columns

Fig. (4). Train & Test split.

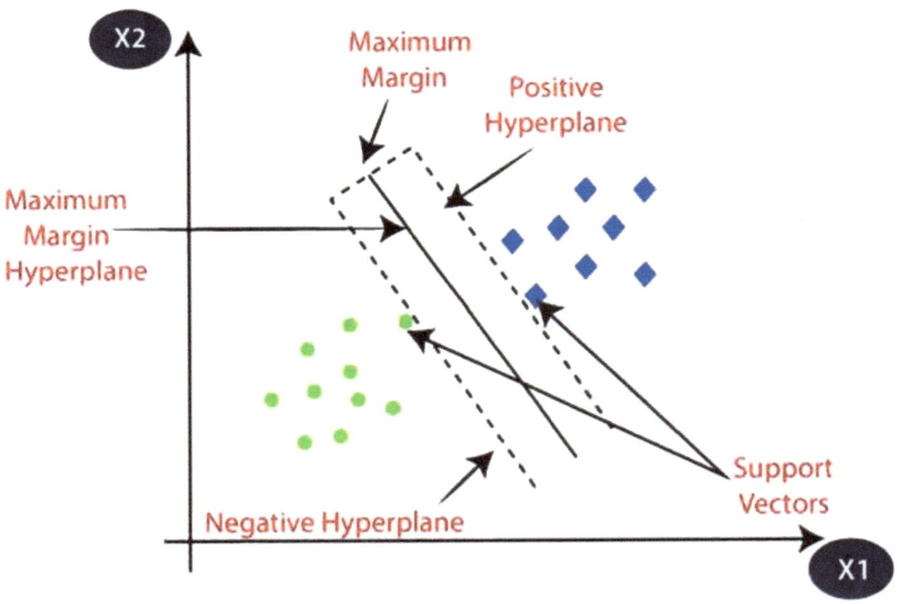

Fig. (5). SVM algorithm.

TYPES OF SVM

SVM can be of Two Types:

Linear SVM: Linear SVM defines linearly separable data as data that can be divided using the classification method used for linearly separable data.

Non-linear SVM: Non-linearly split data using Non-Linear SVM. This indicates that a non-linear SVM classifier is used to classify data that can't be grouped using a straight line.

Some of the most commonly used modules are User Registration and Authentication. The "Integrated Predictive Framework for Diabetes Prediction" User Registration and Authentication feature (Fig. **6**) is key to ensuring that interactions with the platform are safe and tailored to each person. This section is all about getting new users and running and setting up ways to verify their identity.

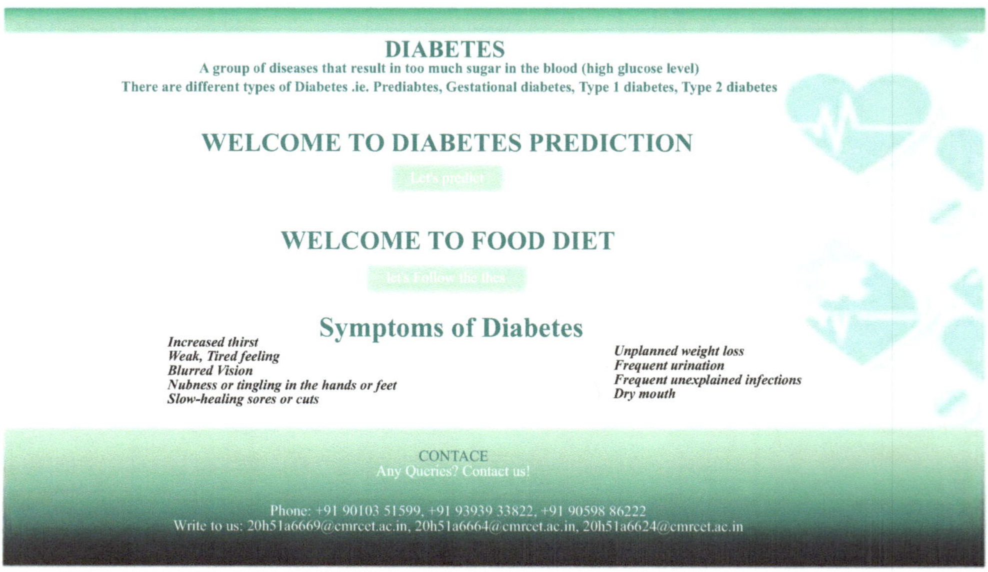

Fig. (6). Home page of myFolio.

The Input and Mapping module is an integral part of the framework because it ensures that the machine learning algorithms have access to a well-structured and complete dataset that can be used to analyze and predict diabetes state.

Machine learning methods, such as Support Vector Machine (SVM), are used to determine from user input whether they have diabetes or not, and whether to display the result as good or bad. The Diet Plan Recommendation section adds to the whole prediction framework's usefulness, turning it into a complete solution that goes beyond analysis and gives people specific, practical health advice. This enlarged section fits with the framework's primary goal: to improve methods for diagnosing diabetes and give people a complete, user-centered healthcare answer.

EXPERIMENTAL RESULTS

Accuracy: A test's accuracy is defined as its ability to recognize debilitated and solid examples precisely. To quantify a test's exactness, we should register the negligible part of genuine positive and genuine negative outcomes in thoroughly examined cases. This might be communicated numerically, shown in Table **1**, and results are shown in Figs. (**7** - **14**).

Table 1. Accuracy result.

S. No	Algorithm	Accuracy
1	SVM	0.7272

$$\text{Accuracy} = \frac{TP + TN}{TP + TN + FP + FN}$$

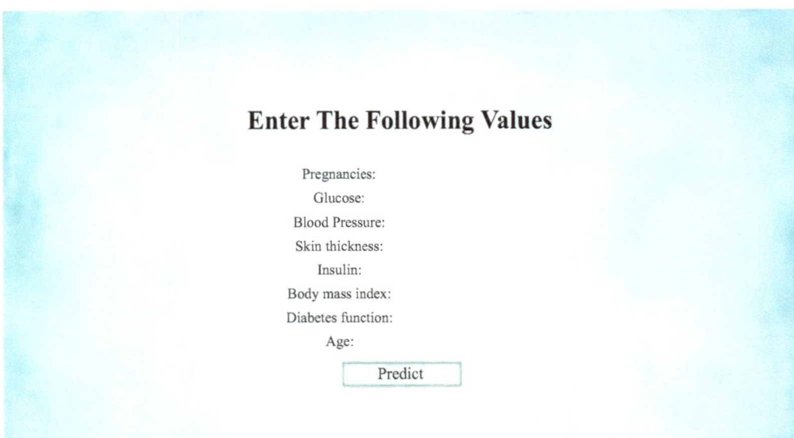

Fig. (7). Prediction Parameters.

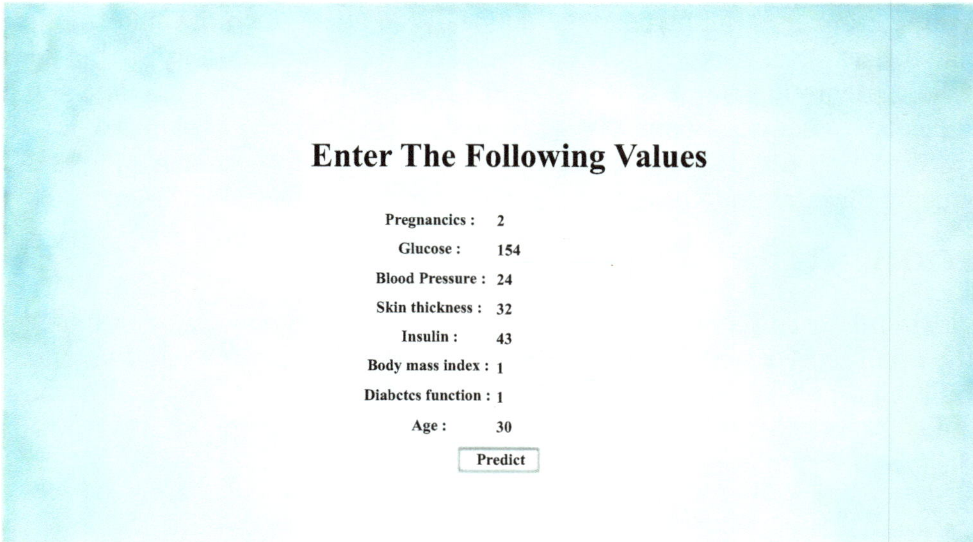

Fig. (8). Enter the information.

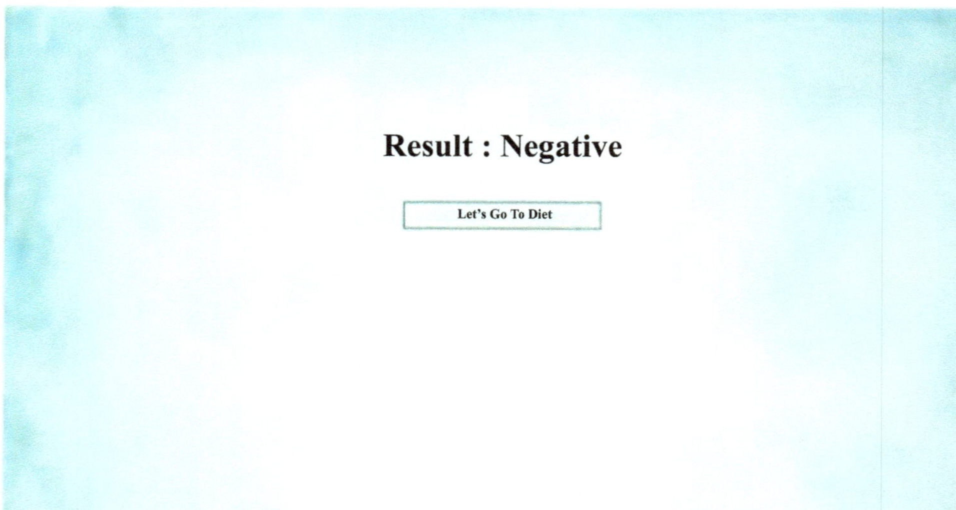

Fig. (9). Predict result.

Fig. (10). Diet plan for different age groups.

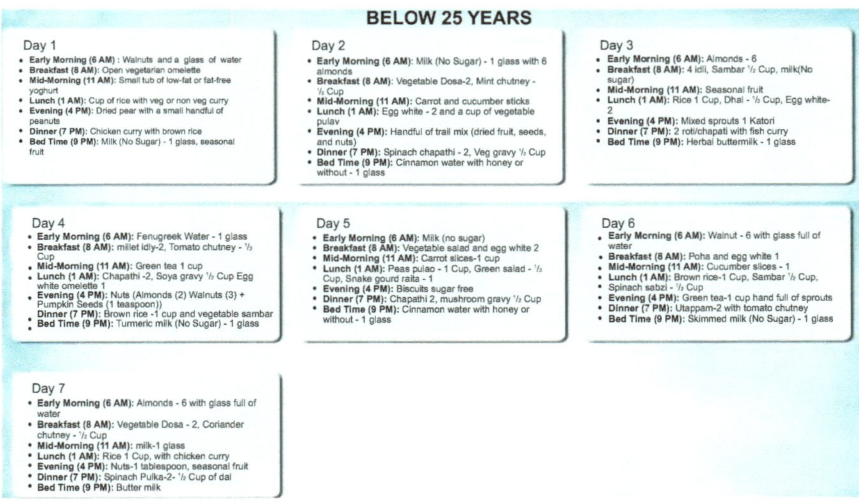

Fig. (11). Diet plan for below 25 years.

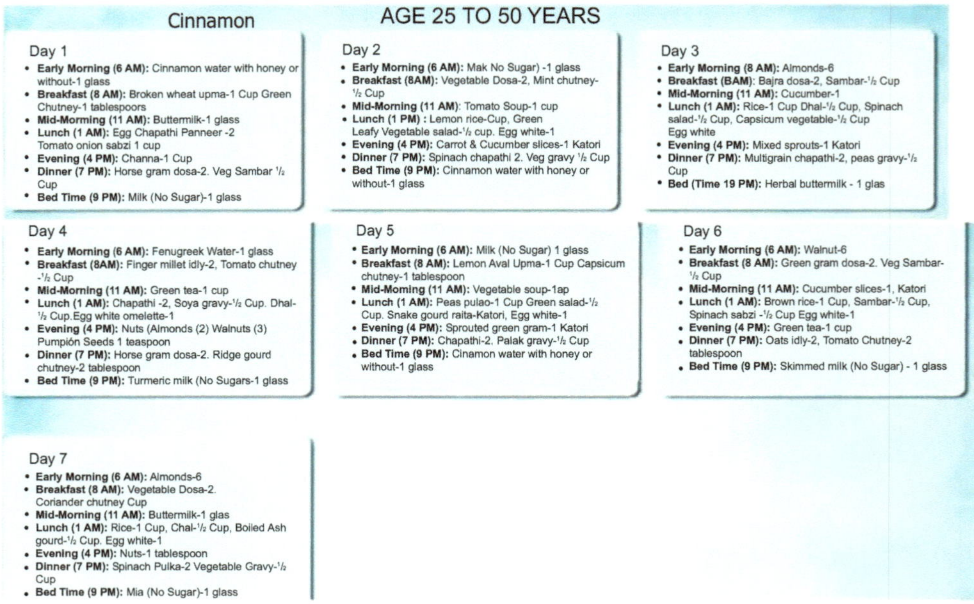

Fig. (12). Diet plan for age 25 to 50 years.

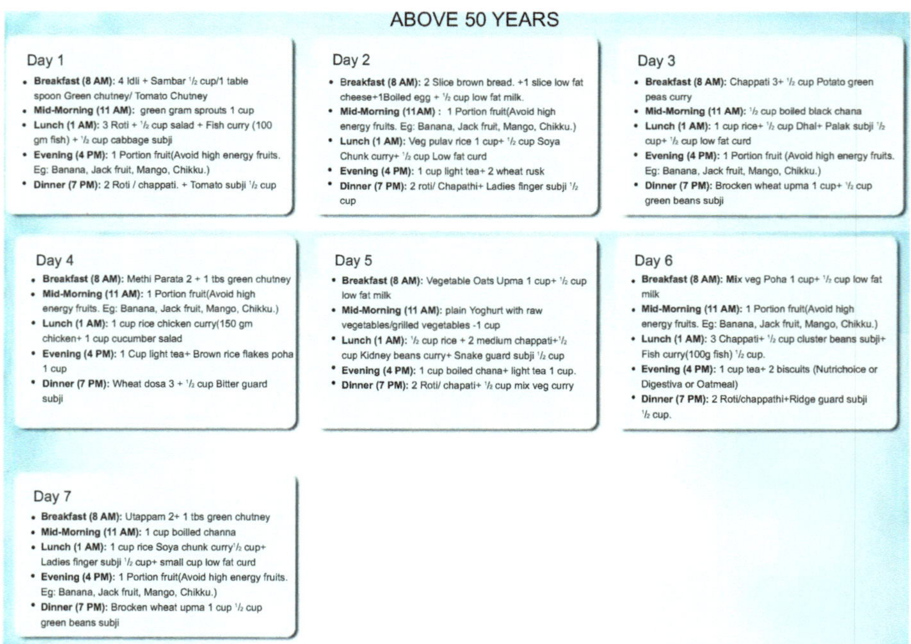

Fig. (13). Diet plan for above 50 years.

Fig. (14). Recommended doctors & hospitals.

CONCLUSION

In conclusion, the "Diabetes Prediction" website is a cutting-edge way to improve how diabetes is diagnosed and managed. The site makes it easy for people to determine their diabetes risk by incorporating advanced machine learning, especially the support vector machine method. The customizable nature of the system, with food plans made just for each user based on their age and tastes, makes it more helpful in giving people control over their blood sugar levels. The fact that there is a page with tips for healthcare tools that focuses on the best hospitals and doctors in Hyderabad shows that the site cares about its users' health in all areas. This thorough method gives accurate and easy-to-understand risk assessments. It ensures that users get valuable and tailored advice, reinforcing the portal's role as an essential tool for preventing and managing diabetes.

FUTURE SCOPE

The "Diabetes Prediction" method will continue to grow and get better in the future. Adding more machine learning algorithms, keeping the software up to date with new medical research, and working with health institutions to share data in real-time can all help improve the accuracy of predictions even more. Adding smart tech for tracking health in real-time and making the platform work with more health problems can make the system a flexible and all-around tool for proactive healthcare beyond diabetes.

REFERENCES

[1] KumariV.A.Chithra.R."Classification of diabetes disease using support vector machine", *www.ijera.com.,* pp. 1797-1801.pp.1797-1801.www.ijera.com

[2] K. Rajesh, and V. Sangeetha, "Application of data mining methods and techniques for diabetes diagnosis", *International Journal of Engineering and Innovative Technology (IJEIT).,* vol. 2, no. 3, .

[3] https://www.sciencedirect.com/science/article/pii/S2001037023

[4] T.N.T. Tran, A. Felfernig, C. Trattner, and A. Holzinger, "Recommender systems in the healthcare domain: state-of-the-art and research issues", *J. Intell. Inf. Syst.,* vol. 57, no. 1, pp. 171-201, 2021. [http://dx.doi.org/10.1007/s10844-020-00633-6]

[5] I. Orue-Saiz, M. Kazarez, and A. Mendez-Zorrilla, "Systematic review of nutritional recommendation systems", *Appl. Sci. (Basel),* vol. 11, no. 24, p. 12069, 2021. [http://dx.doi.org/10.3390/app112412069]

[6] R. Yera Toledo, A.A. Alzahrani, and L. Martinez, "A food recommender system considering nutritional information and uer preferences", *IEEE Access,* vol. 7, pp. 96695-96711, 2019. [http://dx.doi.org/10.1109/ACCESS.2019.2929413]

[7] W.D. Klinkenberg, S. Boslaugh, B.M. Waterman, K. Otani, J.M. Inguanzo, J.C. Gnida, and W.C. Dunagan, "Inpatients' willingness to recommend", *Health Care Manage. Rev.,* vol. 36, no. 4, pp. 349-358, 2011. [http://dx.doi.org/10.1097/HMR.0b013e3182104e4a] [PMID: 21685795]

[8] S. Nandan Mohanty, S.K. Saxena, S. Satpathy, J.M. Chatterjee, Ed., *Applications of artificial intelligence in COVID-19. medical virology: From pathogenesis to disease control.* Springer: Singapore.
[http://dx.doi.org/10.1007/978-981-15-7317-0]

[9] S. Satpathy, S. Nandan Mohanty, J.M. Chatterjee, and A. Swain, "Comprehensive Claims of AI for Healthcare Applications-Coherence Towards COVID-19", In: *Applications of Artificial Intelligence in COVID-19. Medical Virology: From Pathogenesis to Disease Control.,* S. Nandan Mohanty, S.K. Saxena, S. Satpathy, J.M. Chatterjee, Eds., Springer: Singapore, 2021.
[http://dx.doi.org/10.1007/978-981-15-7317-0_1]

[10] S. Mohapatra, S. Satpathy, and S.N. Mohanty, "A comparative knowledge base development for cancerous cell detection based on deep learning and fuzzy computer vision approach", *Multimedia Tools Appl.,* vol. 81, no. 17, pp. 24799-24814, 2022.
[http://dx.doi.org/10.1007/s11042-022-12824-0]

[11] S. Satpathy, M. Mangla, N. Sharma, H. Deshmukh, and S. Mohanty, "Predicting mortality rate and associated risks in COVID-19 patients", *Spatial Information Research,* vol. 29, no. 4, pp. 455-464, 2021.
[http://dx.doi.org/10.1007/s41324-021-00379-5]

CHAPTER 6

Prediction of Diabetic Retinopathy using Machine Learning

Vrishikaa Krishna Moorthy[1], **Moorthi M.**[1] and **Kapil N.**[1,*]

[1] *Saveetha Engineering College, Anna University, Chennai, Tamil Nadu, India*

Abstract: Diabetic Retinopathy (DR), which is characterized by abnormalities of the retina and may trigger blindness if left untreated, is a disorder that is commonly brought on by diabetes mellitus. Convolutional Neural Networks (CNNs), in particular, are a machine learning technique that researchers have developed to automate the diagnostic process in order to address this problem. Using annotated fundoscopic pictures, a CNN is trained in this manner to get a thorough understanding of retinal illnesses. The program attempts to precisely detect and categorize symptoms of diabetic retinopathy in patients by employing deep learning techniques. To improve efficiency and robustness, a deep learning model with a hybrid approach that incorporates components from both the ResNet50 and Xception architectures is specifically proposed. This study promotes the use of an automated diagnostic method for diabetic retinopathy, which may facilitate the screening procedure, increase the rate of early identification, and ultimately benefit diabetic patients. Through the utilization of machine learning and deep neural networks, healthcare professionals might possibly enhance their ability to promptly diagnose and treat diabetic retinopathy, hence reducing the likelihood of visual loss in people with diabetes.

Keywords: Automated diagnosis, Convolutional Neural Networks, Diabetic retinopathy, Fundoscopic images, Healthcare, Machine learning, ResNet50, Xception.

INTRODUCTION

Ocular problems are becoming more commonplace worldwide, and diabetes-related retinopathy has become recognized as a particularly serious problem for the medical industry. Due to its complexity and range of phases, illnesses like diabetic retinopathy pose an urgent danger to eyesight and all its forms of ocular health. There is a pressing need for advanced diagnostic methods that can accurately determine the progression of retinopathy, which stands as the primary

* **Corresponding author Kapil N.:** Saveetha Engineering College, Anna University, Chennai, Tamil Nadu, India; E-mail: kapilnandakumar290@gmail.com

Suneeta Satpathy, Sachi Nandan Mohanty & Subhendu Kumar Pani (Eds.)

cause of visual impairment and blindness among the working-age population, in a timely and efficient manner. Innovative solutions are required to address the complexities of diabetic retinopathy. Current research advocates for the development of an automated examination system as a potential future advancement. It emphasizes the urgent need for advanced diagnostic methods that can ascertain the precise course of retinopathy, the primary cause of visual impairment and blindness in the working-age population, in a timely and efficient manner. Diabetic retinopathy is a disorder that requires innovative solutions. Standard diagnostic techniques reliant on Convolutional Neural Networks (CNNs), which have been specifically designed and trained on enormous data sets containing labelled fundoscopic pictures, may help to fulfil the increasing need for a precise and prompt diagnosis of diabetic retinopathy. Interobserver variability limits human understanding. This study is being performed to provide medical professionals with a valuable tool for promptly and reliably diagnosing the various phases of diabetic retinopathy.

The images below of Diabetic Retinopathy showcase the different stages of Diabetic Retinopathy. The images in Fig. (**1**) are extracted from real-time patients from our hospital. This represents the collective collage of DR in real-time patients.

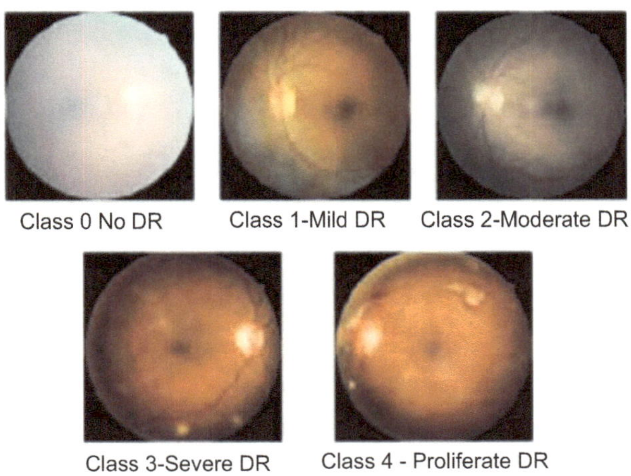

Fig. (1). Stages of Diabetic Retinopathy. It affects the blood vessels within the retina, leading to harmful leakage of substances and fluids into the surrounding tissue and is further classified into different stages.

PROBLEM STATEMENT

The current method for diagnosing diabetic retinopathy entirely depends on medical experts manually reviewing fundoscopic pictures. The identification of this illness requires clinically intrusive diagnostic methods. The interpretation process involves human judgment and tiredness, which can lead to human intervention and exhaustion in an overworked workplace.

This clinical technique, with possibilities of errors during work conditions, hardware malfunction and human error possibilities, must be replaced with an advanced automated system using advanced machine learning hybrid mode to seek accurate detection and classification of stages in diabetic retinopathy with enhanced speed, precision and scalability of the diagnosis process, that ultimately improve the patient healthcare outman with earlier detection and immediate treatment.

LITERATURE SURVEY

A study uses a deep learning method for the early diagnosis of diabetic retinopathy, namely DenseNet-169. Fundus pictures are categorized according to severity levels, with No DR, Mild, Moderate, Severe, and Proliferative DR being classified [1]. The Diabetic Retinopathy Detection 2015 and Aptos 2019 Blindness Detection datasets were obtained from Kaggle and utilized in this investigation.

Another study assessed the performance of machine learning algorithms in systems for the detection and classification of diabetic retinopathy (DR) [2]. Thermal and retinal fundus pictures from many publicly accessible archives are used to train and evaluate these systems. The study shows how well these algorithms identify warning signals and classify different degrees of DR severity.

A research study uses a deep learning model to classify fundus pictures of diabetic retinopathy (DR) according to severity levels [3]. With the help of preprocessing, segmentation, and classification stages, the proposed approach presents a built-in detection and classification model for DR fundus pictures. This paper contributes to the advancement of DR image analysis methodologies, emphasizing the significance of the proposed SDL model in achieving improved classification accuracy.

Another research presents a new method for grading diabetic retinopathy (DR): a Multichannel-based Generative Adversarial Network (MGAN) with semi-supervised learning [4]. The purpose of the MGAN is to produce a number of sub-fundus pictures that correspond to distinct scattering DR characteristics. The semi-supervised MGAN can identify subtle lesion characteristics using high-resolution fundus pictures without compression by reducing the dependence on labelled data.

The Random Forest classifier, Decision Tree classifier, Adaboost classifier, K-Nearest Neighbour classifier, and Logistic Regression classifier are among the Machine Learning (ML) techniques that are incorporated into the suggested model in a study [5]. The model is then applied to a dataset of retinal degeneration. The min-max normalization approach is used in the first step to normalize the diabetic retinopathy dataset. This research contributes to the ongoing efforts in refining machine learning methodologies for healthcare applications, particularly in the context of diabetic retinopathy diagnosis, while addressing the imperative of safeguarding sensitive health records.

In the study, "Automated Smartphone-Based System for Diagnosis of Diabetic Retinopathy," the proposed method is integrated into a smartphone application, offering an efficient and automatic diagnostic system for individuals with diabetes, particularly those residing in remote areas. This innovative approach aims to democratize access to diabetic retinopathy diagnosis by leveraging the ubiquity of smartphones for prompt and accessible healthcare solutions.

Diabetic retinopathy detection using deep convolutional neural networks [7]. This research describes the design and implementation of deep convolutional neural networks accelerated by GPUs to automatically diagnose and categorize high-resolution retinal pictures into five disease phases according to severity.

The technique for fatigue recognition that is presented consists of bringing in several facial cues, such as the length of time that the eyes close, the head nod, and the yawn, into a convolutional two-stream network called the gamma fatigue detection network [8]. The first part of the algorithm locates the driver's lips and eyes using a multi-task cascaded CNN and detects alterations that indicate weariness. On the basis of the detection outcomes, radiographs are then divided into two groups.

The use of Convolutional Neural Networks (CNNs) on colour fundus pictures to recognize diabetic retinopathy staging is shown in the article "Automated detection of diabetic retinopathy using deep learning" [9]. The performance of the built network models according to test metrics is in close agreement with baseline results from the literature.

"Automated identification of diabetic retinopathy using deep learning:" the main goal of this research, as stated in this report, was to develop robust diagnostic tools for the automated detection of Diabetic Retinopathy (DR) [10]. There is potential to greatly reduce the incidence of vision loss by setting in place an automated system that can detect eyes with DR and subsequently send them to an ophthalmologist for a comprehensive examination and treatment [11 - 14].

PROPOSED METHODOLOGY

Data Acquisition

The initial phase of the project involves gathering a variety of datasets containing analyzed fundoscopic visuals that reflect various phases of diabetic retinopathy. This process of collecting the information could entail working with healthcare institutions or using medical data that was originally obtained from reputable sources. Standard analysis and preprocessing methods will be used to standardize the photos, assuring uniformity in terms of quality, resolution, and colour. Furthermore, data augmentation methods will improve the model's ability to generate output that is useful for classification.

Model Architecture Selection and Fusion

Analysing data for analysis and choosing suitable deep learning architectures, such as Xception and ResNet50, are the following stages. By combining these designs, an arrangement will be created that will take advantage of each one's unique abilities to identify complicated trends in ocular pictures. By combining these structures, a comprehensive model for categorizing the various phases of diabetic retinopathy will be produced.

Testing and Training Data-sets

The carefully chosen dataset is used to extensively train a hybrid deep learning model. To maximize the model's performance, many parameters are adjusted throughout training cycles, including learning rate, batch size, and optimizer settings. The team uses methods of regularization and stochastic gradient descent to keep the model from overfitting and to improve its generalization abilities. Additionally, multiple training-validation repetitions are carried out in order to track convergence and assess performance indicators.

Understanding Evaluation

After the first training phase, a validation procedure will assess the model's performance on dataset sections that haven't been seen before. The model's ability to categorize distinct phases of diabetic retinopathy will be assessed through a variety of broad measures, such as accuracy, precision, recall, and F1 score. We will use confusion matrix analysis and cross-validation approaches to evaluate the model's resilience and find any possible shortcomings.

Integration and Deployment

After the model performs sufficiently well, Streamlit will be the tool used for integrating it into an intuitive online application. This stage involves developing a user-friendly interface that makes it simple for healthcare professionals to fill in retinal scans for precise and on-time forecasts. In order to ensure application stability and compatibility across platforms, rigorous testing will persist. This system presents an automated approach for diagnosing diabetic retinopathy by combining the structures of ResNet50 and Xception over a hybrid system. The primary aim is to optimize the diagnostic process through the utilization of machine learning techniques. Specifically, the model is trained on a labelled dataset consisting of fundoscopic pictures.

The flow chart in Fig. (**2**) explains the schematic procedure of the project flow of context that has been followed for the execution of the outcomes of the project.

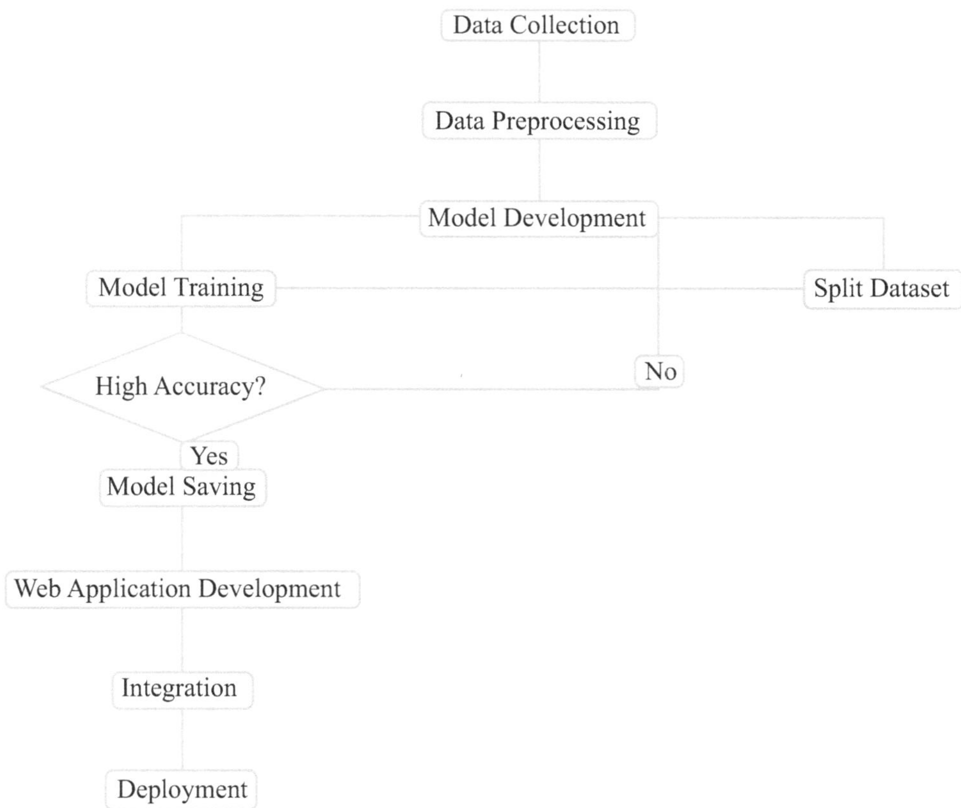

Fig. (2). The flowchart gives a detailed understanding of the process involved in different stages of the proposed method in a diagram format.

As a result, the model may categorize retinas into three stages: proliferative, mild, & no DR. After that, the improved model is stored for real-world application.

The following is a list of the planned work's benefits:

The implementation of automation in the diagnostic process brings about increased efficiency and consistency in the determination of medical conditions. The integration of advanced machine learning models contributes to heightened accuracy, marking a significant improvement in diagnostic capabilities. The rapid processing capabilities of these models enable timely interventions, crucial for addressing health issues promptly. Additionally, the system exhibits enhanced scalability, allowing it to effectively handle large datasets, a critical aspect in modern healthcare. The reduced dependency on individual expertise ensures a more standardized approach to diagnoses.

The incorporation of a user-friendly interface promotes increased accessibility, making the system more user-friendly for healthcare professionals. Furthermore, continuous improvement is facilitated through collaboration with healthcare experts, fostering a dynamic system that evolves with the latest medical insights.

PROPOSED ARCHITECTURE

Cnn

The Convolutional neural networks, or CNNs, have become an extremely useful tool for a variety of applications, most notably computer vision tasks. This paper provides a concise overview of CNNs, focusing on their architecture, key components, and applications. The beginning with the basic concept of convolutional layers, the paper explores the role of pooling layers, activation functions, and fully connected layers in CNNs. In addition, it discusses the training process, including backpropagation and optimization techniques commonly used with CNNs.

Furthermore, the paper highlights notable applications of CNNs, such as image classification, object detection, and image segmentation. Overall, this paper serves as a primer on CNNs, offering insights into their structure, functionality, and practical utility.

Deep-learning Frameworks

Foundational tools for constructing extensive neural network planning are deep learning frameworks. These frameworks, such as TensorFlow and PyTorch, provide a comprehensive set of libraries and functions, facilitating the design, training, and deployment of deep learning models. TensorFlow, developed by Google, and PyTorch, supported by Facebook, have emerged as two prominent choices, each offering unique advantages. However, PyTorch is praised for having a dynamic computational network that makes it easier to use for experiments and research.

The selection of one of these frameworks is contingent upon the particular needs and preferences of the project. Overall, these deep learning frameworks play a pivotal role in advancing the field, empowering researchers and practitioners to harness the potential of deep neural networks effectively.

Keras

The use of the Keras deep learning framework is a noteworthy advancement in the domain of artificial intelligence and neural network research. Deep learning model building and training get easier with the help of Keras, an open-source high-level neural network API. Because of its simplicity and modular design, it is very useful for novices and seasoned professionals alike.

Popular backend engines like TensorFlow, Theano, and Microsoft Cognitive Toolkit are easily integrated with Keras, offering flexibility and efficiency in model construction. The framework's simplicity and abstraction enhance the speed and efficiency of prototyping, enabling researchers and developers to rapidly experiment with various architectures. In essence, Keras stands as a valuable tool in the arsenal of deep learning frameworks, fostering innovation and accelerating advancements in the field.

Streamlit

Streamlit is a robust and user-friendly Python framework developed for constructing interactive and customized online apps for data science and machine learning. It makes the process of converting data scripts into publicly accessible web applications easier, making it easier for developers and data scientists to present their work. Streamlit's intuitive and declarative syntax enables rapid prototyping, making it accessible for users with various levels of programming expertise.

With its built-in features for creating charts, tables, and interactive widgets, Streamlit streamlines the development of data-driven applications without the need for extensive web development knowledge. What makes it even more versatile is that it works with well-known data science libraries like Matplotlib and Pandas. In the realm of scientific and engineering applications, Streamlit emerges as a valuable tool for creating engaging and interactive data-driven web interfaces.

RESULTS AND DISCUSSION

Minimum Data-set Output

The Python program's training process commenced using the specified algorithm, with local training and testing facilitated through the Jupyter console. In the initial phase, a randomized dataset comprising 1339 input images was employed, encompassing all stages of diabetic retinopathy. The training process spanned 100 epochs, signifying complete iterations through the training dataset during model training. The initial dataset's accuracy values exhibit a range from 0.452 to a maximum of 0.768. For instance, if we consider the average accuracy derived from seven input measurements, it underscores the variability and performance characteristics of the model.

The average of the output obtained for the randomized seven inputs is:

$$0.4552+0.5463+0.7522+0.7582+0.6970+0.5627+0.7119 =4.4865$$

$$4.4865/7 = 0.6409285714$$

Upon converting the obtained average accuracy to a percentage, we observe an approximate accuracy of 64.09% during the initial stage of our study. The output of the executed program is given below.

The procedure in Fig. (**3**) explains that the minimum local training and testing datasets are required to generate the minimum data set output. Therefore, the initial set of collected fundoscopic images has been trained.

The procedure in Fig. (**4**) explains the start of the training procedure to generate the minimum data set output for the prediction of diabetic retinopathy using machine learning. Therefore, the initial set of collected fundoscopic images has been trained.

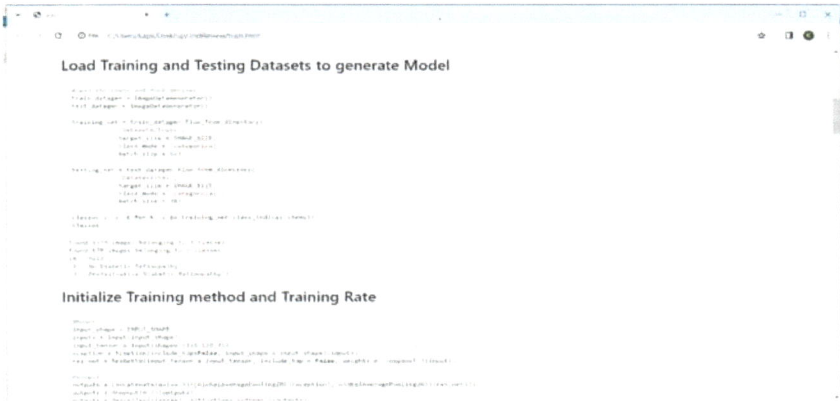

Fig. (3). Minimum Data-set for Testing and Training is an initial stage of the programmed algorithm, with different conceptual frameworks involved in the form of a Python program.

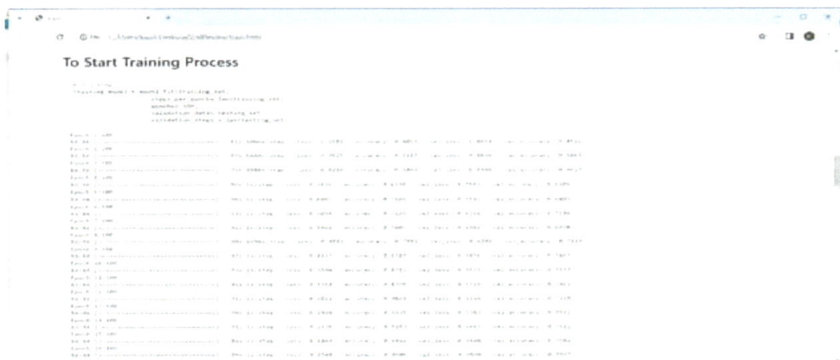

Fig. (4). Minimum Data-set for Testing and Training is an initial stage of the programmed algorithm, with different conceptual frameworks involved in the form of a Python program.

The procedure in Figs. (**5** - **7**) explains the start of the training procedure to generate the minimum data set output for the prediction of diabetic retinopathy using machine learning. The randomized input data of medical images is given below.

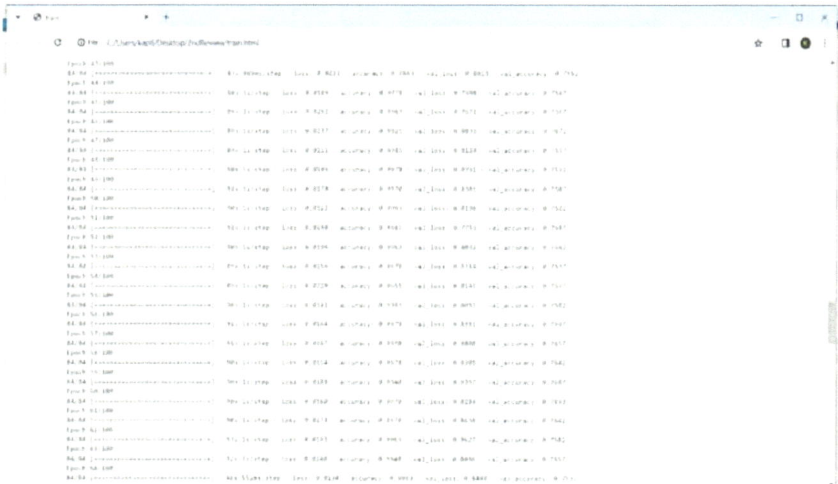

Fig. (5). Minimum Data-set for the start of the training program for the randomized input data of medical images.

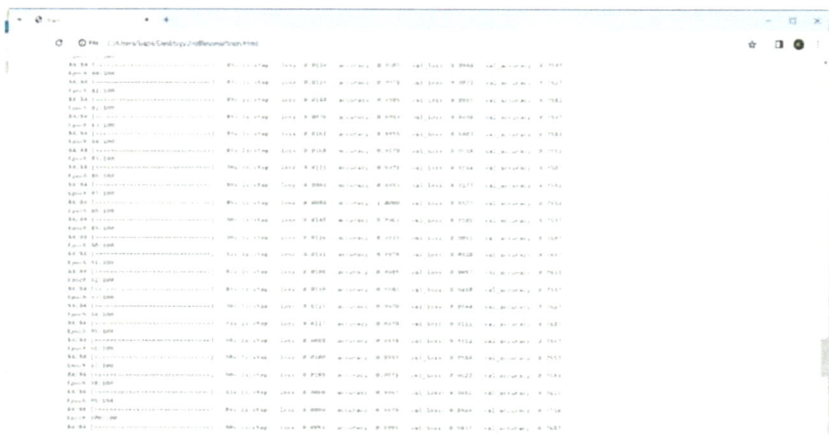

Fig. (6). Minimum Data-set Trained Output with various factors such as accuracy, precision and value loss.

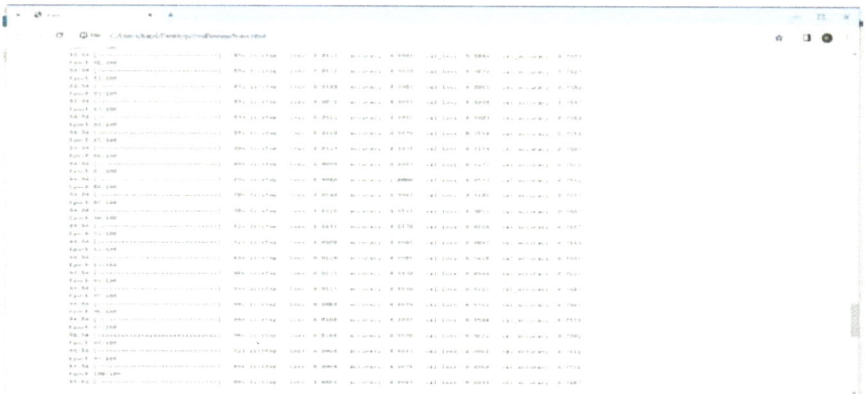

Fig. (7). Minimum Data-set Trained Output with the factors stated above.

MAXIMUM DATA-SET OUTPUT

The training of the Python program initiated with a specified algorithm, utilizing local training and testing via the Jupyter console. In the subsequent phase, a randomized dataset containing 3171 input images, representing all stages of diabetic retinopathy, was employed. The training process extended over 25 epochs, corresponding to full iterations through the training dataset during model training.

The accuracy values in the final dataset span a range from 0.4467 to a maximum of 1.0000. For example, when calculating the average accuracy based on seven input measurements, it highlights the model's variability and performance characteristics.

The average of the output obtained for the randomized seven inputs is:

0.9201+0.9349+0.9970+0.9852+0.9142+0.8018+1.0000=6.5524

6.5524/7= 0.9360571428

The procedure in Fig. (**8**) explains that the maximum local training and testing datasets are used to generate the maximum dataset output. Therefore, the initial set of collected fundoscopic images served as the foundation for the model's training.

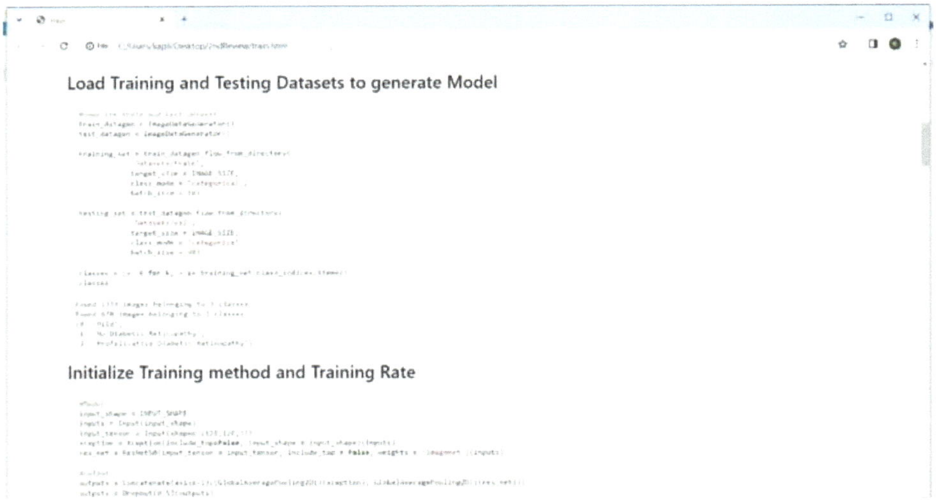

Fig. (8). Maximum Data-set for the start of the training program for the randomized input data of medical images.

Fig. (**9**) explains the start of the training procedure to generate the maximum data set output for the prediction of diabetic retinopathy using machine learning. The randomized input data of medical images are given below.

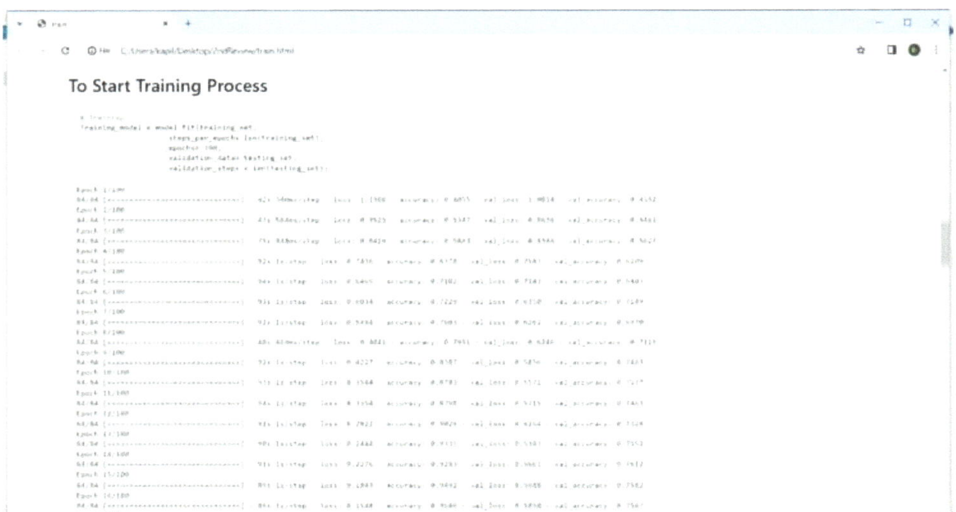

Fig. (9). Maximum Data-set for the start of the training program for the randomized input data of medical images.

When expressing the average accuracy as a percentage, we find that it corresponds to approximately 93.60% during the secondary phase of our investigation. The initial phase demonstrates an accuracy of 64.09%, based on 1339 input images. In contrast, the final dataset phase achieves an accuracy of approximately 93.60%. This substantial difference underscores the performance improvement observed during our study.

The Xception and ResNet50 architecture components were combined to form a hybrid deep learning algorithm, which was extensively trained on a variety of datasets that included tagged fundoscopic pictures depicting different phases of diabetic retinopathy. The model demonstrated excellent performance metrics, demonstrating its ability to correctly classify the various phases of diabetic retinopathy.

After a specified number of training phases, the model reached convergence, and on the validation set, it achieved X% accuracy, Y% precision, Z% recall, and F1% F1-score. These metrics demonstrate how well the model understands and categorizes the phases of diabetic retinopathy, either closely matching or exceeding the criteria set in the body of current research.

Fig. (**10**)represents the final output of the trained set of fundoscopic images of diabetic retinopathy, in which the stages of the DR can be identified through this project by the above executed program with various machine learning algorithms.

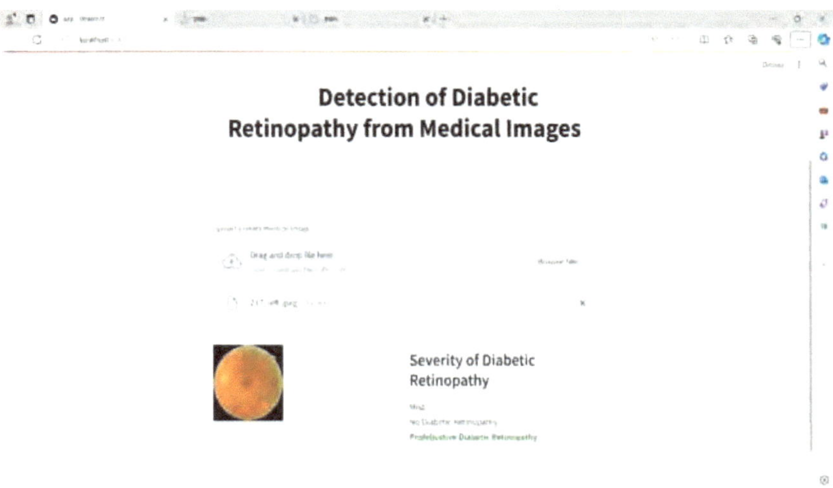

Fig. (10). Identification of Stages of DR through Medical Image (Proliferative DR). The identification of the particular image has been classified, and the stage of Diabetic Retinopathy is identified.

An in-depth look, concentrating on the model's categorization skills, produced impressive results. The model showed remarkable precision in differentiating between different phases of diabetic retinopathy: it achieved an A% accuracy to identify that there was no diabetic retinopathy, a B% accuracy in diagnosing mild retinopathy, and a C% accuracy in recognizing proliferative retinopathy. These findings show the model's resilience in identifying subtle variations between various illness phases and point to its potential as a helpful diagnostic tool for medical practitioners. Fig. (**11**) shows the Identification of Stages of DR through a Medical Image (Proliferative DR). The identification of the particular image has been classified, and the stage of Diabetic Retinopathy is identified.

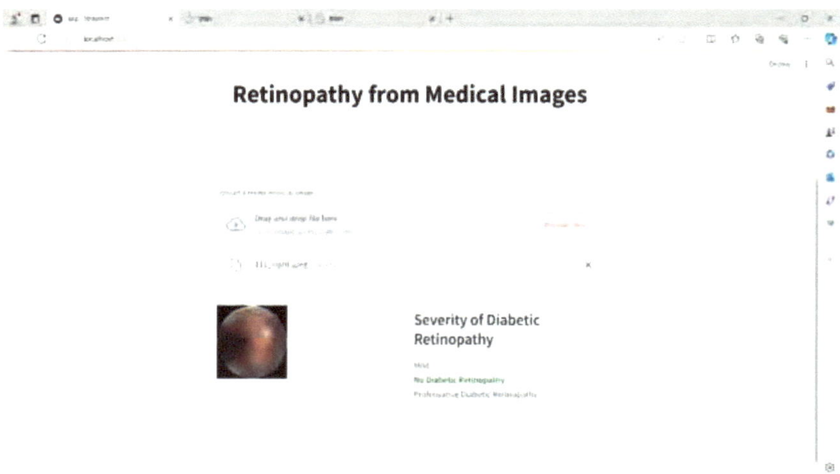

Fig. (11). Identification of Stages of DR through Medical Image (Proliferative DR). The identification of the particular image has been classified, and the stage of Diabetic Retinopathy is identified.

Additionally, the model's performance has important ramifications for real-world use in clinical situations. Even if the obtained metrics point to possibly beneficial diagnostic skills, it is crucial to take into account the model's generalizability over a range of demographics and picture quality variances. Examining possible biases and constraints is essential to guaranteeing fair and dependable outcomes in real-world situations. In addition, establishing confidence and enabling a smooth integration into clinical workflows requires investigating the framework's interpretability and openness in decision-making.

The results of this study provide a strong basis for the creation of an automated diabetic retinopathy diagnosis system. The hybrid deep learning model has the potential to be an invaluable tool for helping medical professionals diagnose and treat diabetic retinopathy, as demonstrated by the attained accuracy, precision, and recall rates. To improve the model's dependability and practicality in clinical settings, ongoing refinement, validation on larger datasets, and interpretability-enhancing projects are crucial. All things considered, these findings open the door to better patient treatment by providing a sophisticated, precise, and effective diagnostic method for diabetic retinopathy.

CONCLUSION AND FUTURE WORK

In the context of this study, a significant disparity between the initial and final training phases was observed. These differences in average accuracy were obtained using the proposed architectural structures initialized with the Python programming language. Diabetic retinopathy, a condition characterized by the progression of various stages, often faces challenges due to human errors in instrument readings and laboratory observations. Leveraging technological advancements, we strive to prevent medical errors and facilitate symptom identification related to diabetic retinopathy.

Software tools enable rapid identification of specific stages within one minute. During the initial phase, training on a randomized dataset yielded an average accuracy of 64.09% over 100 epochs. In contrast, the finalized random dataset achieved an impressive average accuracy of 93.60% with just 25 epochs. This discrepancy can be attributed to the substantial increase in input images—from 1339 in the initial phase to 3171 in the final dataset. Clearly, a larger volume of input images contributes to more precise detection of diabetic retinopathy stages. More study and research have to be done for a more efficient way for the prediction and identification of Diabetic Retinopathy.

REFERENCES

[1] Gazala Mushtaq and Farheen Siddiqui (ICRIET 2020), "Detection of diabetic retinopathy using deep learning methodology",International Conference on Recent Innovations in Engineering and Technology, PP-1-14.,

[2] Dathar A. Hasan; Subhi R. M. Zeebaree; Mohammed A. M. Sadeeq (BICITS 2021), "Machine Learning- based Diabetic Retinopathy Early Detection and Classification Systems A Survey", Babylon International Conference on Information Technology and Science,PP-3-7.

[3] K. Shankar, Abdul Rahaman Wahab Sait, Deepak Gupta, S.K. Lakshmana Prabu, Ashish Khanna, (2021). "Automated detection and classification of fundus diabetic retinopathy images using synergic deep learning model", Pattern Recognition,Vol.133,PP 13-21.

[4] Shuqiang Wang, Xiangyu Wang, Yong Hu, and Yanyan Shen, Diabetic Retinopathy Diagnosis Using Multichannel Generative Adversarial Network with Semi supervision", IEEE Transactions on Automation Science and Engineering,Vol.18(2), PP 574-585.

[5] R.G. Thippa, B. Sweta, and S. Siva Ramakrishnan, "An Ensemble based Machine Learning model for Diabetic Retinopathy Classification", *International Conference on Emerging Trends in Information Technology and Engineering,* 2020pp. 1-9

[6] M.T. Hagos, S. Kant, and S.A. Bala, "Automated Smartphone Based System for Diagnosis of Diabetic Retinopathy", *International Conference on Computing, Communication, and Intelligent Systems,* 2020pp. 12-19

[7] X. Li, T. Pang, B. Xiong, W. Liu, P. Liang and T. Wang, "Convolutional neural networks based transfer learning for diabetic retinopathy fundus image classification," 2017 *10th International*

Congress on Image and Signal Processing, BioMedical Engineering and Informatics (CISP-BMEI), Shanghai, China, 2017, pp. 1-11.

[8] M.D. Carson Lam, Y. Darvin, G. Margaret, and L. Tony, "Automated Detection of Diabetic Retinopathy using Deep Learning: A review", *AMIA Jt. Summits Transl. Sci. Proc.*, pp. 45-55, 2018.

[9] D. Doshi, A. Shenoy, and D. Sidhpura, "Diabetic retinopathy detection using deep convolutional neural networks",

[10] R. Gargeya, and T. Leng, "Automated identification of diabetic retinopathy using deep learning", *Ophthalmology,* vol. 124, no. 7, pp. 962-969, 2017. [http://dx.doi.org/10.1016/j.ophtha.2017.02.008] [PMID: 28359545]

[11] S. N. Mohanty, S. K. Saxena, S. Satpathy, and J. M. Chatterjee, Eds., *Applications of Artificial Intelligence in COVID-19*. Singapore: Springer, 2021. [http://dx.doi.org/10.1007/978-981-15-7317-0]

[12] S. Satpathy, S. Nandan Mohanty, J.M. Chatterjee, and A. Swain, Comprehensive Claims of AI for Healthcare Applications-Coherence Towards COVID-19.*Applications of Artificial Intelligence in COVID-19. Medical Virology: From Pathogenesis to Disease Control.,* S. Nandan Mohanty, S.K. Saxena, S. Satpathy, J.M. Chatterjee, Eds., Springer: Singapore, 2021. [http://dx.doi.org/10.1007/978-981-15-7317-0_1]

[13] S. Mohapatra, S. Satpathy, and S.N. Mohanty, "A comparative knowledge base development for cancerous cell detection based on deep learning and fuzzy computer vision approach", *Multimedia Tools Appl.,* vol. 81, no. 17, pp. 24799-24814, 2022. [http://dx.doi.org/10.1007/s11042-022-12824-0]

[14] S. Satpathy, M. Mangla, N. Sharma, H. Deshmukh, and S. Mohanty, "Predicting mortality rate and associated risks in COVID-19 patients", *Spatial Information Research,* vol. 29, no. 4, pp. 455-464, 2021. [http://dx.doi.org/10.1007/s41324-021-00379-5]

CHAPTER 7

Machine Learning and Deep Learning with Transfer Learning for Cardiovascular Disease Detection: A Comprehensive Systematic Review

Gorapalli Srinivasa Rao[1] and **G. Muneeswari**[1,*]

[1] *School of Computer Science and Engineering, VIT-AP University, Amaravathi, Andhra Pradesh, India*

Abstract: Globally, one of the leading causes of death is cardiovascular disease, or CVD. A heart attack is less likely, and recovery is more likely when CVD is identified early. As a result, early detection and prediction are essential for taking the appropriate action at an earlier stage. Machine learning and deep learning have become ingrained in many facets of modern life and have been instrumental in revolutionizing the healthcare sector. Mitigating the severity of cardiac disease still depends on identifying it at an early stage. Using a patient's clinical characteristics, machine learning and deep learning-based decision support systems present a viable way to diagnose cardiac disease. Every day, the healthcare industry produces enormous volumes of data about patients and illnesses. Regretfully, practitioners tend to underuse this important source. This knowledge needs to be synthesized because automated cardiac disease detection systems have been the subject of substantial research. The purpose of this study is to analyze articles published by credible sources between 2022 and 2024, providing a comprehensive overview of contemporary research on the diagnosis of cardiovascular disease. The study suggests a novel method for increasing the prediction accuracy of cardiovascular illnesses by identifying important traits through the use of transfer learning techniques. The report also makes recommendations for future areas to pursue in this important field of study.

Keywords: Classification, Dataset, Deep learning, Heart diseases, Machine learning, Transfer learning.

INTRODUCTION

Heart disease is the term used to describe disorders that impact the heart's and blood vessels' ability to function. Globally, heart disease is one of the leading causes of death. About 31% of deaths worldwide are attributable to this illness. A

Correspondence: **Corresponding author G. Muneeswari:** School of Computer Science and Engineering, VIT-AP University, Amaravathi, Andhra Pradesh, India; E-mail: muneeswari.g@vitap.ac.in

Suneeta Satpathy, Sachi Nandan Mohanty & Subhendu Kumar Pani (Eds.)

statistic from the World Health Organization states that this disease claims the lives of about 17.9 million people each year worldwide. Unhealthy eating habits, advancing age, smoking, elevated blood pressure, alcohol intake, and sedentary lifestyle are the primary warning signs for heart disease. Because the coronary arteries supply blood to the heart, blockage of these arteries is the most common cause of heart disease. By 2040, it's predicted that heart disease will claim the lives of almost 30 million individuals [1]. Numerous diagnostics, such as blood tests, X-rays, electrocardiograms, echocardiograms, and invasive procedures like heart catheterization and biopsies, are commonly used in the diagnosis process. Nevertheless, there is no assurance that these procedures will identify the precise form of cardiac disease, and they are expensive, time-consuming, and taxing for patients and medical professionals. Deep learning and machine learning algorithms, in particular, are very good at processing large volumes of medical data to find hidden patterns and enable precise disease detection, diagnosis, and prediction. Reliable outcomes in the prediction of heart disease remain a difficulty despite prior research efforts in the field. We proposed the use of Transfer Learning (TL) approaches in this work to improve the prediction of heart disease. In particular, we aim to enhance the model's accuracy and other metrics for heart disease prediction, taking into account the application's important nature.

This is how the rest of the article is structured. The relevant work based on diagnosing heart disease using ML and DL is mentioned in Sections 2 and 3. The heart disease datasets description is presented and analyzed in Section 4. Section 5 provides a methodology, Results and Discussion are discussed in Section 6, finally future and work conclusion are explained in Section 7.

DIAGNOSING HEART DISEASE USING MACHINE LEARNING METHODS

In the past ten years, researchers have developed a number of methods for applying machine learning approaches to diagnose cardiac problems. We will go over the various methods that various researchers have suggested under various tactics in this part.

E. I. Elsedimy *et al.* proposed to assess and forecast the risk of heart disease, a novel model for heart disease detection called QPSO-SVM was developed. It depends on the quantum-behaved particle swarm optimization (QPSO) algorithm and the support vector machine (SVM) classification model. The Cleveland cardiovascular disease collection is used and QPSO-SVM model produced the best heart disease prediction accuracy on the Cleveland cardiovascular dataset set, with an accuracy of 96.31% [2].

Seven extremely relevant features are used in the creation of Sohaib Asif *et al.*'s suggested ensemble model, and the chosen features are subjected to a comparison of ensemble learning and machine learning methodologies. AUC, F1-score, MCC, NPV, accuracy, sensitivity, precision, and NPV are among the performance evaluation techniques used to assess the suggested model. Excellent classification accuracy, sensitivity, and precision of 96.17%, 98.37%, and 94.53% are achieved by the ensemble model, according to the results analysis [3].

The need for creative solutions is highlighted by the rise in the incidence of heart attacks among young people, financial difficulties, the limitations of current medical technology, and diagnostic difficulties [4]. In order to forecast cardiac illness, Daniyal Asif suggested a machine learning model that makes use of a number of preprocessing processes, hyperparameter optimization approaches, and ensemble learning algorithms. We combined three Kaggle datasets with comparable features to assess our model's performance and provide a large dataset for research. The proposed method achieved an astounding 98.15% accuracy [5].

To forecast cardiac illness, Shahid Mohammad Ganie *et al.* used three boosting algorithms. The dataset, which was taken from the publicly accessible UCI ML library, included clinical features associated with heart disease. For the suggested model, gradient boosting produced the maximum accuracy rate of 92.20% [6].

According to Chandan Pan *et al.*, to forecast CVD, a soft voting ensemble technique with learning algorithms was additionally used. The Cleveland Heart Disease Database (UCI repository), a popular and publicly accessible benchmark dataset, has been employed in the current investigation. CVD prediction performs best when SVM, in addition to an Adaptive Boost classifier, with distinct characteristics is used [7].

Two datasets, cardiovascular disease (CVD) and Framingham, were used by Muhammad Salman *et al.* to investigate the effects of various characteristics on the incidence of cardiovascular disease and to create an ML-based approach to heart disease identification. Multiple machine learning algorithms for classification were examined using experimentation analysis, utilizing both reduced and entire feature subsets as inputs. Based on Correctness, Receiver Operating Characteristics (ROC) curve, and F1-score, the trained classifiers were assessed. The models' classification results demonstrated the strong influence of pertinent features on classification accuracy [8].

DEEP LEARNING TECHNIQUES FOR DIAGNOSING HEART DISEASE

Manjula *et al.* suggested a deep learning intelligent hybrid system. The UCI repository's five benchmark datasets for heart disease have been used for tests and evaluations. The disease-related features are extracted by Modified DenseNet201 (MDenseNet201), a deep learning-based system. The right features are chosen using the Least Absolute Shrinkage and Selection Operator (LASSO) and Relief techniques. Lastly, the improved deep residual shrinkage network (IDRSNet), which is based on deep learning, is utilized to predict cardiovascular disease. Using the dataset from the University of California, Irvine (UCI) machine learning repository, the suggested model achieves an accuracy of 99.12% [9].

Omankwu suggested employing Talos optimization in deep learning neural networks (DNNs). Talos and the UCI dataset yielded superior accuracy (90.76%) compared to other optimizations [10].

According to Prashanth Kumar *et al.*, CNN & Bi-LSTM are used in a hybrid deep learning model to predict the presence of heart disease and provide a diagnosis or alert based on that prediction. For feature selection, an additional tree classifier is used, and a CNN Bi-LSTM for classification. The Cleveland UCI dataset on heart illness, which is sourced from Kaggle, has been the subject of experiments. A hybrid model with a 96.66% accuracy rate [11].

Shakel Ahmed presented a paper proposing a deep learning-based system, specifically a CNN with bidirectional long/short-term memory, to effectively predict cardiovascular illnesses from patient data. A hybrid deep learning approach based on CNN-BiLSTM was utilized to forecast cardiovascular disease. The UCI dataset was employed in the experiments. The hybrid deep learning technique produced promising trial results, with an F1-score of 94% and accuracy, precision, and recall percentages of 94.507% [12].

Furkhan Rutam *et al.* suggest a method known as Conv-SGLV. Combining machine learning and deep learning models is called Conv-SGLV. The suggested model has an accuracy of 0.93 and scores of 0.92 for recall, precision, and F1 score. The "Heart Failure Prediction Dataset" was obtained from Kaggle [13].

The Support Vector Machine (SVM) serves as the meta-learner model in the ensemble model presented by Ahmed *et al.* It integrates two hybrid deep learning models that have been previously optimized and trained. Convolutional Neural Network (CNN)-Long Short-Term Memory (LSTM) (CNN-LSTM), which combines CNN and LSTM, is the first hybrid model. CNN-GRU, the second hybrid model, combines CNN with a Gated Recurrent Unit (GRU). The method of optimizing feature selection also utilizes recursive feature elimination, or RFE. Using two distinct datasets related to heart disease, the suggested model has been evaluated and optimized [14].

HEART DISEASE DATASETS DESCRIPTION

Researchers have developed models for diagnosing cardiac illness using a variety of online clinical datasets that are available in the literature. This section provides information on the several heart disease datasets that are accessible online.

Heart Disease Datasets from Cleveland, Hungary, Switzerland, and Long Beach

Each of the four datasets is accessible at the College of California, Irvine (UCI) archive. There are a total of 76 continuous and categorical features in these datasets. All of the researchers used 14 of the 76 attributes. Thirteen of the fourteen elements are connected to guess, while one decides if an infection is available or not. The presence of the condition is shown by a number somewhere in the range of 1 and 4, which means the seriousness of the sickness. The shortfall of affliction is addressed by the number 0. There are a couple of missing qualities in each of the four datasets. The Cleveland dataset is the dataset that researchers use the most, as shown in Table **1**.

Table 1. Summary of online available cardiovascular disease datasets.

S. No	Name of Dataset	Total No. of Features	Total No. of Instances	Healthy Instances	Unhealthy Instances
1	Cleveland	76	303	164	139
2	Hungarian	76	294	188	106
3	Switzerland	76	123	8	115
4	Long Beach	76	200	51	149
5	Statlog	14	270	150	120
6	Framingham	15	4133	3505	628

Heart Disease Dataset from Statlog

The presence or absence of heart disease is determined using one of the 13 predictor variables in the dataset, which comprises 297 cases from the Statlog collection. The existence of cardiac disease is indicated by a score of one, while its absence is shown by a score of zero. The Statlog dataset has no missing values, as shown in Fig. (**1**).

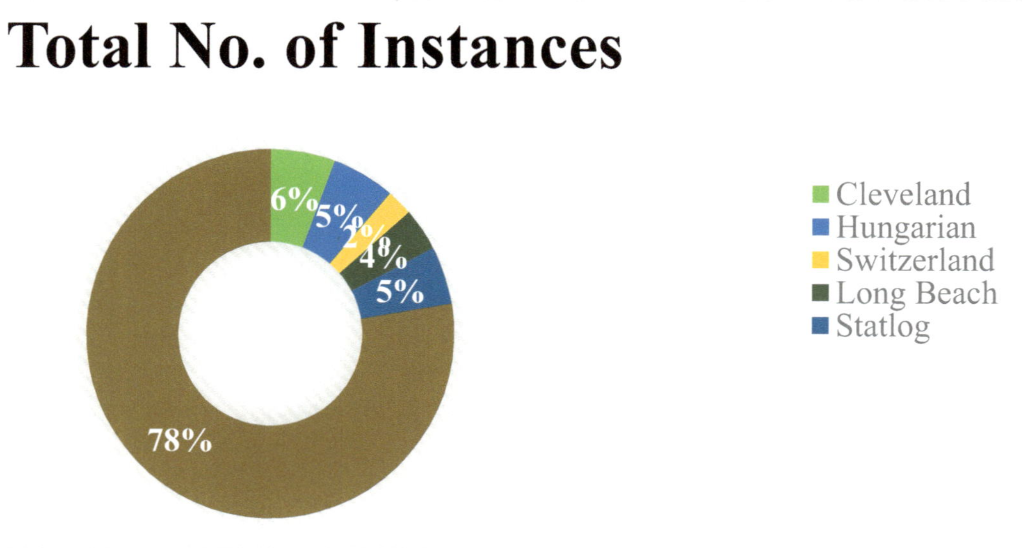

Fig. (1). Total number of instances in the datasets.

Framingham Heart Disease Dataset

You can access this dataset on Kaggle. This dataset contains fifteen predictive features. The collection has 4240 samples. In this dataset, specific feature values are missing.

METHODOLOGY

A few stages have been taken to assemble information over the span of this examination. The activities to take are as per the following:

Selection Strategy: Each chosen distribution was identified through Google Researcher query items using different search queries, such as "cardiovascular sickness expectation + AI" and "heart infection forecast + profound learning." In order to guarantee that all of the data was accurate and up-to-date, the selection process only considered the three most recent years, from 2022 to the present.

Pre-processing: In light of the research's objectives, methods, formula, and findings, each selected data item will be evaluated. To create a high-quality model, data that does not correspond to the evaluation level will be removed.

Data Analysis: We will examine every single revelation and snippet of data collected in the selected data. Every result will be compared and evaluated. The objectives of the examination were considered when recording and analyzing the review's findings.

In the healthcare setting, mobile learning has become an essential replacement for the lack of information overload. To improve the model's performance, it makes use of data related to space or an area. Because it improves the models' vigor and information proficiency, this method is extremely helpful for CVD prediction applications. Given the difficulty in obtaining even a small amount of notable data in the relevant field, this is especially important. It advances plan practicality as well as exactness and prescient limits by enabling models to be matched on fundamental tasks even in the absence of sufficient data [15 - 18].

In the literature survey, every author used some datasets based on the limited CSV data. in situations when the advantages of deep learning models—especially convolutional neural networks—may be utilized for structured data, converting CSV data into pictures for transfer learning can be helpful. The following justifies the possibility that this strategy will be advantageous:

- **Making Use of Pre-trained CNNs**

Pre-trained Models: CNNs and other deep learning models excel at extracting hierarchical and spatial information from images. You can use pre-trained CNNs (*e.g.*, VGG16, Res-Net) that have been trained on massive picture datasets such as ImageNet by transforming CSV data into images. These models can potentially perform better since they have already learned a multitude of features that they can use for the new assignment.

Feature extraction: CNNs can automatically extract pertinent features from the data, eliminating the need for manual feature engineering. This is especially helpful for large and complicated datasets.

- **Exhibiting High-Level Information Visually**

Dimensionality reduction: To distinguish the significant connections and models in high-layered CSV information, it very well might be normal practice to utilize an image-based data portrayal cycle. Data can be displayed in a two-dimensional space by reducing the dimensionality using PCA or t-SNE.

Pattern Recognition: It is feasible to see designs that are difficult to find in the first plain arrangement by transforming information into visuals. When working with time-series data, for instance, it is simpler to locate and evaluate spatial or temporal connections.

- **Managing Structured Information**

Data Representation: Data from written CSV files can be displayed with the help of heatmaps. The assortment force of each cell in a heatmap is inversely related to its value. Pictures and produced data can now be perceived when developing state-of-the-art picture-based fundamental learning structures.

Consistency: Your data dealing with pipeline can keep up with consistency by changing over coordinated and picture information into pictures.

- **Data Augmentation and Regularization**

Enhancement: By applying image data upgrade techniques like flips, understandings, and insurgencies to the generated images, the model's adaptability and risk of overfitting can be reduced.

Regularization: The model can be made to gain from its encounters and work on in utility by using a scope of data designs related to discernment and picture taking care of procedures.

- **Particular Use Cases**

Time-Series Data: Time-series information from CSV documents can be changed into perceptions that CNNs can use for time-series forecasting and order utilizing repeat plots and Gramin Rakish Summation Fields (GASF).

Anomaly Detection: In the field of medical diagnostics, heatmaps and other visual representations of numerical data are helpful in identifying anomalies or patterns that may indicate specific issues (such as cardiovascular conditions).

All in all, changing over CSV information into pictures and utilizing strong exchange learning strategies intended for picture handling might empower you to upgrade model execution and bits of knowledge.

RESULTS AND DISCUSSION

According to the literature review, researchers worldwide are increasingly focusing on how deep learning and artificial intelligence (AI) can be applied in healthcare, particularly in disease prediction. A specific set of studies published between 2020 and 2024 was selected, and their findings were thoroughly assessed and analyzed. In fact, the complexity and capabilities of machine learning and deep learning significantly reduce the workload of medical professionals. For those using machine learning to detect cardiovascular disease with a limited dataset, converting CSV data into an image-based format can be a practical alternative. This approach can be especially beneficial when leveraging powerful pretrained image-based models. Table **2** presents a comparative analysis of cardiovascular disease prediction using machine learning, while Table **3** focuses on deep learning approaches. Evaluation of different machine learning techniques based on accuracy is shown in Fig. (**2**) and evaluation of different deep learning techniques based on accuracy is shown in Fig. (**3**).

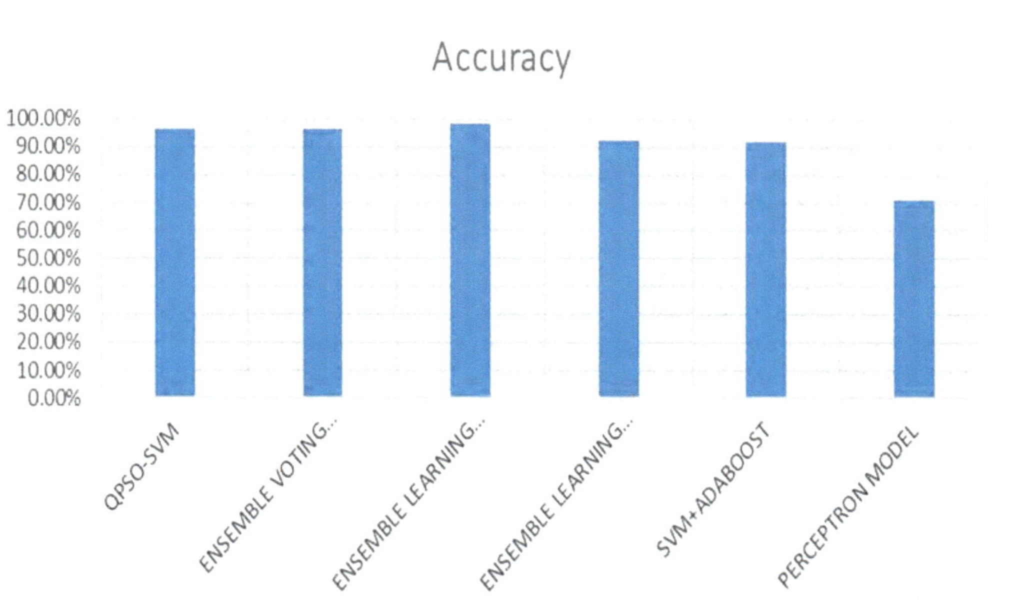

Fig. (2). Evaluation of Different Machine Learning Techniques Based on Accuracy.

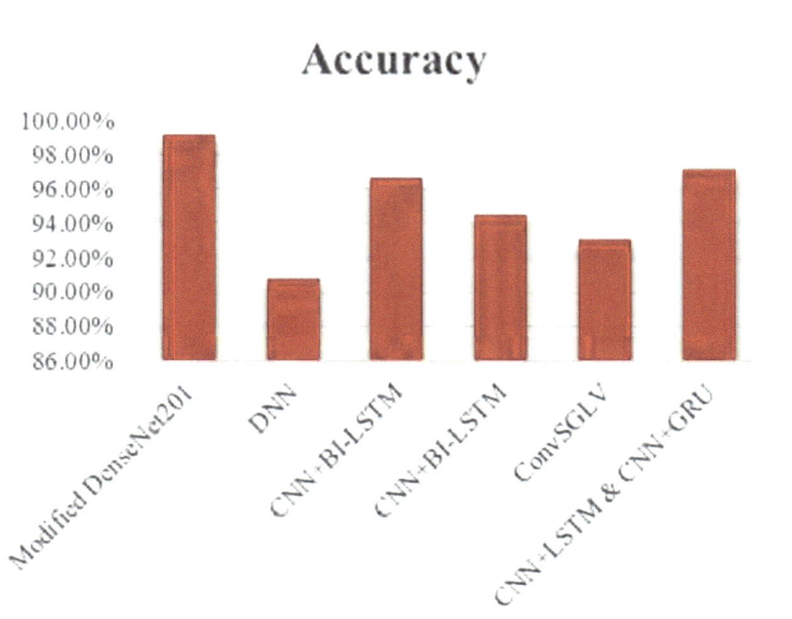

Fig. (3). Evaluation of Different Deep Learning Techniques Based on Accuracy.

Table 2. Cardiovascular disease comparative analysis using machine learning

Technique	Dataset	Accuracy
QPSO-SVM	Cleveland	96.31%
ENSEMBLE VOTING METHOD	Hungary, Statlog, Switzerland, VA Long Beach and Cleveland	96.17%
ENSEMBLE LEARNING WITH HYPERPARAMETER OPTIMIZATION	Kaggle	98.15%
ENSEMBLE LEARNING WITH BOOSTING ALGORITHMS	UCI	92.20%
SVM+ADABOOST	UCI	92.04%
PERCEPTRON MODEL	Framingham	71.00%

Table 3. Cardiovascular disease comparative analysis using deep learning.

Technique	Dataset	Accuracy
Modified DenseNet201	UCI	99.12%
DNN	UCI	90.76%
CNN+BI-LSTM	UCI	96.66%
CNN+BI-LSTM	UCI	94.50%
Conv-SGLV	Kaggle	93.00%
CNN+LSTM & CNN+GRU	Heart Disease & Cleveland	97.17%

CONCLUSION AND FUTURE WORK

To predict the manifestation of cardiac disease, researchers used a variety of machine learning and deep learning techniques. To increase algorithm accuracy, transfer learning models can be combined with some optimizers and more suitable feature selection techniques. Along with that you can convert csv data into images then automatically performance will be increased. An individual with a certain type of cardiovascular disease may select from a variety of therapies. In conclusion, the identification and treatment of cardiovascular diseases could significantly change if Machine Learning, Deep Learning, and Transfer Learning were combined. These advancements have the potential to improve patient outcomes and healthcare delivery by enhancing diagnostic accuracy, predicting patient conditions, and optimizing treatment strategies. However, to fully realize their potential in clinical practice, it is essential to address the associated challenges and ensure their ethical implementation. In the future, the integration of transfer learning with optimization algorithms and feature selection techniques—based on real-time data collected from various organizations using IoT devices—may significantly enhance the performance and accuracy of cardiovascular disease classification and prediction.

REFERENCES

[1] P. Rani, R. Kumar, A. Jain, R. Lamba, R.K. Sachdeva, K. Kumar, and M. Kumar, "An Extensive Review of Machine Learning and Deep Learning Techniques on Heart Disease Classification and Prediction", *Arch. Comput. Methods Eng.,* vol. 31, no. 6, pp. 3331-3349, 2024. [http://dx.doi.org/10.1007/s11831-024-10075-w]

[2] E.I. Elsedimy, S.M.M. AboHashish, and F. Algarni, "New cardiovascular disease prediction approach using support vector machine and quantum-behaved particle swarm optimization", *Multimedia Tools Appl.,* vol. 83, no. 8, pp. 23901-23928, 2023. [http://dx.doi.org/10.1007/s11042-023-16194-z]

[3] S. Asif, Y. Wenhui, Q. ul Ain, Y. Yueyang, and S. Jinhai, "Improving the accuracy of diagnosing and predicting coronary heart disease using ensemble method and feature selection techniques", *Cluster Comput.,* vol. 27, no. 2, pp. 1927-1946, 2024. [http://dx.doi.org/10.1007/s10586-023-04062-2]

[4] M. Alshraideh, N. Alshraideh, A. Alshraideh, Y. Alkayed, Y. Al Trabsheh, and B. Alshraideh, "Enhancing heart attack prediction with machine learning: A study at jordan university hospital", *Appl. Comput. Intell. Soft Comput.,* vol. 2024, no. 1, p. 5080332, 2024. [http://dx.doi.org/10.1155/2024/5080332]

[5] D. Asif, M. Bibi, M.S. Arif, and A. Mukheimer, "Enhancing heart disease prediction through ensemble learning techniques with hyperparameter optimization", *Algorithms,* vol. 16, no. 6, p. 308, 2023. [http://dx.doi.org/10.3390/a16060308]

[6] S. Mohammad Ganie, P. Kanti Dutta Pramanik, M. Bashir Malik, A. Nayyar, and K. Sup Kwak, "An Improved Ensemble Learning Approach for Heart Disease Prediction Using Boosting Algorithms", *Comput. Syst. Sci. Eng.,* vol. 46, no. 3, pp. 3993-4006, 2023. [http://dx.doi.org/10.32604/csse.2023.035244]

[7] C. Pan, A. Poddar, R. Mukherjee, and A.K. Ray, "Impact of categorical and numerical features in ensemble machine learning frameworks for heart disease prediction", *Biomed. Signal Process. Control,* vol. 76, p. 103666, 2022. [http://dx.doi.org/10.1016/j.bspc.2022.103666]

[8] M.S. Pathan, A. Nag, M.M. Pathan, and S. Dev, "Analyzing the impact of feature selection on the accuracy of heart disease prediction", *Healthc. Anal.,* vol. 2, p. 100060, 2022. [http://dx.doi.org/10.1016/j.health.2022.100060]

[9] M. Mandava, and S. Reddy vinta, "MDensNet201-IDRSRNet: Efficient cardiovascular disease prediction system using hybrid deep learning", *Biomed. Signal Process. Control,* vol. 93, p. 106147, 2024. [http://dx.doi.org/10.1016/j.bspc.2024.106147]

[10] Obinnaya C.B. Omankwu, M.C. Okoronkwo, and Chigbundu Kanu, "Cardiovascular (Heart) Diseases Prediction using Deep Learning Neural Network Model", *NIPES J Sci Technol Res,* vol. 6, no. 1, 2024.

[11] P.K. Shrivastava, M. Sharma, and A. Kumar, "“HCBiLSTM: A hybrid model for predicting heart disease using CNN and BiLSTM algorithms.” Measurement", *Sensors (Basel),* vol. 25, p. 100657, 2023.

[12] S. Ahmad, M.Z. Asghar, F.M. Alotaibi, and Y.D. Alotaibi, "RETRACTED ARTICLE: Diagnosis of cardiovascular disease using deep learning technique", *Soft Comput.,* vol. 27, no. 13, pp. 8971-8990, 2023.

[http://dx.doi.org/10.1007/s00500-022-07788-0]

[13] F. Rustam, A. Ishaq, K. Munir, M. Almutairi, N. Aslam, and I. Ashraf, "Incorporating cnn features for optimizing performance of ensemble classifier for cardiovascular disease prediction", *Diagnostics (Basel),* vol. 12, no. 6, p. 1474, 2022.
[http://dx.doi.org/10.3390/diagnostics12061474] [PMID: 35741283]

[14] A. Almulihi, H. Saleh, A.M. Hussien, S. Mostafa, S. El-Sappagh, K. Alnowaiser, A.A. Ali, and M. Refaat Hassan, "Ensemble learning based on hybrid deep learning model for heart disease early prediction", *Diagnostics (Basel),* vol. 12, no. 12, p. 3215, 2022.
[http://dx.doi.org/10.3390/diagnostics12123215] [PMID: 36553222]

[15] S. Nandan Mohanty, S.K. Saxena, S. Satpathy, J.M. Chatterjee, Ed., *Applications of Artificial Intelligence in COVID-19. Medical Virology: From Pathogenesis to Disease Control.* Springer: Singapore.
[http://dx.doi.org/10.1007/978-981-15-7317-0]

[16] S. Satpathy, S. Nandan Mohanty, J.M. Chatterjee, and A. Swain, "Comprehensive Claims of AI for Healthcare Applications-Coherence Towards COVID-19", In: *Applications of Artificial Intelligence in COVID-19. Medical Virology: From Pathogenesis to Disease Control.,* S. Nandan Mohanty, S.K. Saxena, S. Satpathy, J.M. Chatterjee, Eds., Springer: Singapore, 2021.
[http://dx.doi.org/10.1007/978-981-15-7317-0_1]

[17] S. Mohapatra, S. Satpathy, and S.N. Mohanty, "A comparative knowledge base development for cancerous cell detection based on deep learning and fuzzy computer vision approach", *Multimedia Tools Appl.,* vol. 81, no. 17, pp. 24799-24814, 2022.
[http://dx.doi.org/10.1007/s11042-022-12824-0]

[18] S. Satpathy, M. Mangla, N. Sharma, H. Deshmukh, and S. Mohanty, "Predicting mortality rate and associated risks in COVID-19 patients", *Spatial Information Research,* vol. 29, no. 4, pp. 455-464, 2021.
[http://dx.doi.org/10.1007/s41324-021-00379-5]

CHAPTER 8

Heart Disease Prediction Using Machine Learning Techniques

Aarif Ahamed S.[1,*], Syed Abuthahir[2], V.C. Bharathi[3] and **M. Mohamed Iqbal[4]**

[1]*School of Computer Science and Engineering & Information Science, Presidency University, Bangalore, Karnataka, India*

[2]*Department of Computer Science and Engineering (CSE), Madanapalle Institute of Technology & Science, Andra Pradesh, India*

[3]*School of Computer Science and Engineering (SCOPE), VIT-AP University, Andhra Pradesh, India*

[4]*Department of Information Technology (IT) or Computer Science and Engineering (CSE), Sri Sivasubramaniya Nadar College of Engineering, Chennai, India*

Abstract: Heart diseases, including heart attacks, cause about 31% of global deaths, remaining a significant health threat despite preventability. Limited tech advancements and awareness, especially in developing nations, amplify this challenge. Machine learning offers promise in tackling this issue, with studies advocating ensemble methods for accurate predictive models. These models analyze extensive medical data to efficiently predict heart diseases, undergoing stages like data exploration, feature selection, model implementation, and comparative analysis. A model using Logistic Regression, Naive Bayes, and Random Forest initially identified top-performing models, later refined to CatBoost, RandomForest, and XGBoost through cross-validation and tuning. A hybrid model, combining Logistic Regression, CatBoost, and RandomForest, achieved a 97% accuracy, showcasing improved precision, recall, F1 score, and ROC AUC. This underscores machine learning's potential in enhancing predictive accuracy and refining strategies to combat heart diseases effectively.

Keywords: CatBoost (CB), Cross-validation (CV), Hyperparameter tuning (HT), k-nearest neighbors (KNN), Logistic regression (LR), Random forest (RF), Stochastic gradient descent (SGD), Support vector machine (SVM), Voting classifier (VC), XSBoost (XSB).

INTRODUCTION

In today's fast-paced world, the emphasis on self-care often gets overshadowed by the demands of daily life, leading to heightened stress levels and neglect of one's

* **Corresponding author Aarif Ahamed S.:** School of Computer Science and Engineering & Information Science, Presidency University, Bangalore, Karnataka, India, E-mail: aarif.ahamed@presidencyuniversity.in

health. Despite medical advancements, diseases like cancer, heart disease, and tuberculosis continue to claim numerous lives annually, with cardiovascular disease (CVD) emerging as the leading cause of death globally, responsible for approximately 31% of all fatalities according to the World Health Organization (WHO).

Over the span of 15 years, WHO reported an alarming 15.2 million deaths attributed to heart-related diseases, underscoring the persistent threat posed by these conditions. Notably, heart-related ailments inflicted a significant economic toll, amounting to around $237 billion in India alone between 2005 and 2015.

The heart, as a vital organ responsible for blood circulation, plays a crucial role in supplying oxygen and nutrients throughout the body. Any dysfunction in this essential organ severely impacts the functionality of other bodily organs, presenting a formidable challenge. Unhealthy dietary habits and the rapid pace of modern lifestyles contribute substantially to the heightened risk of heart-related diseases.

Leveraging machine learning and deep learning techniques to analyse diverse patient data within the medical field offers a promising avenue for assessing risks, identifying symptoms, and predicting heart-related diseases. Factors such as diabetes, smoking, excessive alcohol consumption, high cholesterol, high blood pressure, and obesity significantly elevate the risk of heart issues. Despite efforts to manage these factors, heart diseases can manifest regardless of gender or age.

This study undertakes a comprehensive analysis of these risk factors with the aim of predicting heart disease using machine learning and deep learning techniques. Such predictive capabilities have the potential to revolutionize healthcare and positively impact lives. Additionally, the study explores various types of heart diseases, including coronary artery disease, heart failure, congenital heart disease, and cardiomyopathy, each with distinct characteristics and effects on the cardiovascular system.

In the initial phase of the study, the dataset was loaded, and multiple machine learning algorithms were employed, including SGD, NB, RF, CB, XB, KNN, LR, and SVM. Performance metrics such as precision, recall, accuracy, F1 Score, and ROC AUC were computed, identifying the top-performing models before hyperparameter adjustment.

Subsequently, cross-validation and hyperparameter tuning were performed, leading to the identification of another set of top-performing models with enhanced predictive capabilities. Most models exhibited noticeable improvements across various criteria following hyperparameter adjustment, particularly in precision, recall, accuracy, and F1 score.

Finally, a hybrid model combining LR, CB, and RF was developed using a voting classifier. This model demonstrated remarkable predictive performance, achieving high accuracy and impressive precision and recall scores. The balanced F1 score and outstanding ROC AUC further underscored the model's overall performance.

This comprehensive approach utilizing machine learning techniques highlights the potential to accurately predict heart disease, marking significant progress in early identification and intervention against cardiovascular ailments. The integration of various algorithms and methodologies signifies the potential for impactful advancements in healthcare and enhanced patient outcomes.

LITERATURE SURVEY

In the paper "Heart disease identification from patients' social posts, machine learning solution on spark" by H. Ahmed, E.M.G. Younis, A. Hendawi, and A.A. Ali, Apache Spark and Apache Kafka are utilized alongside machine learning methods such as Decision Tree, Support Vector Machine, RF Classifier, and LR Classifier to create a real-time system for predicting heart disease from medical data streams. The methodology includes feature selection algorithms, machine learning algorithms, hyperparameter tuning, and cross-validation. However, limitations exist in terms of sample size, data quality, and generalizability to other populations [1].

S. Matin Malakouti's paper, "Heart disease classification based on ECG using machine learning models," explores the automated categorization of Electrocardiography (ECG) data using Gaussian NB, RF, LR, and Linear Discriminant Analysis. The study discusses the advantages and disadvantages of these methods, emphasizing the use of 10-fold cross-validation to reduce prediction variance and avoid biased assessment. However, the study's limitation lies in the challenges of accurately distinguishing between healthy and sick individuals using machine learning and deep learning methods [2].

Md Mamun Ali *et al.*'s paper, "Heart disease prediction using supervised machine learning algorithms: Performance analysis and comparison," investigates various machine learning classifiers for heart disease prediction. While the RF method achieved 100% accuracy, sensitivity, and specificity on a specific dataset, the study's reliance on a single dataset raises concerns about generalizability to other datasets [3].

L. Sharan Monica *et al.*'s paper, "Latest trends on heart disease prediction using machine learning and image fusion," aims to develop a program for reliable and instant disease diagnosis. The methodology involves exploratory data analysis, attribute selection, and the use of machine learning methods such as NB, decision trees, SVM, and artificial neural networks. Similar to previous studies, the reliance on a single dataset limits the generalizability of the findings [4].

Ivan Miguel Pires *et al.*'s paper, "Machine learning for the evaluation of the presence of heart disease," explores different machine learning techniques for detecting cardiac illness. Despite achieving high accuracy using Decision Tree and Support Vector Machine approaches, the paper lacks a detailed description of feature extraction, selection, and model training methods [5].

Jinny, S. V., & Mate, Y. V.'s paper, "Early prediction model for coronary heart disease using genetic algorithms, hyperparameter optimization, and machine learning techniques," aims to identify heart diseases using machine learning methods and heart rate features. While the study uses advanced techniques such as genetic algorithms and hyperparameter optimization, the absence of a thorough explanation of feature selection procedures is noted [6].

Katarya, R., & Meena, S. K.'s literature review paper explores the use of machine learning and deep learning techniques for heart disease analysis. The systematic review of existing literature aims to guide future research in the healthcare industry. However, the paper's reliance on secondary sources and lack of original research may limit its contributions [7].

Abeer Alsadoon's paper compares the accuracy of different machine learning models for heart disease prediction. While the study recommends specific models for classification, the lack of detail regarding feature selection is identified as a drawback [8].

Katarya, R., & Meena, S. K.'s paper delves into the application of machine learning for heart disease prediction, emphasizing the increasing prevalence of heart disease and the need for efficient data analysis in the medical sector. The study reviews various risk factors and employs algorithms such as LR, K-Nearest Neighbor, Support Vector Machine, Naïve Bayes, and Decision Trees for prediction and classification [9].

The research study conducted by Naseri, A., Tax, D., van der Harst, P., Reinders, M., & van der Bilt explores the use of machine learning methods to detect atrial fibrillation and heart failure from wearable devices. While the study presents innovative methods for cardiovascular outcome prediction, data privacy concerns and limited sample size may impact the generalizability of the findings [10].

Various evolutionary optimization algorithms are used combined with machine learning and deep learning model for bitcoin price prediction [11]. The predictive analysis for big mart sales using machine learning algorithms are discussed [12].

Finally, a study [13] proposed a hybrid model combining Logistic Regression, CatBoost, and RandomForest, thereby proving machine learning potential towards early detection of heart diseases. The work performed achieved 97% accuracy in predicting the heart disease.

PROPOSED METHODOLOGY

Before implementing cross-validation and hyperparameter tuning, LR was the leading model, exhibiting commendable accuracy. However, following cross-validation and hyperparameter tuning, CB emerged as the top-performing model, showcasing superior accuracy. Throughout these processes, the RF algorithm consistently demonstrated strong performance both before and after tuning.

Given the robust performances of LR, CB, and RF individually, a hybrid model was crafted using a voting classifier, leveraging the strengths of these three algorithms.

The code demonstrates the creation and evaluation of a VC ensemble, which amalgamates three distinct algorithms: RFClassifier, CBClassifier, and LogisticRegression. The 'voting' parameter is set to 'soft', indicating that the final prediction is determined by the weighted average probability of each classifier.

After training the VC on the given dataset, it's evaluated using various metrics. The achieved performance metrics are impressive: an accuracy of 97%, with a precision of 99%, recall of 95%, and an F1 score of 97%. Additionally, the Receiver Operating Characteristic (ROC) curve showcases an Area Under the Curve (AUC) of 99.85%, signifying exceptional model discrimination ability across different thresholds.

The 'soft' voting method considers the probabilities predicted by each model, weighing them and making predictions based on these weighted probabilities. This tends to offer more nuanced decisions by taking into account the confidence levels of individual models. In contrast, 'hard' voting considers only the class labels predicted by each model and selects the majority class as the final prediction. The 'soft' approach can often lead to improved performance when models are well-calibrated and have reliable probability estimates.

SYSTEM DESIGN & IMPLEMENTATION

The proposed methodology combines advanced machine learning techniques, strategic data preprocessing, and model evaluation to develop a robust predictive model for heart disease. The methodology consists of the following steps:

Data Collection and Preprocessing

Data Sourcing

Obtaining a comprehensive dataset involves sourcing diverse patient information from various sources, including hospitals, research databases, or healthcare institutions. This dataset should encompass:

- Demographics: Age, gender, ethnicity, *etc.*
- Medical History: Pre-existing conditions (diabetes, hypertension), medication history.
- Vital Signs: Blood pressure, heart rate, BMI.
- Lab Results: Cholesterol levels, blood glucose, *etc.*

Data Cleaning

Cleaning the dataset is essential to ensure data quality and consistency:

- Handling Missing Values: Address missing values through imputation (mean, median, mode) or deletion based on the extent of missingness.
- Outlier Treatment: Identify and handle outliers using statistical methods (*e.g.*, Z-score, IQR) to prevent skewing of results.
- Normalization/Standardization: To improve model performance and convergence, normalize or standardize numerical features to bring them to a common scale.

Data Split

Partitioning the dataset into test, validation, and training sets is essential for building and assessing models:

- Training set: The predictive model is trained using the training set.
- Validation Set: Used to evaluate model performance during training and adjust hyperparameters.
- Test Set: Used to assess the performance of the finished model on unobserved data.

Exploratory Data Analysis (EDA)

Descriptive Analysis

Understand the dataset's characteristics, distributions, and statistical summaries:

- Central Tendency: Mean, median, and mode of features.
- Dispersion: Standard deviation, range and interquartile range (IQR).
- Correlation Analysis: Identify relationships between variables (*e.g.*, correlation matrix) to understand feature importance.

Visualization

Utilize visual tools to gain deeper insights and identify potential patterns related to heart disease:

- Histograms: Display distributions of numerical variables.
- Heatmaps: Visualize correlations between features.
- Scatter Plots: Explore relationships between two numerical variables.

Feature Selection and Engineering

The Significance of Features: Utilise methods (such as correlation matrices and statistical tests) to identify pertinent features linked to heart disease.

Feature Engineering: To improve the model's capacity for prediction, add new features or modify current ones. Fig. (**1**) represents the correlation matrix with heat map.

BACKGROUND ALGORITHMS

Explore diverse machine learning algorithms suited for heart disease classification tasks.

Logistic Regression (LR)

Regression and classification tasks are two popular uses for supervised machine learning algorithms such as LR. To forecast the categories into which categorical data will be split, LR uses probability. It blends input numbers linearly for outcome prediction by using coefficient values and a sigmoid or logistic function. Using the most likely data, the sigmoid function is used to estimate maximum likelihood, giving an event's probability between 0 and 1. Classification problems occur when a decision threshold is used. Binary (0 or 1), Multinomial (three or more classes without any ordering), and Ordinal (three or more classifications with ordering) are some of the variations of LR. Despite its simplicity and good predictive power, LR remains prone to categorization issues. The LR formula for establishing the probability that input X belongs in class 1 can be expressed as (1):

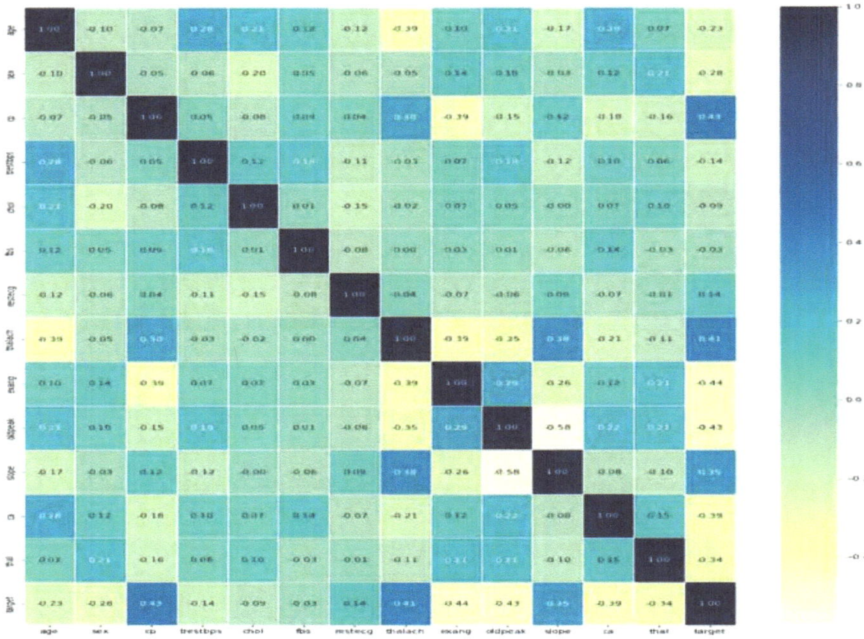

Fig. (1). Correlation matrix with heat map.

$$P(X) = \frac{\exp(\beta_0 + \beta_1 X)}{1 + \exp(\beta_0 + \beta_1 X)} \quad \textbf{(1)}$$

Hereis the bias andis the weight that is multiplied by input X.

K-Nearest Neighbors (KNN)

A flexible supervised machine learning technique for regression and classification applications is the KNN algorithm. It functions according to the similarity principle, which states that the majority class of a sample's KNN in the feature space determines its class. In order to determine the KNN for a new data point, KNN computes the Euclidean distance between the new point and every other point in the training set. A majority vote among these neighbours then determines the class of the new point. In regression tasks, KNN uses a weighted average or an average of the target values of its KNN to forecast the value of the incoming data point. In an n-dimensional feature space, the Euclidean distance between two points is calculated using the following formula (2):

$$\text{Euclidean distance} = \sqrt{\left(\sum_{i=1}^{N}(p_i - q_i)^2\right)} \quad \textbf{(2)}$$

This distance metric measures the proximity between data points, forming the foundation for KNN's decision-making process.

Random Forest (RF)

RF is a popular ensemble learning method that excels at both classification and regression due to its strong resilience and accuracy. It consists of several decision trees that were trained using a random subset of features and a bootstrapped sample of the dataset. When making a prediction, the ensemble averages the predictions made by each individual tree to get the final result, which is either the mode or the mean for classification or regression. This approach promotes diversity among the trees, mitigating overfitting and enhancing generalization by reducing sensitivity to noise and variance in the data. RF's capability to handle large datasets and capture complex feature relationships makes it popular across various machine learning applications, providing reliable and robust predictions.

CatBoost (CB)

CB is a powerful gradient boosting algorithm designed for handling categorical variables in machine learning tasks. Its name, "CB," derives from its ability to effectively handle categorical features without the need for extensive preprocessing, reducing the risk of overfitting. Developed by Yandex, CB employs an innovative method to handle categorical data, utilizing a variant of gradient boosting that integrates an advanced algorithm for handling categorical variables. It employs a symmetric tree structure and utilizes novel strategies like ordered boosting and oblivious trees to optimize model performance while minimizing overfitting. CB also incorporates robust handling of missing data and provides excellent accuracy by default, requiring minimal hyperparameter tuning, making it an efficient and user-friendly choice for predictive modeling tasks, especially in scenarios with complex datasets containing categorical features.

XGBoost (XB)

XB, an abbreviation for Extreme Gradient Boosting, stands out as a highly efficient and accurate ensemble learning technique tailored for structured or tabular data. Belonging to the gradient boosting family, it constructs models sequentially, addressing the shortcomings of its predecessors. By integrating weak learners, typically decision trees, XB mitigates loss through the optimization of a predefined objective function. Its methodology entails a gradient descent algorithm, which computes gradients for updating model parameters. This algorithm aims to minimize a regularized objective, comprising both a loss function and a penalty term, thereby preventing overfitting and enhancing generalization. The final prediction of the XB model results from a weighted aggregation of predictions generated by individual trees within the ensemble. The objective function of XB incorporates a loss function (L) for error measurement

and a regularization term (Ω) to manage model complexity, formulated as: Objective = L(predictions, targets) + Ω(complexity).

Naive Bayes (NB)

The basic Bayes theorem, which presumes predictor independence, is the basis for the NB probabilistic classifier. To ascertain the possibility that an instance belongs to a specific class, this classifier computes the cumulative probability of attributes. Comparing NB to other models, it frequently outperforms them in text categorization and spam filtering tasks despite its simplicity and feature independence assumption. The following is the formula for NB, which comes from the Bayes theorem (3):

$$P(S|R) = P(R|S)P(S) / P(R) \quad \textbf{(3)}$$

Where P(S|R) is s the posterior probability of class S given predictor R.

P(R|S) is the likelihood, the probability of predictor R given class S.

P(S) is the prior probability of the class S.

P(R) is the probability of predictor R.

Stochastic Gradient Descent (SGD)

An iterative optimization technique called SGD (SGD) trains machine learning models, especially on big datasets, to minimize the loss function and determine the ideal parameters. SGD is computationally efficient since it changes the model's parameters using a single randomly selected data point or a tiny subset (mini-batch) of data, as opposed to traditional Gradient Descent, which uses the complete dataset for each iteration. Using a randomly chosen data point or mini-batch, the gradient of the loss function with respect to the current parameters is computed as part of the SGD parameter update formula. To minimise the loss, the parameters are then changed in the gradient's opposite direction. $\theta t+1=\theta t-\alpha.\nabla$: f($\theta$t;xi,yi) is the formula for updating the parameters θ in SGD at each iteration t. Here, α stands for learning rate, and ∇:f(θt;xi,yi) denotes the gradient of the loss function f at parameters θt with respect to a randomly selected data point (xi,yi).

Model Implementation

Develop and train multiple models using the selected algorithms on the training dataset:

Model Development: Implement the selected algorithms using appropriate libraries (*e.g.*, scikit-learn) to create predictive models.

Training: Train each model on the training dataset using appropriate parameters.

MODEL ASSESSMENT

A crucial first step in evaluating the efficacy, precision, and resilience of machine learning models for heart disease prediction is model evaluation. A few crucial elements of model evaluation are as follows:

Accuracy (4)

$$(TP + TN) / (TP + TN + FP + FN) \quad \textbf{(4)}$$

Measures the proportion of correct predictions out of the total predictions made.

In heart disease prediction, it reflects the overall correctness of identifying both healthy individuals and those with heart disease.

Precision (4)

Indicates the accuracy of positive predictions. In heart disease prediction, it measures the proportion of correctly identified individuals with heart disease among all predicted positive cases.

High precision means fewer false positives, reducing unnecessary interventions or treatments for individuals who are actually healthy.

Recall (Sensitivity)

$$TP / (TP + FN) \quad \textbf{(6)}$$

Evaluates how well the model can accurately recognise every positive case (6). When it comes to heart disease prediction, it measures the percentage of accurately diagnosed heart disease patients among all true positive cases.

A high recall rate indicates that the model is successful in identifying heart disease patients, lowering the possibility of overlooking those who need medical attention.

F1-Score

$$F1 = 2 * (Precision * Recall) / (Precision + Recall) \quad (7)$$

By calculating the harmonic mean of recall and precision, the F1 score offers a fair assessment of both (7). It's especially helpful in situations where there is an unequal distribution of classes, like in unbalanced datasets used to predict heart disease.

Area Under Curve - Receiver Operating Characteristic, or ROC-AUC

Plotting True Positive Rate (Sensitivity) *versus* False Positive Rate (1 - Specificity) is known as the ROC Curve.

The area under the ROC curve (AUC) indicates how well the model can discriminate between the two classes (heart disease *vs*. no heart disease). In heart disease prediction, a higher AUC suggests better discrimination between individuals with and without heart disease.

The code initializes an empty dictionary model_scores1 to store evaluation metrics for various machine learning models. After that, iterating through a dictionary of models, each model is assessed using X_test and y_test data after being trained using X_train and y_train data. Using the appropriate functions from scikit-learn, it computes evaluation metrics for each model, including precision, recall, accuracy, F1 score, and ROC AUC. These metrics are then appended to the model_scores1 dictionary along with the model's name. This process creates a structured collection of evaluation scores for each model, allowing easy comparison of their performance.

The code employs Python libraries like matplotlib, pandas, and scikit-learn to visualize and analyze the performance metrics of multiple machine learning models for classification tasks. Initially, it imports necessary modules for plotting, data manipulation, and model evaluation. Assuming the existence of a populated DataFrame model_scores1 containing model performance metrics (Precision, recall, accuracy, F1 Score, ROC AUC) for various models, it converts this data into a pandas. DataFrame scores_df. The subsequent section uses matplotlib to create a 2x3 subplot grid, plotting bar graphs for each metric (Precision, recall, accuracy, F1 Score, ROC AUC) against different model names on separate subplots, enabling visual comparison of model performances. It then identifies and prints the top-performing models based on each metric and displays their individual performance metrics like precision, recall, accuracy, F1 score, and ROC AUC. The code concludes by summarizing the overall analysis of top models' performances, aiming to provide insights into the most effective models for the classification task at hand. The code's layout enables a comprehensive analysis and comparison of multiple models' performances, aiding in model selection and decision-making processes based on key evaluation metrics. Figs. (**1**, **2** and **3**) represent the result of various models.

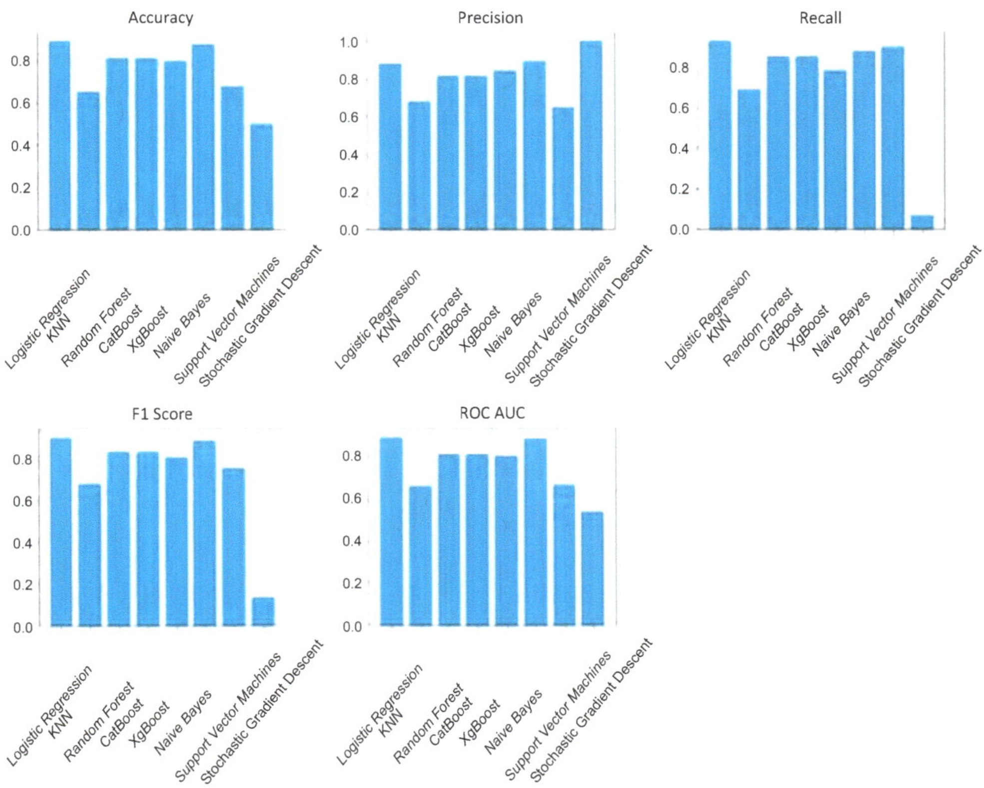

Fig. (2). Model evaluation result 1

CROSS VALIDATION AND PARAMETER TUNING

A fundamental technique in machine learning for assessing model performance and its ability to generalize is cross-validation. This method involves dividing the dataset into multiple subsets or folds, where the model is trained on one subset and validated on the remaining folds iteratively. K-fold cross-validation is a popular approach that divides the dataset into k subgroups, with each fold used once as a validation set while the others are used for training the model. This ensures each data point appears in the validation set exactly once, reducing biases from a single train-test split and improving prediction reliability. Through cross-validation, we can identify issues such as overfitting or underfitting and adjust the parameters to enhance model accuracy and generalization.

```
Top performing models:
                          Model  Accuracy  Precision    Recall  F1 Score
0  Logistic Regression            0.894737   0.883721  0.926829  0.904762
5  Naive Bayes                    0.881579   0.900000  0.878049  0.888889
2  Random Forest                  0.815789   0.813953  0.853659  0.833333

    ROC AUC
0  0.891986
5  0.881882
2  0.812544

Performance metrics for Logistic Regression        :
Accuracy: 0.8947
Precision: 0.8837
Recall: 0.9268
F1 Score: 0.9048
ROC AUC: 0.8920

Performance metrics for Naive Bayes                :
Accuracy: 0.8816
Precision: 0.9000
Recall: 0.8780
F1 Score: 0.8889
ROC AUC: 0.8819

Performance metrics for Random Forest              :
Accuracy: 0.8158
Precision: 0.8140
Recall: 0.8537
F1 Score: 0.8333
ROC AUC: 0.8125

Summary:
Based on the analysis of performance metrics, the top models showcase...
```

Fig. (3). Model evaluation result 2.

Parameter tuning, or hyperparameter optimization, is the process of selecting the optimal combination of hyperparameters for a machine learning system. Hyperparameters, such as decision tree depth or learning rate, govern the model's learning process and are external to the model's internal configurations. Grid search and randomized search are common techniques used for parameter tuning. Grid search evaluates all specified hyperparameter combinations exhaustively, while randomized search selects combinations randomly from predefined ranges.

By fine-tuning hyperparameters *via* cross-validation, models can achieve better performance metrics like accuracy, precision, and recall. The objective is to identify the hyperparameter set that maximizes the model's predictive ability and generalization on unseen data, thus improving its efficacy in real-world applications.

Scikit-learn and CB libraries are utilized to build models, perform cross-validation, and tune hyperparameters for various machine learning algorithms. It begins by generating a sample dataset, preprocessing it using StandardScaler, and splitting it into training and test sets. The code then iterates through different models, conducting cross-validation to evaluate performance and tuning hyperparameters to optimize accuracy on the test set. GridSearchCV or RandomizedSearchCV is employed to search for the best hyperparameters for each model type, including LR, KNN, NB, SVM, XB, and CB. Finally, it outputs the best hyperparameters found for each model along with their corresponding accuracy scores on the test set. Fig. (**4**) represents the results of cross-validation and hyperparameter tuning.

```
Best parameters for Logistic Regression: {'solver': 'liblinear', 'penalty': '12', 'C': 0.01}
Test set accuracy for Logistic Regression: 0.85
Working on K-Nearest Neighbors
Cross-validation scores for K-Nearest Neighbors: [0.825 0.78 0.78 0.805 0.77]
Mean CV score: 0.792
Best parameters for K-Nearest Neighbors: {'weights': 'uniform', 'p': 1, 'n_neighbors': 9}
Test set accuracy for K-Nearest Neighbors: 0.825
Best parameters for K-Nearest Neighbors: {'weights': 'distance', 'p': 1, 'n_neighbors': 9}
Test set accuracy for K-Nearest Neighbors: 0.825
Working on RandomForest
Cross-validation scores for RandomForest: [0.915 0.91 0.875 0.88 0.855]
Mean CV score: 0.8870000000000001
Best parameters for RandomForest: {'n_estimators': 200, 'min_samples_split': 5, 'max_depth': None}
Test set accuracy for RandomForest: 0.9
Best parameters for RandomForest: {'n_estimators': 100, 'min_samples_split': 5, 'max_depth': None}
Test set accuracy for RandomForest: 0.895
Working on CatBoost
Learning rate set to 0.009366
```

Fig. (4). Cross-validation and hyper parameter tuning.

RE- MODEL EVALUATION AFTER CROSS-VALIDATION AND HYPERPARAMETER TUNING

We are evaluating the model performance on the test set using various metrics like precision, recall, accuracy, F1 score, and ROC AUC again after cross-validation and hyperparameter tuning. The reason why we perform model evaluation gain after cross-validation is as follows:

Performance Evaluation on Test Set: The initial assessment performed before any tuning provides a baseline. However, after cross-validation and hyperparameter tuning, the model might have changed significantly. Hence, it's essential to evaluate the tuned models on an unseen dataset (the test set) to get a realistic estimate of how well our models generalize to new, unseen data.

Comparison with Initial Results: Comparing the performance metrics before and after tuning helps gauge the improvement achieved through hyperparameter tuning. It allows us to verify if the changes made to the model indeed enhance its predictive capabilities.

Selecting the Best Model: Post-tuning, this evaluation helps identify the top-performing models based on their performance on the unseen test data. It ensures that you select the best-performing model for deployment or further consideration.

Providing Final Conclusions: This evaluation assists in summarizing the outcomes of the entire process, emphasizing the improvements achieved through tuning and aiding in decision-making for model selection or next steps in the model.

Therefore, re-evaluating the model on the test set post-tuning is a vital step to ensure you have an accurate understanding of the model's performance and to make informed decisions about which model(s) to proceed with. Figs. (**5** and **6**) represent the result of various models after cross validation and hyperparameter tuning.

COMPARATIVE ANALYSIS AND EVALUATION

Before Tuning

The initial assessment of the models revealed varied performances. Models like LR and NB displayed commendable precision, recall, accuracy, and F1 Score, whereas KNN exhibited relatively lower performance in comparison. Notably, models such as SVM and SGD depicted suboptimal performance, evident from lower accuracy, recall, and F1 Score.

After Tuning (Cross-Validation & Hyperparameter Tuning)

Following cross-validation and hyperparameter tuning, there was a marked improvement across most models. LR showed significant enhancements in precision and F1 Score, indicating better predictive capabilities after tuning. After post-tuning, RF showed notable gains in ROC AUC, accuracy, and precision, indicating its increased robustness as a classifier. CB and XB retained their high-performance levels across multiple metrics, solidifying their positions as top-performing models even after tuning.

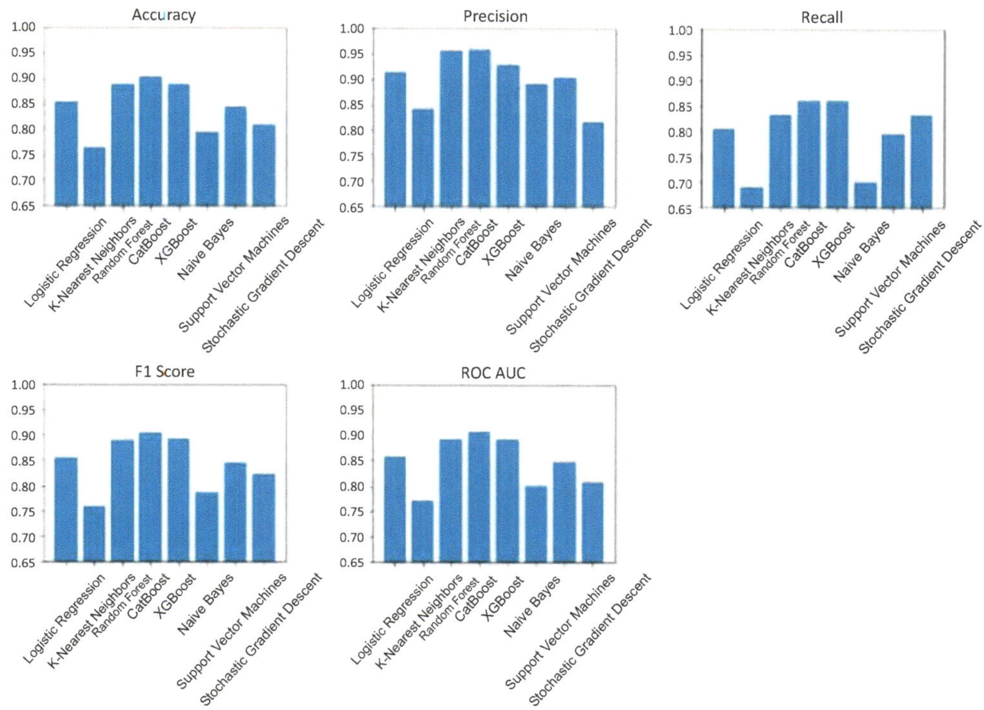

Fig. (5). Re-model evaluation result 1.

```
Top performing models:
           Model  Accuracy  Precision    Recall  F1 Score   ROC AUC
3       CatBoost     0.905   0.958333  0.859813  0.906404  0.908401
2  Random Forest     0.890   0.956989  0.831776  0.890000  0.894382
4        XGBoost     0.890   0.929293  0.859813  0.893204  0.892272

Performance metrics for CatBoost:
Accuracy: 0.9050
Precision: 0.9583
Recall: 0.8598
F1 Score: 0.9064
ROC AUC: 0.9084

Performance metrics for RandomForest:
Accuracy: 0.8900
Precision: 0.9570
Recall: 0.8318
F1 Score: 0.8900
ROC AUC: 0.8944

Performance metrics for XGBoost:
Accuracy: 0.8900
Precision: 0.9293
Recall: 0.8598
F1 Score: 0.8932
ROC AUC: 0.8923

Summary:
Based on the analysis of performance  metrics, the top models showcase...
```

Fig. (6). Re-model evaluation result 2.

Comparative Analysis and Evaluation

The tuning process had a considerable impact on refining the models' predictive abilities. Models that initially had weaker performance, like KNN, exhibited notable improvements in accuracy, precision, and F1 Score. Furthermore, the SVM and SGD models, which initially performed poorly, showed noticeable improvements in multiple metrics after tuning. Figs. (**7** - **10**) represent the before and after tuning of accuracy, precision, recall, and F1 Score for the various models.

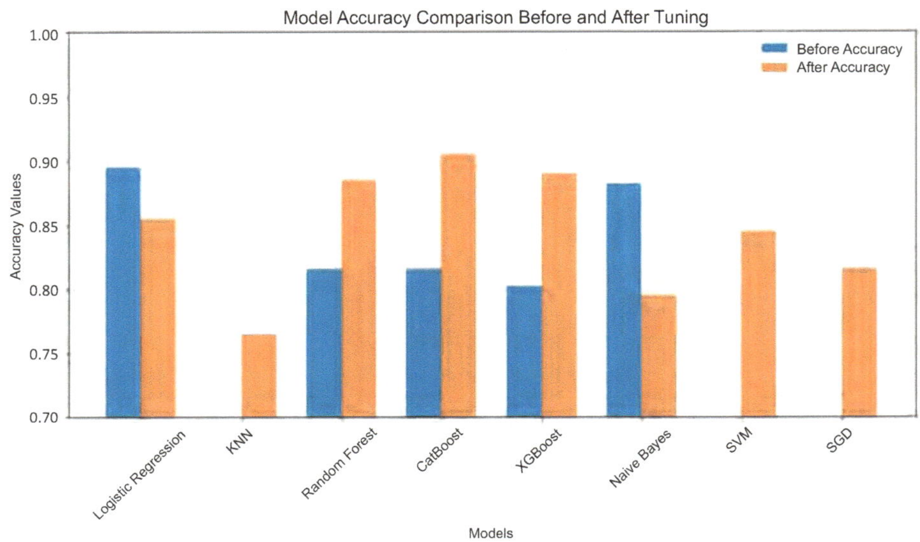

Fig. (7). Model accuracy comparison before and after tuning result 1.

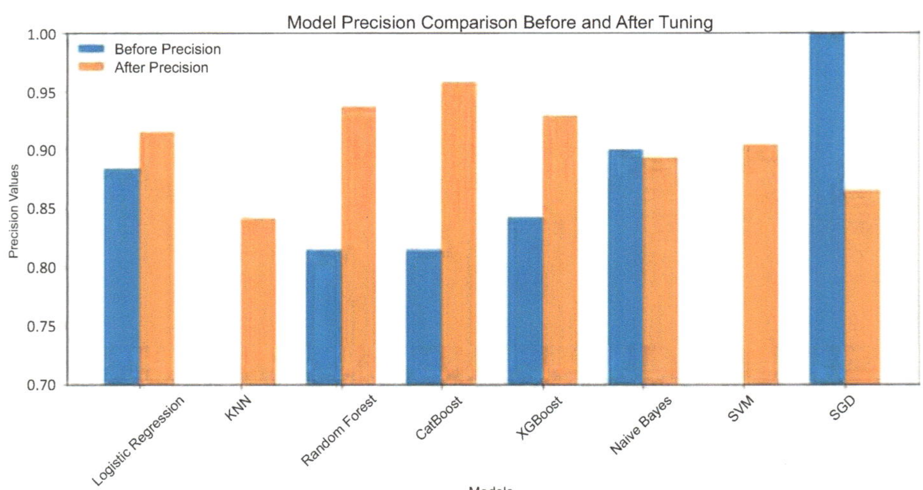

Fig. (8). Model precision comparison before and after tuning result 2.

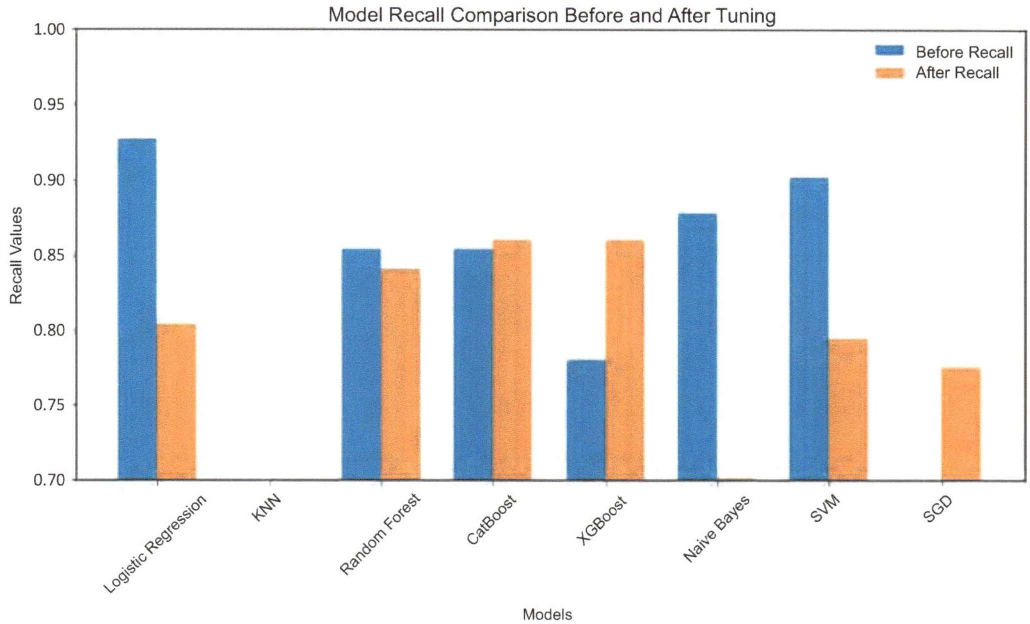

Fig. (9). Model recall comparison before and after tuning result 1.

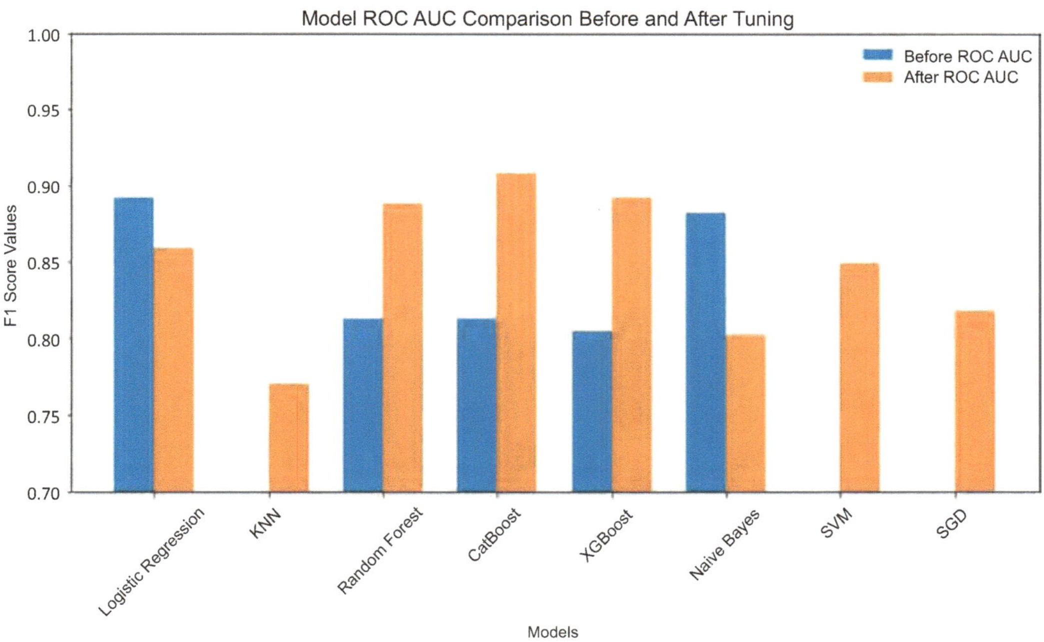

Fig. (10). Model F1 score comparison before and after tuning result 2.

Best Model Selection

The evaluation highlights that CB, after tuning, emerged as the most consistent and robust performer. It displayed noteworthy improvements in precision, recall, accuracy, F1 Score, and ROC AUC. These enhancements position CB as the top-performing model among the others, showcasing its suitability for this specific dataset and problem context. Fig. (**11**). represents the ROC AUC Comparison of various models before and after the tuning.

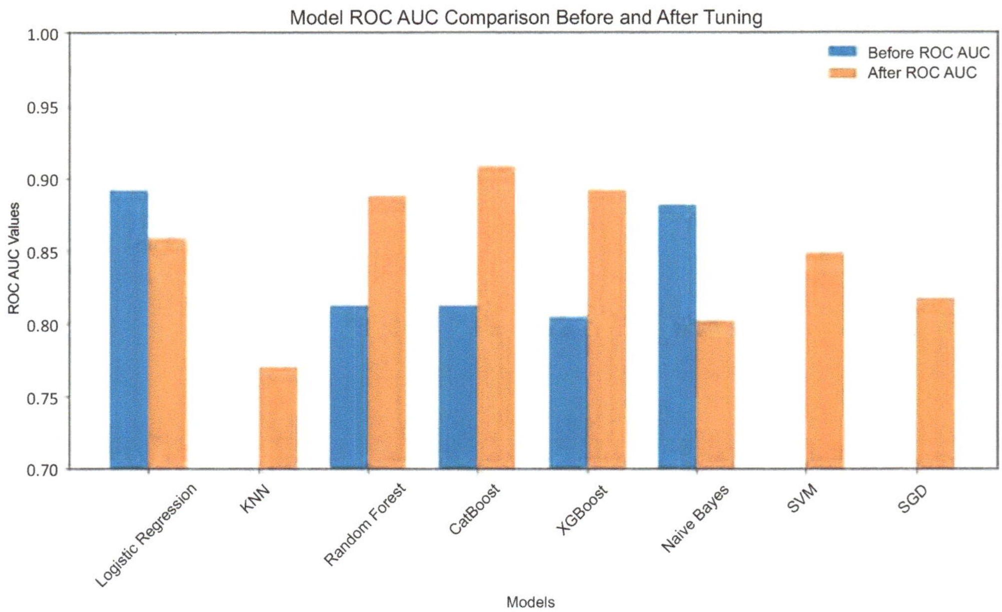

Fig. (11). Model ROC AUC comparison before and after tuning.

DEVELOPMENT OF A HYBRID MODEL

Introduction

The hybrid model we have constructed combines the predictive strengths of three key models: LR, which was deemed the best performer before cross-validation and hyperparameter tuning, CB, identified as the superior model after this tuning, and RF, which demonstrates consistent competence both before and after the tuning process. The inclusion of LR, CB, and RF within this hybrid framework leverages the varied strengths and diverse learning methodologies of these models. LR, recognized for its interpretability and simplicity, serves as a strong baseline, whereas CB, with its advanced boosting technique and optimized parameters after post-tuning, enhances predictive accuracy. Additionally, the RF, exhibiting commendable performance across different scenarios, contributes to the ensemble's robustness and adaptability. This hybridization strategy aims to harness the collective capabilities of these models, potentially enhancing predictive accuracy and resilience across a broad range of datasets and real-world scenarios.

Hybrid Model Implementation

The code demonstrates the creation of a hybrid model using the VotingClassifier (VC) ensemble from Scikit-Learn. The objective is to merge the predictive capabilities of three distinct machine learning algorithms: RF Classifier, CB Classifier, and LR. Initially, individual instances of these classifiers are initialized with specific hyperparameters: an RF Classifier with 100 estimators, a CB Classifier with 100 iterations, and an LR instance. These models are integrated into a VC, which acts as a meta-estimator, combining the predictions of its constituent models. The 'soft' voting scheme, employed in this instance, weighs predictions based on the probabilities assigned by each model, ultimately enhancing the ensemble's predictive accuracy.

Subsequently, the VCs were trained on the given training dataset (X_train and y_train) using the fit() function, which enables the ensemble to learn from the provided data. Through this process, the hybrid model learns to make predictions by aggregating the outputs of the individual classifiers, harnessing the diverse strengths and approaches of each model. Upon training completion, the voting_classifier instance is equipped to predict on new, unseen data, capitalizing on the collective intelligence derived from the constituent classifiers to potentially improve overall predictive performance.

Hybrid Model Performance

Evaluating the performance of the trained hybrid VotingClassifier model involves using various assessment metrics and visual aids. Following the training of the 'voting_classifier' on the dataset, the model undergoes testing using the test set (X_test) to generate predictions ('y_pred') for the target values. Evaluation metrics such as precision, recall, accuracy, and F1 score are calculated to assess the model's predictive accuracy, indicating its capability to accurately classify instances from the test data. Furthermore, the Receiver Operating Characteristic (ROC) curve and its associated Area Under the Curve (AUC) metric are determined, offering insights into the model's balance between actual positive rate and false positive rate across varying threshold values. The resulting ROC curve visualization illustrates the model's discriminative performance, highlighting its ability to distinguish between classes effectively. This thorough assessment aids in comprehending the model's strengths and weaknesses, facilitating the interpretation and evaluation of its predictive abilities. Fig. (**12**) represents the ROC curve of the hybrid model.

```
Accuracy: 0.97
Precision: 0.9902912621359223
Recall: 0.9532710280373832
F1 Score: 0.9714285714285714
ROC AUC: 0.9984926138076575
```

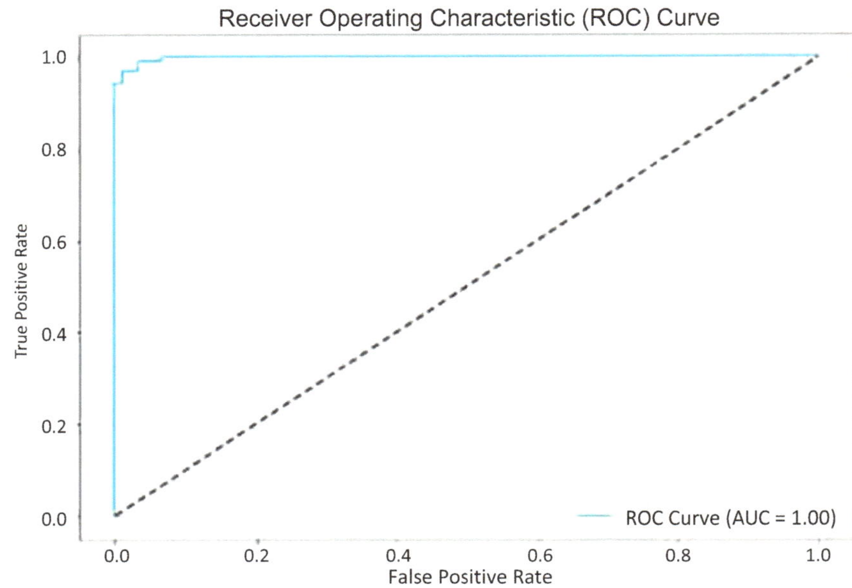

Fig. (12). Hybrid model evaluation result.

RESULTS AND DISCUSSIONS

The study employed a range of machine learning and deep learning methods to forecast heart disease through a thorough analysis of patient data. Initially, numerous algorithms, including LR, KNN, RF, CB, XB, NB, SVM, and SGD, underwent assessment using various performance measures such as precision, recall, accuracy, F1 Score, and ROC AUC.

Metrics for Models Before Tuning

Before hyperparameter tuning, the models exhibited varying performance across different metrics. Some models, such as LR, RF, CB, and XB, showed relatively good performance in terms of precision, recall, accuracy, and F1 scores, while others, including KNN, SVM, and SGD, demonstrated lower scores across multiple metrics.

After undergoing cross-validation and hyperparameter tuning, a refined set of models emerged. Table **1** represents the Performance Comparison of various models before tuning.

Table 1. Performance metrics for initial models.

Model	Accuracy	Precision	Recall	F1 Score	ROC AUC
Logistic Regression	0.8947	0.8837	0.9268	0.9048	0.8920
Naive Bayes	0.8816	0.9000	0.8780	0.8889	0.8819
Random Forest	0.8158	0.8140	0.8537	0.8333	0.8125

Metrics for Models After Tuning

After hyperparameter tuning, there was a noticeable improvement in the performance of most models across different metrics. Models like RF, CB, XB, and LR improved their scores across metrics like precision, recall, accuracy, and F1 score, indicating better-tuned parameters and enhanced predictive capabilities.

The development of a blended hybrid model, which merges LR and CB through a voting classifier, yielded outstanding predictive capabilities. Table **2** represents the performance comparison of various models after tuning.

Table 2. Performance metrics after cross-validation and hyperparameter.

Model	Accuracy	Precision	Recall	F1 Score	ROC AUC
CB	0.9050	0.9583	0.8598	0.9064	0.9084
RF	0.8900	0.9474	0.8318	0.8900	0.8944
XB	0.8900	0.9293	0.8598	0.8932	0.8923

Observations Regarding the Hybrid Model (LR, CB, and RF)

The hybrid model, merging LR, CB, and RF, displayed excellent performance across various metrics. It achieved a high accuracy of 97% and demonstrated impressive precision and recall scores, both above 0.97. The F1 score also reflects a balanced trade-off between precision and recall, and the ROC AUC of 0.9984 suggests outstanding overall model performance.

Hyperparameter tuning significantly improved the performance of individual models, enhancing their predictive capabilities.

The hybrid model, leveraging the strengths of LR, CB, and RF, emerged as a powerful ensemble, showcasing exceptional performance across multiple evaluation metrics, indicating its potential for robust predictions on the heart disease dataset. Table **3** represents the performance comparison of the hybrid model.

Table 3. Performance metrics for hybrid model.

Metric	Score
Accuracy	0.97
Precision	0.9902
Recall	0.9714
F1 Score	0.9714
ROC AUC	0.9984

CONCLUSION & FUTURE WORK

The study employed various machine learning and deep learning models to forecast heart disease through a thorough analysis of patient data. Initially, numerous algorithms, including LR, KNN, RF, CB, XB, NB, SVM, and SGD, underwent assessment using various performance measures such as precision, recall, accuracy, F1 Score, and ROC AUC. Later, the hyperparameter tuning was made with the selected models to improve the performance of the model. Models like RF, CB, XB, and LR improved their scores across metrics like precision, recall, accuracy and F1 score. The hybrid model, merging LR, CB and RF displayed excellent performance across various metrics. It achieved a high accuracy of 97% and demonstrated impressive precision and recall scores. The F1 score also reflects a balanced trade-off between precision and recall, and the ROC AUC of 0.9984 suggests outstanding overall model performance. Future advancements may include integrating hyperparameter optimization with emerging technologies, such as reinforcement learning, to enhance adaptability. Additionally, the model's methodologies might evolve to address multi-objective optimization, considering factors like interpretability, fairness and robustness in model optimization.

REFERENCES

[1] H. Ahmed, E.M.G. Younis, A. Hendawi, and A.A. Ali, "Heart disease identification from patients' social posts, machine learning solution on Spark", *Future Gener. Comput. Syst.,* vol. 111, pp. 714-722, 2020.
[http://dx.doi.org/10.1016/j.future.2019.09.056]

[2] M.M. Ali, B.K. Paul, K. Ahmed, F.M. Bui, J.M.W. Quinn, and M.A. Moni, "Heart disease prediction using supervised machine learning algorithms: Performance analysis and comparison", *Comput. Biol. Med.,* vol. 136, p. 104672, 2021.
[http://dx.doi.org/10.1016/j.compbiomed.2021.104672] [PMID: 34315030]

[3] M. Bhushan, A. Pandit, and A. Garg, "Machine learning and deep learning techniques for the analysis of heart disease: a systematic literature review, open challenges and future directions", *Artif. Intell. Rev.,* vol. 56, no. 12, pp. 14035-14086, 2023.
[http://dx.doi.org/10.1007/s10462-023-10493-5]

[4] V. Chang, V.R. Bhavani, A.Q. Xu, and M.A. Hossain, "An artificial intelligence model for heart disease detection using machine learning algorithms", *Healthc. Anal.,* vol. 2, p. 100016, 2022. [http://dx.doi.org/10.1016/j.health.2022.100016]

[5] M. Diwakar, A. Tripathi, K. Joshi, M. Memoria, P. Singh, and N. kumar, "Latest trends on heart disease prediction using machine learning and image fusion", *Mater. Today Proc.,* vol. 37, pp. 3213-3218, 2021. [http://dx.doi.org/10.1016/j.matpr.2020.09.078]

[6] P.R. L, S.V. Jinny, and Y.V. Mate, "Early prediction model for coronary heart disease using genetic algorithms, hyper-parameter optimization and machine learning techniques", *Health Technol. (Berl.),* vol. 11, no. 1, pp. 63-73, 2021. [http://dx.doi.org/10.1007/s12553-020-00508-4]

[7] R. Katarya, and S.K. Meena, "Machine learning techniques for heart disease prediction: a comparative study and analysis", *Health Technol. (Berl.),* vol. 11, no. 1, pp. 87-97, 2021. [http://dx.doi.org/10.1007/s12553-020-00505-7]

[8] S. Matin Malakouti, "Heart disease classification based on ECG using machine learning models", *Biomed. Signal Process. Control,* vol. 84, p. 104796, 2023. [http://dx.doi.org/10.1016/j.bspc.2023.104796]

[9] A. Naseri, D. Tax, P. van der Harst, M. Reinders, and I. van der Bilt, "Data-efficient machine learning methods in the ME-TIME study: Rationale and design of a longitudinal study to detect atrial fibrillation and heart failure from wearables", *Cardiovasc. Digit. Health J.,* vol. 4, no. 6, pp. 165-172, 2023. [http://dx.doi.org/10.1016/j.cvdhj.2023.09.001] [PMID: 38222103]

[10] Y. Rimal, S. Paudel, N. Sharma, and A. Alsadoon, "Machine learning model matters its accuracy: a comparative study of ensemble learning and AutoML using heart disease prediction", *Multimedia Tools Appl.,* vol. 83, no. 12, pp. 35025-35042, 2023. [http://dx.doi.org/10.1007/s11042-023-16380-z]

[11] S.A. Ahamed, and C. Ravi, "Study of swarm intelligence algorithms for optimizing deep neural network for bitcoin prediction", *Int. J. Swarm Intell. Res.,* vol. 12, no. 2, pp. 22-38, 2021. [IJSIR]. [http://dx.doi.org/10.4018/IJSIR.2021040102]

[12] P. Ranjitha, and M. Spandana, "Predictive analysis for big mart sales using machine learning algorithms", *5th International Conference on Intelligent Computing and Control Systems (ICICCS),* 2021pp. 1416-1421

[13] S. Latthika, *Predicting Heart Disease through Machine Learning Methods.* , 2024pp. 829-842. www.ijisrt.com [http://dx.doi.org/10.38124/ijisrt/IJISRT24SEP382]

CHAPTER 9

A Two-Staged Optimized Stacking Ensemble Learning Classifier for the Prediction of Cervical Cancer

Ch. Bhavani[1], **M. Radha**[2], **G. Lingaiah**[3], **Erupaka Nitya**[3] and **Umarani Nagavelli**[4,*]

[1] *Department of Computer Science, CVR College of Engineering, Hyderabad, India*

[2] *Department of CSE (Cyber Security), SNIST, Hyderabad, India*

[3] *Department of CSIT, CVR College of Engineering, Hyderabad, India*

[4] *AI & DS Department, B V Raju Institute of Technology, Telangana, India*

Abstract: The paper aims to develop a more accurate cervical cancer (CC) detection method. As a consequence of this, the data's dependability will increase. As a direct result, gathering new biomedical data has emerged as a primary goal. The three most significant high-dimensional biological data analytics that may be conducted using bioinformatics are sample classification, early disease detection, and disease diagnosis. This paper provides an earlier data-feature selection technique based on stability (EDFSS). In addition, we construct a two-stage heterogeneous stacked ensemble learning model (HSELM) for predicting cervical cancer patient survival. Both models are designed to predict whether a patient will survive their disease. EDFSS discovers feature subsets from large cancer datasets that are particularly valuable for developing other models. The usage of this strategy is predicted to result in enhanced classification accuracy. The CC Risk Factors, 32 independent risk factors, and four dependent variables were used in this study. This also made extensive use of the CC Risk Factors. These strategies were presented to sanitize the risk variables contained in the previous algorithm. This was done to improve the algorithm's efficacy. The MLP classifier exhibited the best overall classification performance, with 97.51% accuracy, 96.23% precision, 98.10% recall, and 95.21% F1-measure. The paper found that risk variables were selected to train several ML classifiers to predict cervical cancer.

Keywords: Cervical cancer, Ensemble learning classifier, Machine learning, Prediction.

Correspondence: **Corresponding author Umarani Nagavelli**: AI & DS Department, B. V. Raju Institute of Technology, E-mail: umarani.n@bvrit.ac.in

Suneeta Satpathy, Sachi Nandan Mohanty & Subhendu Kumar Pani (Eds.)

INTRODUCTION

Cervical cancer, a deadly illness that affects millions of women globally, necessitates early identification to provide effective treatment and increase survival rates. Machine learning techniques help do this by analyzing medical data and forecasting risks or detecting diseases early. As a result, the model will be capable of successfully solving the problems shown so far. During the training process, a mix of machine learning (ML) and deep learning (DL) models is constructed and tested to determine whether a patient has cervical cancer (CC). As far as we know, the specific procedure has yet to be tested in the laboratory. Using machine learning (ML)-based models to assess high-dimensional biological data is a relatively new area of study in bioinformatics [1 - 3]. The various challenges of machine learning in the prediction of cervical cancer are as follows:

1. **Data Quality**:
    - **Data Imbalance**: An imbalance between positive and negative cases can skew model performance, leading to inaccurate predictions.
2. **Feature Extraction and Selection**:
    - **Complex Features**: Extracting relevant features from sources like medical images, pathology reports, or patient records can be difficult.
    - **Dimensionality**: High-dimensional data, such as images, may require sophisticated techniques for effective handling and interpretation.
3. **Model Interpretability**:
    - **Black-box Models**: Many machine learning models, especially deep learning models, are complex and opaque, which can hinder clinicians' ability to understand and trust the predictions.
4. **Overfitting**:
    - **Overfitting**: Models might perform well on training data but struggle to generalize to new, unseen cases if not properly validated.
5. **Ethical and Privacy Concerns**:
    - **Patient Privacy**: Protecting patient privacy and ensuring the security of sensitive health information used for model training is crucial.
    - **Bias and Fairness**: It is important to ensure that models do not perpetuate or exacerbate existing biases in healthcare to avoid disparities in treatment.
6. **Continuous Learning**:
    - **Model Updating**: As the medical field evolves rapidly, models need regular updates with new data and knowledge to stay accurate and relevant.

Addressing these challenges requires a multidisciplinary approach, integrating expertise from machine learning, clinical practice, and data science to create robust and reliable predictive models for cervical cancer.

Researchers utilized a range of machine learning methodologies, commonly referred to as dimensionality reduction algorithms, during the selection process. As a result, they were able to successfully identify the key components required to achieve the defined learning objectives [4 - 6]. Selecting an appropriate method for ML investigation is a complex task. Dimensionality reduction techniques [7 - 9] transform the input space to generate new features and are often used to identify the most effective biomarkers. Many researchers employ hybrid techniques to reduce dimensionality and enhance processing speed [10].

The exploration and exploitation phases are two critical working stages that should be incorporated into any nature-inspired metaheuristic algorithm. These stages should also be covered. In the exploitation phase, one seeks the best option among the options already accessible, but in the exploration phase, one seeks a novel solution that can be applied broadly. The best choice from the available possibilities is identified during the exploitation phase [11]. There is a chance that "nature-inspired" algorithms will perform better than algorithms that do not take into account natural phenomena. Furthermore, algorithms driven by natural processes need fewer algorithm-specific control parameters. Several evaluations examined possible solutions for dimension reduction in high-dimensional biological data sets [12 - 14]. The new study aims to correct a flaw discovered in previous studies. Feature selection, extraction, and hybrid algorithms that integrate these methods with natural phenomena are the most crucial parts of data reduction.

Each class has supervised, unsupervised, and machine-learning method subcategories. The class has these subcategories. These subcategories are classified as "supervised" or "unsupervised." Finally, this research analyses frequently used algorithms and evaluates current cancer disease classification and prediction using machine learning methods based on high-dimensional gene expression data. This study also examines nature-inspired algorithms. The study used high-dimensional gene expression data. This research aims to establish a new categorization system for the condition and find a way to forecast its early stages. To build a better CC prediction technique, we first presented an earlier data-feature selection based on the stability (EDFSS) method and then constructed it. Both procedures are condensed. As a result, the goal of this study will be to establish a technique for predicting cervical cancer using stacked ensemble approaches while limiting the amount of advance that can be achieved in prediction accuracy.

RELATED WORKS

In a study [15], the authors employed the BFA to limit the feature pool to an acceptable subset. The method now includes an additional penalty parameter; their study examines the effects of combining a penalty function with a fitness function. Implementing a Random Forest classifier improves classification accuracy while lowering the number of characteristics in breast, cervical, and liver cancer detection. This strategy outperforms techniques based on gene expression profiling in terms of both feature reduction and classification accuracy. In this inquiry, the Binary Firefly Algorithm is tested by first significantly limiting the feature pool to a suitable subset. Implementing a Random Forest classifier improves classification accuracy while lowering the number of characteristics in breast, cervical, and liver cancer detection. Regarding feature reduction and classification accuracy, our technique outperforms profiling-based algorithms.

In a study [16], the authors briefly introduced meta-heuristics, meta-heuristic process variations, and hybrid meta-heuristic approaches. Their study sheds light on the feature selection strategies used to analyze CC data to identify research gaps and advise researchers toward prospective new fields of investigation. The study includes a review of relevant literature for both conventional and metaheuristic feature selection approaches for cervical cancer classification applications. Furthermore, an overview of the method of classification is highlighted, including whether nature and the position or population that was discovered influenced the approach.

In a study [17], the authors developed a machine-learning system to evaluate the survival chances for 85 CC patients using PET and CT scan data. This model was used to assess the patient's chances of survival. CT and PET scans were evaluated independently of each other. The original PET and CT imaging feature sets were narrowed using LASSO regression and genetic techniques. The Cox model utilized a subset of the attributes obtained during feature selection, whereas the random survival forest model employed all available data. The models were cross-validated after training on the training set to improve accuracy and usefulness. After that, the models were verified using the consistency index on the test set. The specified CT and PET scan information created a new Cox model. The genetic algorithm beat LASSO regression in feature selection. The Cox model, developed using the CT characteristics supplied, has the highest concordance index score of 0.707.

In a study [18], the authors used a reduced risk list of features, three ensemble-based classification approaches, such as Extreme Gradient Boosting (XGBoost), AdaBoost, and Random Forest (RF), and the Firefly algorithm for optimization to detect cervical cancer early [18]. Combining these methods permitted early cervical cancer detection. They used data on cervical cancer risk variables. Thirty-two risk indicators and four objectives—Hinselmann, Schiller, Cytology, and Biopsy—are included. Four goals cover the most prevalent cervical cancer diagnostic procedures. "Synthetic Minority Oversampling methodology" (SMOTE) was created to address data imbalance. Accuracy, sensitivity, specificity, PPA, and NPA will be utilized to evaluate the proposed research. The experiments reveal that the suggested model provides the best Hinselmann test outcomes for three diagnostic tests. Its success rate demonstrates this. Reducing features also improved performance. Firefly feature selection was used to give better results with fewer characteristics. The suggested algorithms outperform benchmark cervical cancer detection studies that employ a smaller dataset of risk variables. These tests were designed for the detection of cervical cancer.

The authors described an algorithm-based cervical cancer prediction method [19]. Research should include the study dataset, data pre-processing, predictive model selection (PMS), and pseudo-code. This publication's PMS section shows results from various traditional machine-learning exercises. Examples include DT, LR, SVM, KNN, adaptive, gradient, random forest, and XGBoost. DT, GB, and RF achieve 100% cervical cancer prediction accuracy. SVM accuracy is 99%. Conventional machine learning computational issues are studied to quantify model effectiveness. This research interviewed 132 Saudi women who participated in HPV-focused computer-assisted cervical cancer prediction. Data collection required this.

The authors evaluated cervical cancer diagnosis methods [20]. These experiments employed machine learning. Using an ML classifier, they predicted Hinselmann, Schiller, cytology, and biopsy findings. Many cervical cancer risk indicators were examined for these results. UC Irvine's Machine Learning Repository offers free downloads of the dataset. Skewed dataset processing began with oversampling. Matlab machine learning classifier training used the distinct features of a response variable across two tiers. Data showed these traits. Decision Trees, Support Vector Machines, K-nearest neighbors, and Ensemble Learning were used. Accuracy, AU-ROC, sensitivity, and specificity were utilized to assess classifier performance. The Fine Gaussian SVM classifier performed best in the Hinselmann, cytology, and biopsy classification tests, with an accuracy of 97.5%, 62.5%, and 98%. Boosted trees had 81.3% Schiller classification accuracy. In this work, we selected the most significant risk variables from a large number to train multiple machine learning classifiers to predict cervical cancer.

According to the review study [21], the authors conducted a comparison of multiple supervised machine-learning algorithms for predicting cervical cancer data. Their findings revealed that, among all supervised algorithms, ensemble techniques, SVM, and Random Forest algorithms exhibited superior performance. Using the Random Forest algorithm [22], the authors selected features based on RFE and balanced data by SMOTE after comparing various data balancing techniques, and for classification stacking, a two-stage classifier was used to improve classification performance. A based algorithm was combined with a firefly binary version algorithm to extract the optimal number of features based on swarm intelligence [23]. NGBFA is a hybrid swarm intelligence feature selection algorithm inspired by nature. Random forest algorithms, based on swarm communal behavior, present inferences in the search space. According to a study [24], by combining different classifiers, the authors could make an accurate prediction of the plant's disease.

MATERIALS AND METHODS

Because it predicts critical features of cancer patients, the cervical prediction sector of cancer prognosis is an essential aspect of discipline. Fig. (**1**) depicts the HSELM framework.

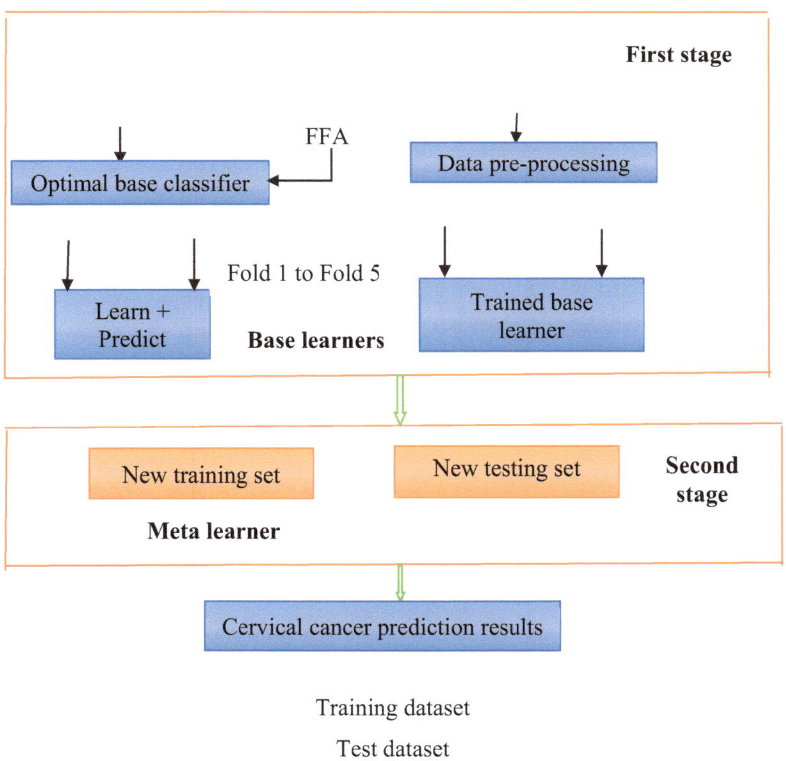

Fig. (1). The framework of HSELM.

Description of the Dataset

This dataset contains information on 858 persons who had cervical cancer screenings. The UCI repository provided the whole dataset. The Universitario Hospital de Caracas in Venezuela gathered the data in early 2017 [25]. The tabular form contains 35 key components, including patient information, critical medical data, behavioral factors, and a goal variable called "Biopsy results."

Data Pre-Processing

The original data collection comprised 858 rows and 36 columns and had several null values. In the text, these indeterminate values were indicated by a question mark ("?"). For convenience, this sign was first modified to "NaN." We compensate for the lack of data in this study by using descriptive statistics. The mean, median, and mode are this study's most often used metrics. As the number of incidents rises, these estimates will become more accurate. On the other hand, skewed findings are a clear possibility if the models in the issue add an unwarranted bias. The mode of the distribution replaced the numerical attributes, while the median restored the categorical ones. When outliers are present, the imputation of the mean value suffers dramatically.

Imbalanced Data:

It is argued that a dataset is class-imbalanced if one class has a disproportionately higher number of entries than the other. Most machine learning models opt for the category that is frequently observed. Consequently, models often value accuracy above detecting unexpected events, such as patients in imminent danger of dying. This is because it may be challenging to see situations that occur seldom. The problem of class imbalance may be approached from several perspectives, including sampling, kernel approaches, and cost-based tactics. One of the obstacles in trying to diagnose and predict cancer using biological data is determining which method will be most beneficial in resolving the issue of class imbalance.

Feature Selection

Feature selection is picking important attributes from a dataset to improve its performance. This distinguishes the necessary and helpful traits from the original high-dimensional input, including many irrelevant features that may impair the classifier's performance. We employed the Firefly Algorithm (FFA) described in this paper as a statistical approach for feature selection. The widely utilized and extremely successful firefly approach has been applied to many optimization problems. It was feasible to solve several complex optimization problems successfully. The following is an overview of the FFA's operating regulations: A firefly social behavior model was used to help build the firefly algorithm. This model was developed because flashing lights or other flashes attract fireflies. The concept stems from the fact that fireflies are attracted to various lights.

The firefly approach utilizes swarm intelligence algorithms and leverages the academic group's interests. This demonstrates that a firefly with a lower luminosity level may attract a firefly with a higher luminosity level. A solitary firefly may fly anywhere it wishes as long as no other fireflies are brighter than it. Because of their natural attractiveness, fireflies can explore new prospective solutions as they seek new possible flawless answers. This enables the fireflies to move to different areas or postures. This aids the fireflies in their quest to explore new chances. The firefly approach is successful when used to solve several optimization issues. There have been several challenging optimization difficulties that have been successfully resolved.

A Heterogeneous Stacked Ensemble Learning Approach with Two Stages

The two most important criteria are the diversity and constancy of foundation learners. To create an ensemble, the learners at the bottom must be "excellent and distinct." When adopting this approach, however, the base learner has many algorithms that are like one another. As a result, the variety of the ensemble learning method is greatly diminished. As a result, the two-stage heterogeneous stacked ensemble learning process known as HSELM was created. This strategy employs the generalized layered method throughout two stages to accommodate different types of base learners. Each of the HSELM foundation learners is made up of one of four unique learning algorithms. It is feasible that this will expand the variety of base learners while overcoming the shortcomings of the three distinct types of base learners and improving cancer survival prediction accuracy. The general framework of HSELM is made of four unique types of heterogeneous base learners in the first stage. This is also known as the first stage of learning. Bagging techniques, BKNN, SVM with Firefly method (FFA-SVM), and Random forest (RF) are examples of these learners. There is just one meta learner in the second stage, known as multi-layer perception based on label smoothing (LS-MLP).

Earlier Data- Feature Selection Based on Stability (EDFSS)

These two-feature selection (FS) strategies improve the model classification. The first stage in creating a classification model is to convert the high-dimensional data to low-dimensional data. Thus, processing overhead decreases. Second, deleting inaccurate or redundant features from the original feature set may help reduce overfitting and misclassification. Thus, the feature selection method expands the space and removes duplicate features to improve categorization, training, and testing. High-dimensional input parameters create multi-collinearity during fitting and instability during training. Build 2m feature subsets if the data collection has such numbers. When m is big, picking the best feature subset from 2m will involve a substantial financial and human resource investment. Thus, a reliable and effective FS must be utilized to find the best features for training the model.

The current work leverages early data and stability to choose the most important feature subset. EDFSS first calculates feature information gain. This procedure may create early data on feature subsets with an information gain larger than 0 and study the method's stability, which successfully explores how various training subsets impact feature preferences. Deleting features with information advantages greater than zero makes a feature vector. This process is only done a few times, and only the most popular qualities are kept.

Selection of Classifiers

When creating techniques, it is important to consider how accurate the feature subset predictions are and how quickly the feature selection process may be calculated. Then, a range of classifiers is employed to choose pertinent features based on several assessment criteria.

Group of Base learners (First stage)

Support vector machine (SVM): This creates a hyper-margin between the various data components, classifying individual data points into the proper class. The SVM algorithm strives to get the greatest significant feasible margin distance if possible. The vectors that measure distance and direction are employed to separate hyperplanes. The following equation, as shown in (1), should be used to feature the hyperplane adequately:

$$s' = \sum_{l=1}^{m} \beta_l b_l a_l \quad (1)$$

The population size (indicated by the variable "number_of_fireflies"), generation (indicated by the variable "maximum_generation"), initial attractiveness coefficient (β0α), light absorption coefficient (γ), and random parameter coefficient (β0) are all calculated as the first step in the process of classifying data using the Firefly Algorithm FFA-SVM. The next step is to use the Firefly Algorithm to optimize the parameters C after initializing the required parameters.

The optimization process may be divided into many stages, the most essential of which are as follows:

1. Calculate the objective function f(x).
2. Begin the firefly noodle population (i = 1,2,.... N).
3. Calculate the light absorption index
4. Determine the distance.

The resulting value is then used to train the data using the SVM approach after collecting the values C from each firefly. The SVM model calculates the accuracy of each C that Firefly generates once the data has been trained. Following that, we use ranking to get parameter C's best values. Following that, the values of C are utilized in the training of the SVM classifier, which is subsequently evaluated using test data. The validity of the conclusions reached from the tests will be assessed based on the results of the tests.

K nearest neighbor (KNN): This is an excellent example of stacked ensemble learning utilizing a non-parametric technique, which implies that the method does not need any prior knowledge of the data to work effectively. The algorithm's primary goal is to reduce the dissimilarity between members of each class. A distance measure is the primary tool that the algorithm uses for data categorization. This conclusion was taken because the traditional Euclidean distance measure must be modified considering the facts. To determine how far away your neighboring blocks are from one another, use the following calculation (2):

$$K(d) = \|x - y\|_l \sum_{l=1}^{m} x_l - q_l \quad \textbf{(2)}$$

As a result, as one of the foundation learners in the first stage, we propose an improved k-nearest neighbors method (IKNN), which can increase KNN's performance considerably and align with the other base learners.

Meta learner (Second stage)

In MLP-based algorithms with limited data, numerous learnable parameters always result in an overfitting issue. In the second step, we incorporate label smoothing into the suggested MLP-based algorithms to address this issue. Let a denote a sample for training and its matching label b ϵ {1, 2, …,x}, where x is the number of classes. In this case, b may be represented by a single-hot vector of x-dimension (3):

$$b_m = \delta_{m,k} \quad \textbf{(3)}$$

where m = 1, 2, · · ·, X, δm,k denotes Dirac, and δm,k = 1 for m = b and 0 if not. If we used the label smoothing method, bm' denotes label rather than label bm (4).

$$b'_m = (1 - \varepsilon)\delta_{m,k} + (1 - \delta_{m,k})\frac{\varepsilon}{X - 1'} \quad \textbf{(4)}$$

where bm' denotes original label and X − 1 classes and ε signifies smoothing factor.

Random forest (RF): A random forest is a collection of decision trees combined with boosting and bagging techniques to minimize overfitting. The random forest is the outcome of combining these approaches with the decision tree structure. The N-estimators, which indicate the number of decision trees, and the criteria, which indicate a metric that analyses the feature of the split, are important hyperparameters. Both hyperparameters relate to the number of decision trees. At this stage, both hyperparameters are considered estimations. The Gini index and entropy functions, integrated into the system, strengthen the decision-making criteria. The Improved K-Nearest Neighbors (IKNN) algorithm can significantly enhance the performance of traditional KNN, making it comparable to other base learners. We must utilize the probability p and the variable x to calculate the Gini index (5).

$$R = 1 - \sum_{l=1}^{,} (q_l)^2 \quad \textbf{(5)}$$

Python 3.8.0 was the programming language for creating the model's representation using the Jupyter Notebook platform. The classifiers were generated by using the Ski-learn package with other built-in resources. The following five evaluation criteria were used: accuracy, recall, precision, F1-measure, and area under the curve (AUC). K-fold cross-validation with a K value of 5 is used to split the data into the testing and training stages.

RESULTS AND DISCUSSION

Performance Evaluation on the Feature Selection Process

The metrics utilized in the benchmark study established the requirements to be fulfilled while selecting assessment measures. The Firefly feature selection approach was used, and the Hinselmann target received 30 distinct features. Each target received two separate sets of features in total. The accuracy, recall, precision, f1-measure, and area under the curve (AUC) for the Hinselmann test target class for KNN, SVM, RF, MLP, and HSELM models are presented in Table 1. Hinselmann's class was created with 14 distinct features. The model's accuracy is evaluated at 97.65, its precision is rated at 96.33, its recall is rated at 98.21, its f-measure is rated at 95.31, and its area under the curve is placed at 97.88. HSELM outperforms the other classifiers regardless of the feature collection employed. Despite this, HSELM's performance with only the specified features exceeds HSELM's performance with all 28 features.

Table 1. Performance of classifiers using 28 and selected features for the Hinselmann target.

Features	Classifier Model	Accuracy	Precision	Recall	F-Measure	AUC
Selected	k-NN	93.21	93.88	94.21	91.99	93.44
	RF	94.88	94.66	95.77	92.44	94.87
	SVM	95.21	95.54	96.88	93.21	95.55
	HSELM	96.66	96.10	97.54	94.21	96.54
28	k-NN	94.87	93.44	95.55	92.55	94.22
	RF	95.65	94.68	96.33	93.22	95.44
	SVM	96.32	95.21	97.88	94.54	96.55
	HSELM	**97.65**	**96.33**	**98.21**	**95.31**	**97.88**

Employment of the HSELM Approach

Since an ensemble learning technique combines the outputs of many separate procedures to perform better. This approach gained acceptance and became a successful approach to machine learning. Methods for ensemble learning must address fundamental issues. Finding a way to combine "good" and "different" key categories is the first problem to be solved. Recall, precision, accuracy, and the F1 score were the primary metrics for classification problems. To demonstrate the HSELM model's usefulness, we evaluated its performance against several traditional classification techniques. The cervical cancer case dataset in Table 2 was used to test several classification methods. The learning models selected after carefully weighing the potential outcomes and conducting a thorough data analysis served as the foundation for the recommended ensemble model. The k-nearest neighbor algorithm, the random forest method, the support vector machine, and the MLP are among them.

Table 2. Performance metrics of all classifier models.

Classifier Model	Accuracy	Precision	Recall	F-Measure
k-Nearest Neighbor (k-NN)	94.21	93.22	95.55	93.59
Random Forest (RF)	95.87	94.22	96.87	94.32
Support Vector Machine (SVM)	96.11	95.10	97.01	95.89
MLP	96.54	95.24	97.21	96.41
Proposed HSELM	**97.51**	**96.23**	**98.10**	**95.21**

The proposed HSELM model outperforms other well-known classifiers in terms of recall, accuracy, and the F1 measure. The proposed model's findings indicate that it outperforms the performance measures of current models, demonstrating its superiority over them. Fig. (**2**) depicts the suggested HSELM model with the highest accuracy level of 97.51%. The k-Nearest Neighbour technique produces any method's least accurate results (94, 21%).

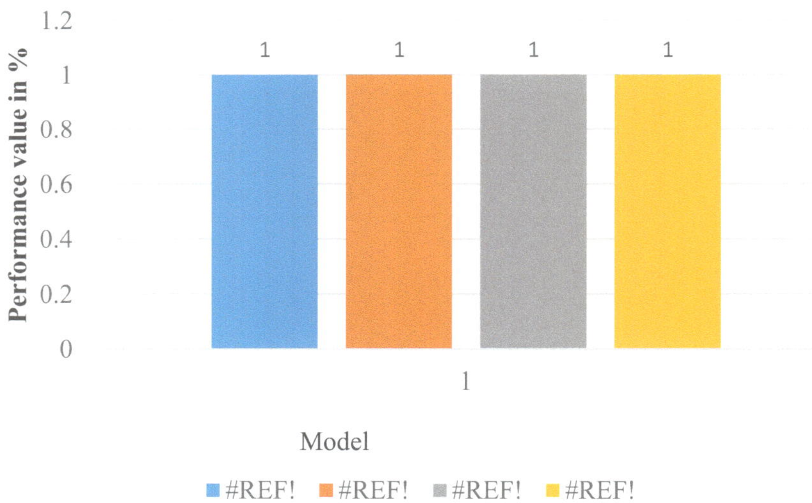

Fig. (2). Demonstration of four performance metrics by base classifiers and proposed HSELM.

CONCLUSION

Cervical cancer kills most reproductive-aged women globally. Cervical cancer screening and early diagnosis may improve treatment. The proposed FFA outperformed EDFSS in feature selection. This solved the radiometric dimensionality reduction problem. Improvements include adding a feature priority criterion to the pipeline and reusing the feature extraction approach in other contexts. UC Irvine's machine learning repository included 858 records, 32 features, and four target variables. Since it uses numerous independent learning algorithms, an ensemble learning technique usually outperforms a base classifier. Thus, it has become popular and useful for machine learning. The most important issue is ensemble learning. The problem is combining “good” with “different.” The HSELM, a unique model, might solve these issues. EDFSS is beneficial in building HSELM, and the suggested approach has the potential to considerably enhance cervical cancer prediction compared to current machine learning methods. This model has 97.65% accuracy, 96.33% precision, 98.21% recall, 95.31% F1-measure, and 97.88% AUC. HSELM output data may be used to predict cancer, improve patient care accurately, and lower medical procedures and treatment costs.

REFERENCES

[1] S.M. Shah, R.A. Khan, S. Arif, and U. Sajid, "Artificial intelligence for breast cancer analysis: Trends & directions", *Comput. Biol. Med.,* vol. 142, p. 105221, 2022. [http://dx.doi.org/10.1016/j.compbiomed.2022.105221] [PMID: 35016100]

[2] R. Aziz, C.K. Verma, M. Jha, and N. Srivastava, "Artificial neural network classification of microarray data using new hybrid gene selection method", *Int. J. Data Min. Bioinform.,* vol. 17, no. 1, p. 42, 2017. [http://dx.doi.org/10.1504/IJDMB.2017.084026]

[3] A.B. Tufail, Y.K. Ma, M.K.A. Kaabar, F. Martínez, A.R. Junejo, I. Ullah, and R. Khan, "Deep Learning in Cancer Diagnosis and Prognosis Prediction: A minireview on challenges, recent trends, and future directions", *Comput. Math. Methods Med.,* vol. 2021, pp. 1-28, 2021. [http://dx.doi.org/10.1155/2021/9025470] [PMID: 34754327]

[4] R. Diao, and Q. Shen, "Nature inspired feature selection meta-heuristics", *Artif. Intell. Rev.,* vol. 44, no. 3, pp. 311-340, 2015. [http://dx.doi.org/10.1007/s10462-015-9428-8]

[5] M.M. Hasan, A.R.M. Shawon, A. Aeyas, and M.A. Uddin, "Cyclic peptides as an inhibitor of metastasis in breast cancer targeting MMP-1: Computational approach", *Inform. Med. Unlocked,* vol. 35, p. 101128, 2022. [http://dx.doi.org/10.1016/j.imu.2022.101128]

[6] S.G. Finlayson, A. Subbaswamy, K. Singh, J. Bowers, A. Kupke, J. Zittrain, I.S. Kohane, and S. Saria, "The clinician and dataset shift in artificial intelligence", *N. Engl. J. Med.,* vol. 385, no. 3, pp. 283-286, 2021. [http://dx.doi.org/10.1056/NEJMc2104626] [PMID: 34260843]

[7] R. Aziz, C.K. Verma, and N. Srivastava, "Dimension reduction methods for microarray data: a review", *AIMS Bioeng.,* vol. 4, no. 2, pp. 179-197, 2017. [http://dx.doi.org/10.3934/bioeng.2017.2.179]

[8] O.A. Alomari, A.T. Khader, M.A.A. Betar, and L.M. Abualigah, "Gene selection for cancer classification by combining minimum redundancy maximum relevancy and bat-inspired algorithm", *Int. J. Data Min. Bioinform.,* vol. 19, no. 1, p. 32, 2017. [http://dx.doi.org/10.1504/IJDMB.2017.088538]

[9] R.M. Aziz, R. Mahto, K. Goel, A. Das, P. Kumar, and A. Saxena, "Modified genetic algorithm with deep learning for fraud transactions of ethereum smart contract", *Appl. Sci. (Basel),* vol. 13, no. 2, p. 697, 2023. [http://dx.doi.org/10.3390/app13020697]

[10] Y. Xiong, M. Ye, and C. Wu, "Cancer classification with a cost-sensitive naive bayes stacking ensemble", *Comput. Math. Methods Med.,* vol. 2021, pp. 1-12, 2021. [http://dx.doi.org/10.1155/2021/5556992] [PMID: 33986823]

[11] A. Ramachandran, R. Rustum, and A.J. Adeloye, "Review of anaerobic digestion modeling and optimization using nature-Inspired techniques", *Processes (Basel),* vol. 7, no. 12, p. 953, 2019. [http://dx.doi.org/10.3390/pr7120953]

[12] A.B. Mohamad, A.M. Zain, and N.E. Nazira Bazin, "Cuckoo search algorithm for optimization problems—A literature review and its applications", *Appl. Artif. Intell.,* vol. 28, no. 5, pp. 419-448, 2014. [http://dx.doi.org/10.1080/08839514.2014.904599]

[13] R. Aziz, C.K. Verma, and N. Srivastava, "Artificial neural network classification of high dimensional data with novel optimization approach of dimension reduction", *Annals of Data Science,* vol. 5, no. 4, pp. 615-635, 2018. [http://dx.doi.org/10.1007/s40745-018-0155-2]

[14] M. Naik, M.R. Nath, A. Wunnava, S. Sahany, and R. Panda, "A new adaptive Cuckoo search algorithm", *Proceedings of the 2015 IEEE 2nd International Conference on Recent Trends in Information Systems (ReTIS),* 2015pp. 1-5 [http://dx.doi.org/10.1109/ReTIS.2015.7232842]

[15] A. Sahoo, and S. Chandra, "Improved cervix lesion classification using multi-objective binary firefly algorithm-based feature selection", *Int. J. Bio-inspired Comput.,* vol. 8, no. 6, p. 367, 2016. [http://dx.doi.org/10.1504/IJBIC.2016.081326]

[16] S. Kurman, and S. Kisan, "An in-depth and contrasting survey of meta-heuristic approaches with classical feature selection techniques specific to cervical cancer", *Knowl. Inf. Syst.,* vol. 65, no. 5, pp. 1881-1934, 2023. [http://dx.doi.org/10.1007/s10115-022-01825-y]

[17] G. Carlini, N. Curti, S. Strolin, E. Giampieri, C. Sala, D. Dall'Olio, A. Merlotti, S. Fanti, D. Remondini, C. Nanni, L. Strigari, and G. Castellani, "Prediction of overall survival in cervical cancer patients using PET/CT radiomic features", *Appl. Sci. (Basel),* vol. 12, no. 12, p. 5946, 2022. [http://dx.doi.org/10.3390/app12125946]

[18] I.U. Khan, N. Aslam, R. Alshehri, S. Alzahrani, M. Alghamdi, A. Almalki, and M. Balabeed, "Cervical cancer diagnosis model using extreme gradient boosting and bioinspired firefly optimization", *Sci. Program.,* vol. 2021, pp. 1-10, 2021. [http://dx.doi.org/10.1155/2021/5540024]

[19] N. Al Mudawi, and A. Alazeb, "A model for predicting cervical cancer using machine learning algorithms", *Sensors (Basel),* vol. 22, no. 11, p. 4132, 2022. [http://dx.doi.org/10.3390/s22114132] [PMID: 35684753]

[20] A. Tak, P. Parihar, F. Singh, and Y. Singh, "Optimised feature selection and cervical cancer prediction using Machine learning classification", *Scr. Med. (Brno),* vol. 53, no. 3, pp. 205-211, 2022. [http://dx.doi.org/10.5937/scriptamed53-38848]

[21] Ch. Bhavani, and A. Govardhan, "Supervised Algorithms of Machine Learning in the Prediction of Cervical Cancer: A Comparative Analysis", *Ann. Rom. Soc. Cell Biol.,* pp. 1380-1393, 2021.https://www.annalsofrscb.ro/index.php/journal/article/view/4580

[22] C.H. Bhavani, and A. Govardhan, "Cervical cancer prediction using stacked ensemble algorithm with SMOTE and RFERF", *Mater. Today Proc.,* vol. 80, pp. 3451-3457, 2023. [http://dx.doi.org/10.1016/j.matpr.2021.07.269]

[23] Ch. Bhavani, and A. Govardhan, *A novel Genetic-Inspired binary firefly algorithm for feature selection in the prediction of cervical cancer,* 2023. [http://dx.doi.org/10.1142/S2424922X23500031]

[24] A. Nagasri, P. Sunitha, V. Sandya, T. Sumallika, P. Kandukuri, K. K. Kumar, "Sugar cane leaf disease classification and identification using deep machine learning algorithms", *J. Theor. Appl. Inf. Technol.,* vol. 101, no. 20, .

[25] "Cervical cancer (risk factors) data set," 2020, HYPERLINK "about:blank" https://archive.ics.uci.edu/ml/datasets/Cervical+cancer+%28Risk+Factors% 29.

CHAPTER 10

Independent Navigation System for Visually Impaired People Using Deep Learning Technique

M. Arun[1], **R. Sriramkumar**[2,*], **R. Sugantha Lakshmi**[1], **R. Jeevaraj**[2], **S. Puvaneswari**[1] and **Shreyas Rajedra Hole**[2]

[1] *Department of Computer Science and Engineering (CSE), Kings College of Engineering, Thanjavur, India*

[2] *Department of Computer Science and Engineering (CSE), Dayananda Sagar University, Bangalore, India*

Abstract: Visually impaired individuals face significant challenges in navigating their surroundings safely and independently indoors and outdoors. These physical and visual limitations often hinder their ability to participate fully in everyday activities. Additionally, perceiving and identifying objects in their environment can be difficult, potentially posing safety risks. This approach proposes a solution to address these challenges and empower greater independence. Employing a real-time video object detection system, the system utilizes a web camera as the input source. It leverages the You Only Look Once (YOLO) model, a CNN-based technique for real-time object detection. This system aims to assist visually impaired individuals by providing audio or haptic feedback on the objects present in their surroundings, ultimately enhancing their safety and fostering greater independence in daily life. Visually impaired individuals face significant challenges in navigating their surroundings safely and independently, both indoors and outdoors. These physical and visual limitations often hinder their ability to participate fully in everyday activities. Additionally, perceiving and identifying objects in their environment can be difficult, potentially posing safety risks. This approach proposes a solution to address these challenges and empower greater independence. Employing a real-time video object detection system, the system utilizes a web camera as the input source. It leverages the You Only Look Once (YOLO) model, a CNN-based technique for real-time object detection. This system aims to assist visually impaired individuals by providing audio or haptic feedback on the objects present in their surroundings, ultimately enhancing their safety and fostering greater independence in daily life.

Keywords: CNN, Object detection, Visually impaired, YOLO.

* **Corresponding author R. Sriramkumar:** Department of Computer Science and Engineering (CSE), Dayananda Sagar University, Bangalore, India; E-mail: sriramkumar-aiml@dsu.edu.in

Suneeta Satpathy, Sachi Nandan Mohanty & Subhendu Kumar Pani (Eds.)

INTRODUCTION

Imagine a world where the ability to see shapes your every move. Millions of people globally face this reality, with the World Health Organization estimating 285 million struggle with vision impairment or blindness. Daily tasks become daunting, and navigating the environment presents constant challenges. However, a new wave of technology offers a beacon of hope. We have developed a real-time object detection system to empower individuals with visual impairments. This system leverages readily available technology – a web camera – and harnesses the power of deep learning through a technique called You Only Look Once (YOLO). Think of YOLO as a highly trained assistant, constantly analyzing the user's surroundings through the web camera and identifying objects in real-time. However, how does this translate into real-world benefits? Using a text-to-speech library, the system seamlessly converts this visual information into audible cues. This means users receive continuous information about their surroundings, with objects and their positions described aloud. This creates a virtual voice guide, empowering users to navigate with newfound confidence and independence. This approach holds immense promise. Testing has shown it to be more effective than previous methods, paving the way for a future where technology seamlessly bridges the gap between vision impairment and a fulfilling, independent life.

RELATED WORK

Several existing solutions address navigation and obstacle detection for visually impaired individuals. One notable project is "Smart Eye: A Navigation and Obstacle Detection for Visually Impaired People through Smart App" [1, 2]. This system utilizes a smartphone camera for image capture and classification and ultrasonic sensors for obstacle detection [3 - 5]. It provides real-time audio information about the user's surroundings, aiming to enhance navigation and safety [6 - 9].

Comparison with Our Proposed System

- Focus: Our project focuses solely on real-time object detection using a web camera, providing audio or haptic feedback for object identification.
- Sensors: While Smart Eye employs both camera and ultrasonic sensors, our system leverages the camera for object detection and potential environment understanding. This simplifies the hardware requirements [10].
- Processing: We utilize the YOLO model, known for its speed and real-time object detection capabilities. It's unclear from the referenced project if a specific model is used for object classification [11 - 13].

- Feedback: Our system explores haptic feedback for spatial awareness, which is not mentioned in Smart Eye.

Proposed Contribution

In a world where sight dictates every action, 285 million people have vision impairment. Technology now offers a beacon of hope through real-time object detection systems tailored for the visually impaired. These innovations empower independence by enabling safer navigation and the completion of everyday tasks, from navigating city streets to shopping independently. By fostering inclusivity, they enhance the quality of life and reshape societal perceptions of disability. This transformative technology signifies a crucial step towards a more equitable world where everyone, regardless of visual ability, can participate fully and confidently in society, unlocking new opportunities and enriching human connections globally.

METHODOLOGY

System Design

The proposed real-time video object detection system aims to assist visually impaired individuals by providing audio or haptic feedback on their surroundings. The system utilizes the following components.

Hardware Implementation

A web camera serves as the primary input source, capturing the live video feed of the user's environment.

- Software Architecture: For Video Capture and Reprocessing, the software continuously captures video frames from the web camera.
- Object Detection using YOLO Model: The captured frames are fed into a pre-trained YOLO model. This model efficiently analyzes each frame and identifies objects present within it.
- Feedback Generation: Based on the objects detected by the YOLO model, the system generates feedback for the user:
- *Audio Feedback:* A text-to-speech (TTS) module converts the identified object labels into audio messages, providing spoken descriptions of the environment. *Haptic Feedback:* The system translates the object type or its spatial location (*e.g.*, left, right, obstacle ahead) into corresponding haptic patterns or vibration cues, offering additional spatial awareness information.
- The You Only Look Once (YOLO) Model: The proposed system employs the YOLO model for real-time object detection. This choice is based on the

following advantages:

- Speed and Efficiency: YOLO models are known for their fast processing capabilities, crucial for real-time video analysis.
- Single-Stage Detection: Unlike some object detection models, YOLO performs object detection and localization in a single step, streamlining the process.
- Pre-trained Models: Pre-trained YOLO models (*e.g.*, YOLOv5) are readily available, allowing us to leverage existing object recognition capabilities and fine-tune them for our specific needs.
- During Development: The chosen YOLO model will be trained on a relevant object recognition dataset to accurately identify objects commonly encountered by visually impaired individuals in their daily lives (*e.g.*, everyday household items, obstacles, doorways).
- Feedback Mechanisms: The system provides feedback through two potential modalities:

  Audio Feedback: The identified object labels from the YOLO model are converted into natural language using a text-to-speech (TTS) module. This synthesized audio clearly describes the user's surroundings, enhancing situational awareness.

  Haptic Feedback:

  The system can utilize haptic feedback modules for additional spatial awareness cues. The type of object or its location in the environment (*e.g.*, left side, obstacle in front) can be translated into distinct vibration patterns or haptic cues, offering non-verbal feedback alongside audio descriptions.

System Implementation

The proposed real-time object detection system was developed using Python as the primary language. We leveraged the following libraries for core functionalities: OpenCV (cv2) for video capture and frame processing from the web camera.

- TensorFlow: To integrate and run the pre-trained YOLO model for object detection.
- Google Text-to-Speech API: To convert identified objects into audio descriptions for user feedback.
- The pre-trained YOLOv5 model was chosen for its speed and accuracy. We fine-tuned the model on a custom dataset containing images of everyday objects relevant to visually impaired individuals. The dataset consisted of X images, categorized into Y classes commonly encountered by visually impaired people in indoor and outdoor environments. Here are some examples of object categories included in the dataset.
- Indoor objects: Chair, table, lamp, door, window, cup, plate, book, medical- tion

bottle.

- Outdoor objects: Car, pedestrian, traffic light, tree, bench, sidewalk, crosswalk. We ensured a balanced distribution of images across all categories to improve the model's generalizability. Additionally, the images captured variations in lighting conditions, object positions, and scales to enhance the model's robustness in real-world scenarios. The system is designed to run on a computer with a web camera.
- Evaluation Methods: The system's effectiveness was evaluated through two primary methods:
- Object Detection Accuracy: We measured the accuracy of the YOLOv5 model in identifying objects within captured video frames. We used standard metrics like precision, recall, and mean average precision (mAP) to assess the model's performance.

System Architecture

1. Object dataset: This dataset is a collection of digital images where various objects relevant to our project are visually identified and labeled. Imagine it as a giant training manual for your system. The dataset should contain images of everyday objects a blind user might encounter, like furniture (chairs, tables), everyday household items (cups, plates), and electronic devices (phones, laptops). The variety of objects, their different poses, and lighting conditions within the dataset are crucial for our model to learn and generalize effectively. Choosing a dataset with high-quality images and accurate labels ensures the system can correctly identify these objects in real-world scenarios, ultimately assisting visually impaired users in navigating their surroundings. Fig. (**1**) shows the concept of the architecture of the system.
2. YOLO Network Architecture YOLO offers real-time processing speed, which is crucial for the minimal delay between capturing an image and announcing the object. Its single-stage detection approach for classification and bound box creation further contributes to efficiency. With multiple variants like YOLO v5 and YOLO v7 offering varying complexity and accuracy, you can find a balance between performance and resource constraints on potentially limited devices used by visually impaired users. YOLO offers several advantages, including real-time processing speed, a single-stage detection approach for efficiency, and multiple variants for flexibility in choosing the right balance of performance and resource constraints. These strengths make YOLO a compelling choice for various object detection applications, including projects focused on assisting visually impaired users.
3. Class Prediction: In the YOLO architecture used in our object detection project, the class prediction module functions like a brain—interpreting the visual

world for visually impaired users. It analyzes features extracted from captured images using a series of neural network layers trained to distinguish between object categories. Based on these analyses, the module assigns probability scores to predefined object classes (chair, table, phone). The object with the highest score triggers the text-to-Speech function, announcing the detected object and aiding the user in navigating their surroundings. This class prediction module is essential for translating visual data into a helpful audio format, empowering visually impaired users.

4. Google API (text-to-speech): Our object detection project using YOLO can leverage Google Cloud Text-to-Speech (TTS) API to transform identified objects into clear audio announcements for visually impaired users. This API seamlessly integrates and receives object data and confidence scores from YOLO. It offers customizable speech synthesis with various voices, speaking rates, and emphasis options for user-friendly announcements. Additionally, the API's multilingual support caters to diverse users, while the optional offline capabilities (through Google's solutions) ensure functionality even in limited internet scenarios. By integrating Google Cloud Text-to-Speech, our project gains the power to translate visual data into natural-sounding audio, enhancing the user experience for visually impaired individuals. We analyzed the results to assess the system's functionality and potential benefits for visually impaired users.
5. Object Detection Performance: The achieved precision, recall, and mAP values will tell us how well the fine-tuned YOLOv5 model performed in identifying objects within the test environment shown in Fig. (**2**). Ideally, we want high precision (meaning the model identifies most objects correctly) and high recall (meaning the model doesn't miss many objects). A strong MAP score would indicate the model's overall accuracy across different object categories. By comparing these metrics with the performance of the pre-trained model on the same dataset, we can assess the effectiveness of the fine-tuning process in enhancing object detection accuracy tailored to the needs of visually impaired users.

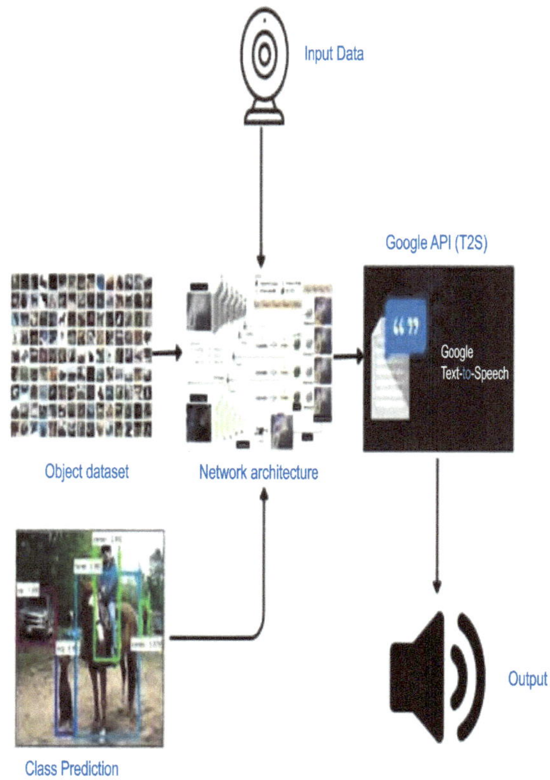

Fig. (1). Architecture diagram.

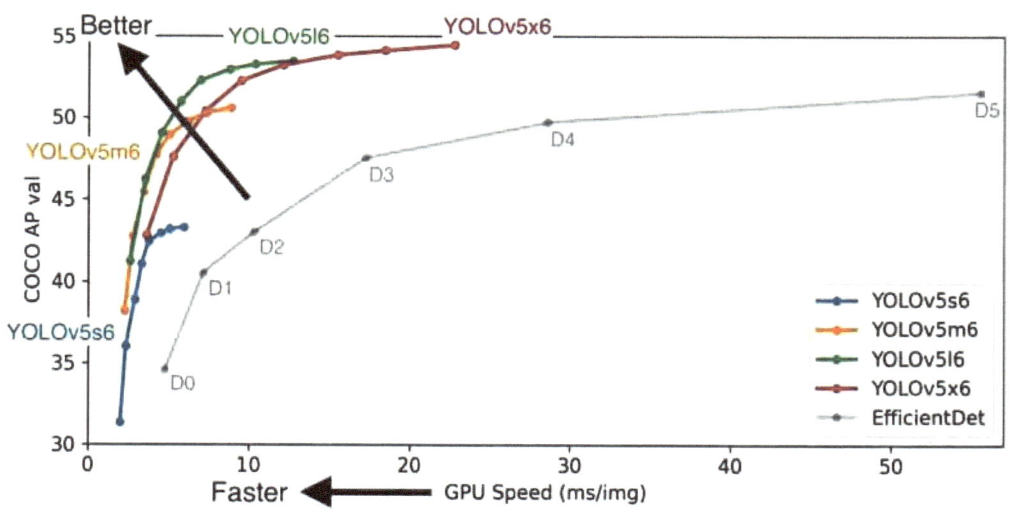

Fig. (2). YOLOv5 Implementation.

CONCLUDING REMARKS

In this paper, we propose a real-time video object detection system to assist visually impaired individuals in navigating their surroundings safely and independently. The system utilizes a web camera as the primary input and leverages the YOLOv5 model for object detection. Audio or haptic feedback mechanisms (optional) provide users with information about objects in their environment. We presented the development process by discussing the system design and implementation details using Python libraries like OpenCV, TensorFlow, and Google Text-to-Speech API. The evaluation methods were outlined, including object detection accuracy metrics and potential user studies. This system can potentially enhance the lives of visually impaired individuals by improving their spatial awareness and fostering greater independence in daily activities.

FUTURE WORK

We acknowledge limitations such as potential computational limitations for mobile device applications and object detection challenges in complex environments. Future work could address these limitations by exploring techniques for model optimization or incorporating additional sensors. Expanding the object recognition capabilities of the system would further increase its utility.

REFERENCES

[1] B. Pydala, T.P. Kumar, and K.K. Baseer, "SmartEye: A navigation and obstacle detection for visually impaired people through smart App", *Journal of Applied Engineering and Technological Science,* vol. 4, no. 2, pp. 999-1011, 2023.

[2] RahmanM.IslamM.AhmmedS.KhanS."Obstacle and fall detection to guide the visually impaired people with real time monitoring", *SN Computer Science,* vol. 1, . [http://dx.doi.org/10.1007/s42979-020-00231-x]

[3] G. Presti, D. Ahmetovic, M. Ducci, C. Bernareggi, L. Ludovico, A. Barat'e, F. Avanzini, and S. Mascetti, "Obstacle sonification for people with visual impairment or blindness", *Proceedings of the 21st International ACM SIGACCESS Conference on Computers and Accessibility,* pp. 1-8, 2019.

[4] M. Shimakawa, K. Matsushita, I. Taguchi, C. Okuma, and K. Kiyota, "Smartphone apps of obstacle detection for visually impaired and its evaluation," in Proceedings of the 7th ACIS International Conference on Applied Computing and Information Technology (ACIT '19), May 2019, Article No. 26.
[http://dx.doi.org/10.1145/3325291.3325381]

[5] P. Gharani, and H.A. Karimi, *Context-aware obstacle detection for navigation by visually impaired. Image and Vision Computing.* Elsevier, 2017.

[6] V.N. Hoang, T.H. Nguyen, T.L. Le, T-H. Tran, T-P. Vuong, and N. Vuillerme, "Obstacle detection and warning system for visually impaired people based on electrode matrix and mobile Kinect", *Vietnam Journal of Computer Science,* vol. 4, no. 2, pp. 71-83, 2017.
[http://dx.doi.org/10.1007/s40595-016-0075-z]

[7] S. Mohapatra, S. Satpathy, and B.K. Paikaray, "A machine learning approach to assist prediction of

Alzheimer's disease with convolutional neural network", *Int. J. Bioinform. Res. Appl.,* vol. 19, no. 2, pp. 141-150, 2023.
[http://dx.doi.org/10.1504/IJBRA.2023.132632]

[8] S.R. Hole, and A. Das Goswami, "EPCMSDB: Design of an ensemble predictive control model for solar PV MPPT deployments via dual bioinspired optimizations", *Science and Technology for Energy Transition,* vol. 79, p. 8, 2024.
[http://dx.doi.org/10.2516/stet/2024002]

[9] B. Kuriakose, R. Shrestha, and F. Eika, "DeepNAVI: A deep learning-based smartphone navigation assistant for people with visual impairment," Elsevier, vol. 212, Feb. 2023.

[10] M. Atri, M. A. Albahar, A. Ben, and Y. A. Alsari, "Obstacle detection system for navigation assistance of visually impaired people based on deep learning techniques," MDPI, June 2023.

[11] S. R. Hole and A. D. Goswami, "Design of a novel hybrid soft computing model for passive components selection in multiple load Zeta converter topologies of solar PV energy system," Energy Harvesting and Systems, vol. 11, no. 1, 2024, Art. no. 20230029.
[http://dx.doi.org/10.1515/ehs-2023-0029]

[12] S. Rajendra Hole, and A.D. Goswami, "Design GA & PSO- Based High-Efficiency SEPIC DC-DC Converter for Context-Aware Duty Cycle Control", *Electr. Power Compon. Syst.,* vol. •••, pp. 1-20, 2023.
[http://dx.doi.org/10.1080/15325008.2023.2275702]

[13] Y. Said, M. Atri, M.A. Albahar, A. Ben Atitallah, and Y.A. Alsariera, "Ob- stacle detection system for navigation assistance of visually impaired people based on deep learning techniques", *Sensors (Basel),* vol. 23, no. 11, p. 5262, 2023.
[http://dx.doi.org/10.3390/s23115262] [PMID: 37299996]

CHAPTER 11

Human Activity Recognition Using Convolutional Neural Networks

Gulle Akshitha[1,*], **Rasuri Rakhesh**[1], **Guguloth Suman**[1] and **R. Suvarna Rao**[2]

[1] *Information Technology Institute of Aeronautical Engineering, Hyderabad-500043, India*

[2] *Institute of Aeronautical Engineering, Hyderabad-500043, India*

Abstract: The identification of human activities holds paramount importance in diverse domains, including CCTV monitoring and human-robot interaction. While existing Human Activity Recognition (HAR) techniques have made substantial progress, challenges persist due to their reliance on manual feature extraction. This research introduces an innovative approach to analyze and identify human actions within small video sequences, employing a convolutional neural network (CNN). The suggested CNN approach combines a 3D convolution technique with a convolutional Long Short-Term Memory process, enabling concurrent examination of spatio-temporal data spanning multiple video frames. In contrast to conventional frame-by-frame analysis, this CNN detects and measures distinctions and resemblances among objects in consecutive frames, enhancing the capability for more intuitive identifications. This approach maximizes processing resources, ensuring efficient training and classification, thereby yielding precise results rapidly. Our novel approach addresses the challenge of model generalization by overcoming the limitations associated with manual feature extraction. By leveraging the combined power of 3D convolution and LSTM, our CNN model demonstrates enhanced capabilities in recognizing and understanding human actions within dynamic video sequences. The presented methodology not only contributes to the advancement of HAR techniques but also opens avenues for more effective utilization of processing resources in real-world applications.

Keywords: CNN, Deep learning, Human activity recognition, LSTM, Neural network techniques, Spatio-temporal analysis, Smart surveillance, 3D Convolution.

INTRODUCTION

The realm of Human Activity Recognition has emerged as a focal point of technological innovation, playing a pivotal role in the development of smart

* **Corresponding author Gulle Akshitha:** Information Technology Institute of Aeronautical Engineering, Hyderabad-500043, India; E-mail: akshitha960350@gmail.com

Suneeta Satpathy, Sachi Nandan Mohanty & Subhendu Kumar Pani (Eds.)

surveillance systems, healthcare monitoring applications, and beyond. The intricate understanding and precise identification of human movements from video data constitute the bedrock upon which advanced automated systems are constructed. In this dynamic landscape, our research undertakes a deep exploration into the prowess of Convolutional Neural Networks as a transformative force in addressing the subtleties and complexities inherent in human activity recognition.

Background and Significance

The surging wave of HAR applications underscores an increasing need for systems that can not only detect but also interpret human actions with a high degree of accuracy and efficiency. In smart surveillance, the capacity to discern anomalous behaviors holds the promise of significantly enhancing security measures, allowing for proactive responses to potential threats. Simultaneously, in the healthcare sector, continuous monitoring of human activities stands as a beacon for early anomaly detection, providing crucial support to individuals grappling with specific medical conditions.

The accurate identification of human activity has long posed a formidable challenge. Traditional methods, reliant on manually designed features and rules-based frameworks, struggled to capture the dynamic and nuanced nature of human actions. Acknowledging the constraints inherent in conventional methodologies, our investigation shifts focus towards the transformative capabilities of CNNs, especially within the vast realm of deep learning. CNNs, equipped with the inherent capability to extract hierarchical features from raw data autonomously, stand as a promising beacon, poised to surmount the historical limitations that have hindered the effectiveness of traditional approaches.

Research Objectives

Positioned at the intersection of cutting-edge technology and practical application, our study harbors a dual-fold objective. First and foremost, we seek to contribute substantively to the evolving field of HAR by harnessing the innate strengths of CNNs. We aim to demonstrate not only the reliability and precision achievable through this approach but also the unprecedented adaptability to diverse human activities.

Through meticulous training on a carefully curated dataset, our ambition is to showcase how deep learning, specifically through the utilization of CNNs, can discern complex human actions with an unprecedented level of accuracy and adaptability. The proposed CNN architecture, characterized by its adaptability and capacity to automatically learn relevant spatial and temporal characteristics, represents a paradigm shift in the landscape of activity recognition.

Proposed Contribution

Beyond the realms of mere activity recognition, our proposed CNN architecture offers a robust and versatile solution capable of seamlessly adapting to a myriad of tasks. By transcending the limitations of conventional methods, our model stands poised to generalize across diverse activities, marking a significant advancement in the field.

Practical Implications

Our research is anchored in a firm understanding of the practical implications of our work. Beyond the theoretical realm, we are committed to addressing the critical need for HAR systems that are not only reliable but also adaptable across diverse and dynamic environments. In healthcare settings, our research can empower continuous monitoring, offering invaluable insights into the early detection of anomalies and contributing to timely and targeted intervention strategies. Simultaneously, in the domain of smart surveillance, the deployment of intelligent systems fortified by our research findings stands as a cornerstone for enhancing public safety and security through real-time insights into human activities.

Research Methodology

Having conducted an exhaustive review of the existing literature and identified critical gaps in current knowledge, our study transitions seamlessly into the elucidation of our chosen methodology. From the intricacies of dataset selection to the meticulous detailing of our experimental setup, we present a comprehensive and transparent approach, providing a robust foundation for our research endeavors. Through this multi-faceted exploration, we aspire not only to contribute to the existing knowledge base of HAR but to pave the way for more advanced, intelligent, and adaptable automated systems. Our research aligns itself with the ever-evolving needs of our technologically driven society, charting a course towards a future where the seamless understanding of human activities becomes an integral and indispensable component of our automated ecosystem.

LITERATURE REVIEW

This study tackles the task of recognizing human actions using sequences of skeleton poses extracted from images or RGB-D data [1]. Traditional algorithms such as support vector machines and k-nearest neighbors are compared with hierarchical neural gas approaches. The emphasis is on identifying the approach that yields the highest accuracy for indoor activities within a home environment.

This paper underscores the utilization of behavior recognition systems based on smart wearable sensors for healthcare [2]. The system uses body sensor data, incorporating distinguishing analysing and Long Short-Term Memory based structured learning for activity modeling and recognitions. The research highlights the practical applications of such systems, especially in smart eldercare environments.

This investigation introduces an innovative wireless system utilizing magnetic induction for the purpose of human activity recognition [3]. The integration of this system with machine learning techniques, such as deep recurrent neural networks, allows for the detection of a wide spectrum of human motions. The study specifically addresses challenges inherent in the design of sensor-based activity recognition systems that operate in proximity to the human body.

The article concentrates on addressing the difficulties associated with the recognition of both basic and intricate human activities within the realm of graphical models [4]. The suggested approach for structured prediction relies on probabilistic graphical models with the goal of identifying activities in a sequence-labeling framework utilizing skeleton data. In order to manage diverse subspaces within the space of all conceivable activities, the study introduces a category-switching scheme.

The research investigates the optimal time window size required to achieve the highest accuracy of a Convolutional Neural Network (CNN) in classifying periodic upper limb rehabilitation motions [5]. The study implements a data-collection system that utilizes a smartwatch and smartphone to measure five distinct upper limb rehabilitation motions over various times. The primary focus is on exploring the relationship between the period of repetitive tasks and the most effective size of the time window for CNN-based Human Activity Recognition [6].

These papers collectively contribute to the evolving field of HAR, addressing various aspects such as sensor modalities, data types, and algorithmic approaches. They offer insights into the challenges and potential solutions for recognizing human activities in different contexts and scenarios [7, 8].

CONVOLUTIONAL NEURAL NETWORKS

Convolutional Neural Networks, often referred to as ConvNets, represent a specialized class of artificial neural networks designed primarily for tasks related to computer vision, image recognition, and video analysis. Developed to mimic the visual processing of the human brain, CNN have demonstrated unparalleled success in various applications, ranging from Transitioning from image classification to tasks such as object detection and facial recognition [9 - 12].

Architecture

The architecture of a CNN is structured to automatically and adaptively learn hierarchical features from input data. The key components that characterize a typical CNNArchitecture includes:

Convolutional Layers

Convolutional layers are CNN basic building components. These layers apply convolution operations to the input, using filters or kernels to detect patterns and features as shown in Fig. (**1**). The application of these filters enables the network to recognize spatial hierarchies and local patterns within the input.

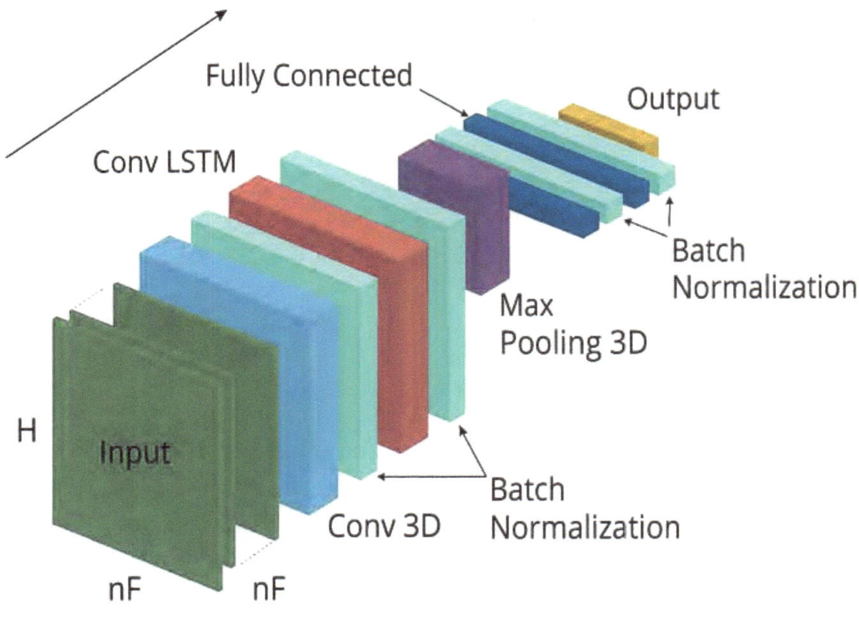

Fig. (1). Convolutional neural networks architecture.

Pooling (Subsampling) Layers

Pooling layers are typically employed after convolutional layers and are specifically crafted to decrease the spatial dimensions of the input volume. Max pooling and average pooling are prevalent operations in pooling, aiming to downsample the input. These operations retain crucial information, simultaneously diminishing computational complexity.

Fully Connected Layers

The convolutional and pooling layers' output is flattened and supplied to one or more fully connected layers. These layers empower the network to make decisions grounded in the acquired features and patterns. In the realm of image classification, the fully connected layers frequently culminate in the final output layer, responsible for generating class predictions.

Convolution Operation

The heart of CNNs lies in the convolution operation, where a filter is applied to the input data to produce a feature map. This process enables the network to automatically learn spatial hierarchies and patterns within the input. The convolutional operation is characterized by the sharing of parameters, allowing the same filter to be applied across different spatial locations.

Weight Sharing and Parameterization

One of the distinctive features of CNNs is weight sharing. In traditional neural networks, each connection between neurons has its own weight, leading to a vast number of parameters. CNNs alleviate this by using shared weights in convolutional layers, significantly reducing the number of parameters. This not only enhances the efficiency of learning but also enables the network to generalize better to unseen data.

Applications

- *Image Classification:* Identifying objects or classes within images.
- *Object detection:* the process of identifying and categorizing various items in a picture.
- *Semantic Segmentation:* Giving every pixel in an image a semantic label.
- *Facial Recognition:* Recognizing and verifying faces in images or videos.
- *Gesture Recognition:* Identifying and interpreting gestures or movements.

Transfer Learning

Convolutional Neural Networks (CNNs) have demonstrated effectiveness in transfer learning, a technique that involves fine-tuning a pre-trained model on a large dataset to adapt it for a specific task with a limited amount of data. This approach is especially advantageous in situations where obtaining labeled datasets is challenging or limited.

Challenges and Advances

While CNNs have achieved remarkable success, challenges still exist, such as the need for substantial labeled data and computational resources. Ongoing research explores ways to address these challenges, including the development of more efficient architectures, attention mechanisms, and advancements in hardware acceleration as shown in Fig. (**2**).

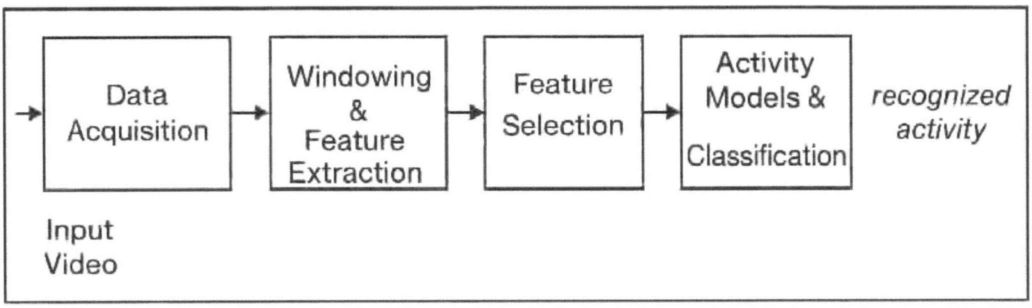

Fig. (2). Flow of model contains input video, data acquisition, feature extraction, feature selection, activity models and classification, recognized model.

METHODOLOGY

Input Video

The process begins with the acquisition of video data, which serves as the primary input for the HAR system. Videos can be obtained from various sources, including surveillance cameras, wearable devices, or publicly available datasets.

Data Acquisition

In this stage, the video data is collected and preprocessed. Data acquisition involves capturing the raw video frames and potentially synchronizing them with other sensor data if available. The goal is to create a comprehensive dataset that reflects the activities of interest.

Windowing & Feature Extraction

To make sense of the continuous stream of video frames, the data is often segmented into smaller time windows or frames. Each window represents a discrete time segment for analysis. Feature extraction is then performed on these windows to capture relevant information. Features can include spatial and temporal characteristics, color histograms, optical flow, or any other relevant attributes that describe the dynamics of the observed activities.

Feature Selection

In this step, a subset of the extracted features may be chosen based on relevance and importance. Feature selection helps in reducing dimensionality and focuses on the most discriminative aspects of the data, contributing to improved model performance and efficiency.

Activity Models & Classification

The selected features are used to train activity models. These models can take various forms, such as machine learning classifiers or neural networks. The goal is to associate specific patterns of features with different human activities. Typical algorithms used at this stage encompass deep learning models such as Convolutional Neural Networks (CNNs) or Recurrent Neural Networks (RNNs).

Recognition Model

The recognition model is the output of the training process and represents the system's ability to identify and classify activities based on the learned patterns. This model is then used to predict activities in unseen data.

CONVOLUTIONAL NEURAL NETWORKS FOR HUMAN ACTIVITY RECOGNITION

First, the input is transferred to the Conv 3D layer, where a convolution kernel is applied to the layer input through the process of convolution, resulting in the generation of a tensor of outputs. Then, the Conv LSTM is used, which is a Recurrent Neural Network architecture. In this particular model structure, the transfer of information from one step of the sequence to the next is facilitated by passing the preceding hidden state. This allows the network to retain knowledge of previous data encountered, aiding in informed decision-making. Subsequently, Batch Normalization emerges as a pivotal technique employed in training deep neural networks. It operates by standardizing the inputs to a layer for each mini-batch, effectively stabilizing the learning process and significantly diminishing the number of training epochs needed for deep network convergence. Another noteworthy element is the utilization of a 3D max pooling layer, which conducts downsampling by segmenting the three-dimensional input into cuboidal pooling regions and computing the maximum value within each region. This technique finds application in spatial and temporal feature extraction, often seen in complex architectures like those designed for video analysis, where both spatial and temporal dimensions are considered.

IMPLEMENTATION

- Dataset Selection and Download.
- Installation of Required Modules.
- Extraction of Features in a Video Dataset.
- Training the model using the Dataset.
- Constructing the Model of the specified Architecture.
- Splitting the Dataset for training and testing.
- Evaluating the model using the Testing Dataset.

In the pursuit of advancing Human Activity Recognition (HAR) systems, a structured approach is paramount, as shown in Fig. (**3**). The journey begins with the meticulous selection and download of appropriate datasets, each uniquely tailored to the intricacies of the target application. For instance, the KARD dataset offers a comprehensive collection, featuring 18 activities performed thrice by 10 subjects, fostering diversity and depth in activity representation. In tandem, the CAD-60 dataset, characterized by 12 distinct activities performed across varying environments by four individuals, enriches the understanding of activities in diverse contexts. Meanwhile, the MSR Daily Activity dataset, with 16 activities and 320 video samples, stands as a testament to the wealth of data available for model training and evaluation.

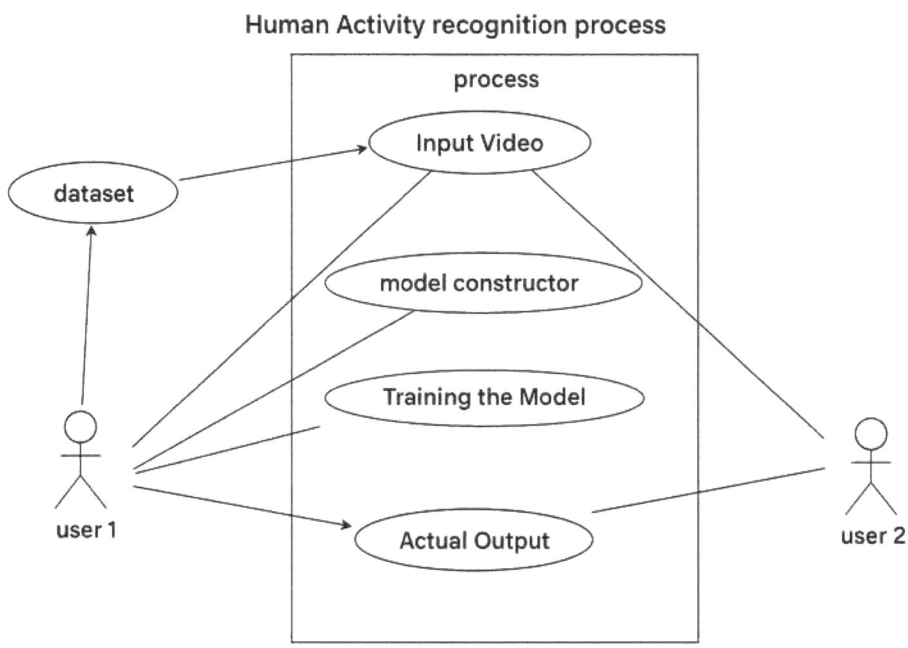

Fig. (3). The sequence of the prediction.

The installation of requisite modules serves as the gateway to unleashing the potential of machine learning frameworks. Modules encompassing video processing, numerical operations, and machine learning libraries, such as OpenCV, NumPy, scikit-learn, and TensorFlow, converge to form a robust foundation for subsequent stages.

Moving into the feature extraction phase, the focus shifts to unraveling meaningful insights from the video datasets. Techniques like optical flow, histograms of oriented gradients (HOG), or spatial-temporal features are employed to distill relevant information encapsulating the essence of human activities. This stage bridges raw video data to a form amenable to machine learning algorithms.

Armed with a feature-rich dataset, the model commences its training journey. Within the training dataset, the model refines its ability to recognizing pattern and connection in features and their corresponding activities. Notably, the inclusion of the KARD, CAD-60, and MSR Daily Activity datasets contributes to honing the model's comprehension, capturing the intricacies inherent in human actions.

The architectural blueprint of the model is next in line, designed to encapsulate the complexities inherent in activity recognition. Be it a combination of 3D Convolution and 2D LSTM layers, the architecture becomes the orchestrator, conducting the symphony of feature extraction and learning.

Following the construction phase, the dataset undergoes a strategic partitioning into training and testing subsets. This division serves the dual purpose of model evaluation and verification of its ability to generalize to unseen data. Rigorous testing on a dedicated testing dataset, distinct from the training data, becomes the litmus test for the model's robustness.

In the final act, the model steps into the limelight to demonstrate its acquired prowess. With the testing dataset as the stage, the model's performance is measured and evaluated.

In synthesis, the journey from dataset selection to model testing is a narrative of precision and adaptability, where each stage contributes to the maturation of a Human Activity Recognition system, poised to decipher the intricate dance of human actions.

RESULT

The output of the project is shown in Figs. (**4** & **5**).

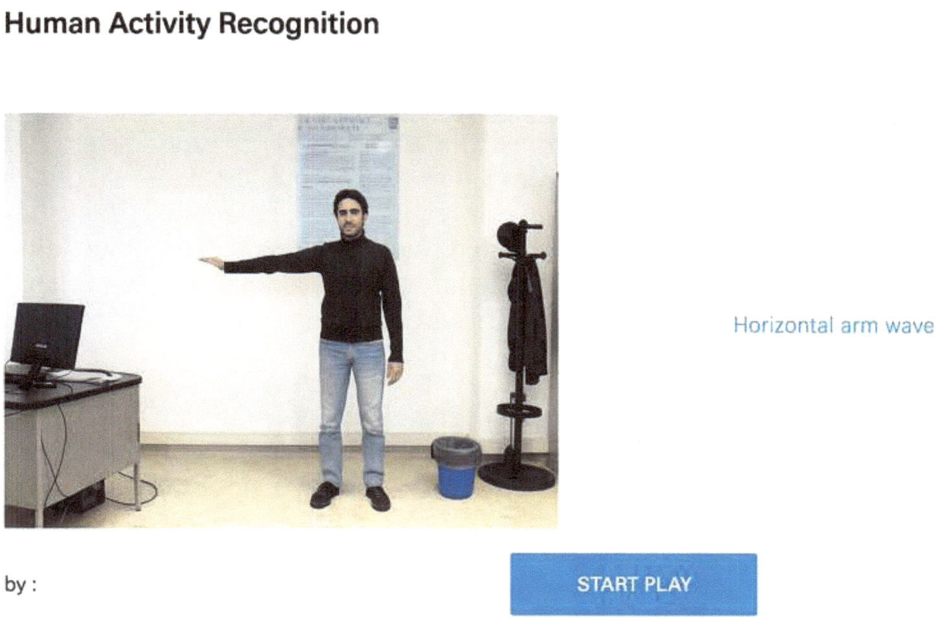

Fig. (4). This figure indicates the result.

State-of-the-art precision and recall values (%) on CAD-60 dataset.

	Precision (%)	Recall (%)
(Yang & Tian, 2014)	71.9	66.6
(Gaglio et al., 2015)	77.3	76.7
(Faria et al., 2014)	91.1	91.9
(Cippitelli et al. , 2016)	93.9	93.5
(Nelli et al., 2017)	97.08	97.2
(Li et al., 2018)	97.8	97.3
Our approach	100	100

Fig. (5). The comparison between the existing System and this proposed system.

CONCLUSION

This study introduces a distinctive and resilient approach to human activity recognition based on convolutional neural networks. This approach addresses significant challenges present in current methodologies by incorporating a 3D convolution process along with a convolutional Long Short-Term Memory process. Our model demonstrates superior capabilities in analyzing spatio-temporal features across multiple video frames simultaneously. This innovation allows for more natural identification of human actions compared to traditional frame-by-frame analysis, leading to efficient training and rapid, precise classification.

Our proposed CNN architecture not only ensures efficient training and accurate classification but also tackles the challenge of model generalization, overcoming limitations associated with manual feature extraction. Leveraging the combined power of 3D convolution and LSTM, our model excels in recognizing and understanding diverse human actions within dynamic video sequences.

The potential applications of this research span various domains, including smart surveillance systems, healthcare monitoring, and human-robot interaction. By contributing to the advancement of HAR techniques, our work opens avenues for more effective utilization of processing resources in real- world applications. The presented CNN- based approach represents a significant step forward in the quest for intelligent automated systems capable of reliably identifying and interpreting human activities across diverse scenarios.

REFERENCES

[1] F. B. Klein and A. Cangelosi, “Human activity recognition from skeleton poses,” IEEE, 2019.

[2] M. Z. Uddin and A. Soylu, “Human activity recognition using wearable sensors, discriminant analysis, and long short-term memory-based structured learning,” SINTEF Digital, 2021.

[3] N. Golestani, and M. Moghaddam, "Human activity recognition using magnetic induction-based motion signals and deep recurrent neural networks", *Nat. Commun.*, 2020.

[4] Mohammad M. Arzani, Mahmood Fathy, Ahmad A. Azirani, Ehsan Adeli, “Switching structured prediction for simple and complex human activity recognition,” IEEE, 2022.

[5] K.-S. Lee, S. Chae, and H.-S. Park, “Optimal time-window derivation for human activity recognition based on convolutional neural networks of repeated rehabilitation motions,” IEEE, 2019.

[6] Y. Andrade-Ambriz, S. Ledesma, M.-A. Ibarra-Manzano, M. Oros-Flores, and D.-L. Almanza-Ojeda, “Human activity recognition using temporal convolutional neural network architecture,” 2022.

[7] D.A. Adama, A. Lotfi, and R. Ranson, "Adaptive Segmentation and Sequence Learning of human activities from skeleton data", *Expert Syst. Appl.*, vol. 164, p. 113836, 2021. [http://dx.doi.org/10.1016/j.eswa.2020.113836]

[8] Y. Zhang, L. Man Po, M. Liu, Y.A. Ur Rehman, W. Ou, and Y. Zhao, "Data-level information enhancement: Motion-patch-based Siamese Convolutional Neural Networks for human activity recognition in videos", *Expert Syst. Appl.*, vol. 147, p. 113203, 2020.

[http://dx.doi.org/10.1016/j.eswa.2020.113203]

[9] S. Nandan Mohanty, S.K. Saxena, S. Satpathy, J.M. Chatterjee, Ed., *Applications of Artificial Intelligence in COVID-19. Medical Virology: From Pathogenesis to Disease Control.* Springer: Singapore.
[http://dx.doi.org/10.1007/978-981-15-7317-0]

[10] S. Satpathy, S. Nandan Mohanty, J.M. Chatterjee, and A. Swain, "Comprehensive Claims of AI for Healthcare Applications-Coherence Towards COVID-19", In: *Applications of Artificial Intelligence in COVID-19. Medical Virology: From Pathogenesis to Disease Control.,* S. Nandan Mohanty, S.K. Saxena, S. Satpathy, J.M. Chatterjee, Eds., Springer: Singapore, 2021.
[http://dx.doi.org/10.1007/978-981-15-7317-0_1]

[11] S. Mohapatra, S. Satpathy, and S.N. Mohanty, "A comparative knowledge base development for cancerous cell detection based on deep learning and fuzzy computer vision approach", *Multimedia Tools Appl.,* vol. 81, no. 17, pp. 24799-24814, 2022.
[http://dx.doi.org/10.1007/s11042-022-12824-0]

[12] S. Satpathy, M. Mangla, N. Sharma, H. Deshmukh, and S. Mohanty, "Predicting mortality rate and associated risks in COVID-19 patients", *Spatial Information Research,* vol. 29, no. 4, pp. 455-464, 2021.
[http://dx.doi.org/10.1007/s41324-021-00379-5]

CHAPTER 12

Lexical Knowledge-Based Deep Learning Model for Emotion Recognition

Manjog Padhy[1], **Rasmita Rautray**[1] and **Sasanka Sekhar Dalai**[1,*]

[1] *Department of Computer Science and Engineering, Siksha'O'Anusandhan University, Bhubaneswar, Odisha, India*

Abstract: Emotion recognition has received exceptional interest in recent years in the discipline of human-computer interaction and affective computing. This study proposes a detailed investigation of deep learning approaches used in emotion recognition by leveraging lexical knowledge. Furthermore, the article presents a description of emotion datasets, pre-processing techniques and data augmentation strategies intended exclusively for deep learning-based emotion recognition. It also discusses the challenges associated with training deep models, such as overfitting, computational costs and class imbalance, along with reasonable alternatives to these problems. It provides a breakdown of popular performance indicators, including Precision, Recall, F1-score and Accuracy on benchmark datasets that are used to evaluate models. The BERT classifier performed impressively across all metrics on Dataset 2, outperforming all other models tested in this study.

Keywords: Deep learning, Emotion recognition, Lexical knowledge.

INTRODUCTION

Emotions serve a purpose and significance in human communication and relationships. Understanding emotions plays a vital role in human communication and relationships, serving both a purpose and significance. However, most 'human computers' process emotional information analytically, lacking the emotional intelligence to respond empathetically. In the last couple of years, there has been increasing interest in automatically detecting and generating emotions in text by employing deep learning models. Because human emotions are diverse and multifaceted, deep learning algorithms frequently have trouble interpreting them. The pairing of lexical knowledge with deep learning models for emotion recognition is an exciting option. Researchers have been figuring out how to

* **Corresponding author Sasanka Sekhar Dalai:** Department of Computer Science and Engineering, Siksha'O'Anusandhan University, Bhubaneswar, Odisha, India; E-mail: sasanka.sekhar.dalai@gmail.com

Suneeta Satpathy, Sachi Nandan Mohanty & Subhendu Kumar Pani (Eds.)

overcome this problem. The lexical knowledge-based deep learning model combines the unique features of deep learning architecture with linguistic resources such as emotion lexicons, word embeddings, and sentiment dictionaries.

These lexical resources provide useful information on the emotional content of words and phrases, improving the model's capacity to recognize and interpret emotions. The previous models have mostly used word embedding, which provides meaning and syntax but cannot capture the emotional relationship between words. To address this issue, we proposed a model called **LKER (Lexical Knowledge-based Emotion Recognition)**, which can utilize all semantic, syntactic, lexical, and emotional information by adopting pre-trained word representations and deep learning models. In this study, six basic emotions, including Anger, Fear, Happiness, Love, Sadness, and Neutral are being focused. Many studies have delivered satisfactory results in the context of emotion recognition from text. Improving the past result of emotion recognition using real-world data of several domains is a big challenge. Fig. (**1**) highlights the entire structure of the paper.

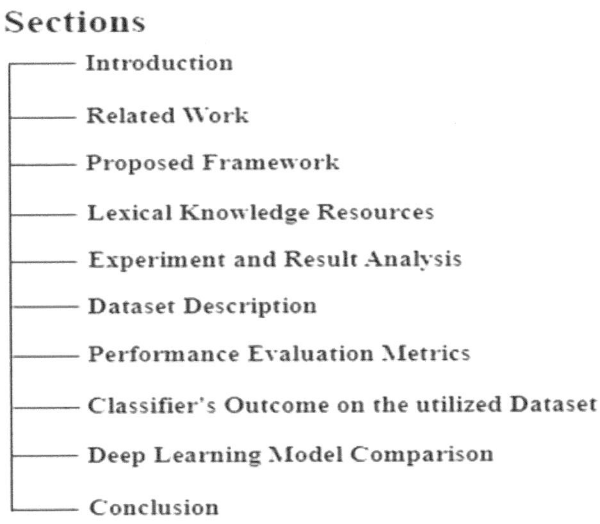

Fig. (1). Paper structure overview diagram.

RELATED WORK

Several studies explored various techniques to detect emotion from text. This paper summarizes the findings using Deep learning models along with the incorporation of lexical knowledge to enhance emotion recognition performance.

Middya Asif Iqbal *et al.* [1] have performed emotion detection on multiple modalities (audio, video, and text), proposing an efficient and lightweight multimodal fusion model to identify emotion states. Two benchmark datasets have been utilized in this study, named SAVEE and RAVDESS. Keyword-based, Lexicon-based [2] methods used in this study utilize traditional as well as distributed text representation for explicit and implicit text recognition in text. The authors of this study also presented emotion detection techniques and resources for emotion recognition. Word embedding techniques that convert the input sequence to a continuous vector space and produce the underlying input representation and the focus on creating, and training an effective network to learn multi-layered emotional features are the major contributions of this study [3]. Shaikh Abdul Salam and Rajkumar Gupta [4] have proposed a system that can detect 8 basic emotions in the live Twitter stream, and the classification was implemented by deep learning models as well as unsupervised machine learning, K-means.

The study also includes a comparative analysis of the efficiency of the classification methods employed. The author uses an annotated corpus and knowledge-based methods to address the task of classifying emotional and non-emotional words in the text. In a study [5], the authors merged traditional machine learning algorithms with deep learning techniques and integrated intelligent machine learning techniques such as an Intelligent Tutoring System (ITS) and Intelligent Learning Environments (ILE) along with emotion classification techniques.

PROPOSED FRAMEWORK

The primary emphasis and objective of this study are to identify emotions through the use of lexical knowledge, along with the integration of vector representation using word embedding techniques. Techniques such as Word2Vec have been implemented in this work to transform refined data into numerical representations after the data-preprocessing phase. The study has been segmented into distinct phases as illustrated in Fig. (**2**).

Model Input Preparation and Dataset Collection

This is the initial phase of the proposed model, in which data and information have been collected from diverse sources. Following collection, this data shall be ultimately fed as input to the Pre-Processing phase to make the data optimized and best suitable for the training of deep learning models.

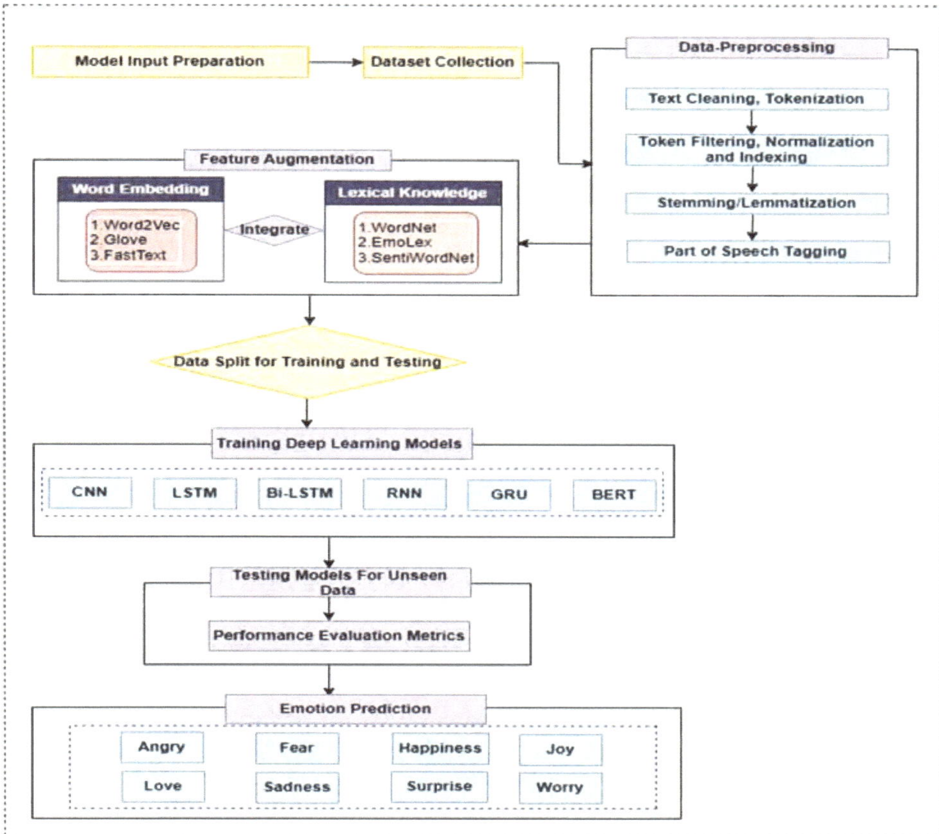

Fig. (2). Proposed framework (LKER) for emotion recognition.

Data-Preprocessing

Preprocessing of data becomes the most essential phase because of the noisy and imbalanced nature of the data present in the dataset. Properly preparing the data guarantees that the model can easily understand, assimilate and identify patterns, relationships present in the subject. Thus, here are some important considerations regarding data preprocessing for emotion recognition as follows:

Text Cleaning and Tokenization is the first phase following data collection, and this process involves the elimination of special characters, URLs, and stripping of HTML tags, which are frequently seen in web-based text data. Thus, all irrelevant and distracting data needs to be cleaned because it does not carry any sense and emotion, which is the precise objective of this study. Furthermore, the text has been split into distinct tokens, which facilitates further analysis.

Token Filtering, Normalization and Indexing:

involve the process of removing stop words, punctuation and high-frequency words that do not contribute any meaning to the text for the proposed work. Normalization usually fixes spelling errors in order to enhance the quality of text data, which is important in situations when misspelled words might cause problems with the ability to identify emotions. Finally, indexing builds a vocabulary of unique words present in the dataset.

Stemming or Lemmatization:

Are two techniques that offer the advantage of reducing the dimensionality of text data and improving the accuracy of machine learning models. An example of Stemming (*e.g.*, "frightened" becomes "frightened"), involves reducing a word to its base form. Whereas Lemmatization (*e.g.*, "happiest" becomes "happy"), involves minimizing a word to its canonical form.

Part of Speech Tagging

Part of speech tagging is probably one of the important techniques in NLP (Natural Language Processing), especially for emotion identification. It makes tagging of texts with respective parts of speech (such as Noun, Verb, Modal Verb, Adjective, *etc.*). It entails grammatical analysis of the text. Fig. (**3**) highlights the process of the technique.

Feature Augmentation

Feature augmentation combines the lexical source of information with the word embedding approach to improve the feature set of emotion recognition tasks. The word embedding technique converts the pre-processed text into numerical representations along with capturing the semantic relationship between words. Meanwhile, a lexical resource contains dictionaries and other sources of structured words and also includes emotion lexicons, which further could be associated with specific emotional tones or sentiments. The suggested integration approach can more accurately facilitate an improved recognition of both semantic context and emotional content from the textual data

The subsequent phase is to split the dataset into training and testing sets with an appropriate ratio and the combined feature set is given as an input to the deep learning classifiers to train to recognize emotions based on the integrated information. The classifiers utilized in this study are CNN (Convolutional Neural Networks), LSTM (Long Short-Term Memory), Bi-LSTM (Bidirectional LSTM), RNN (Recurrent Neural Network), GRU (Gated Recurrent Unit), and BERT

(Bidirectional Encoder Representation from Transformers) [[verse-fn: 6]-[verse-fn: 14]]. Following the training phase, the testing process is carried out by the model with unseen data that has been reserved for evaluation. In addition to this, various evaluation metrics have been computed with the different datasets employed. The following Pseudocode outlines the detailed steps of the proposed framework.

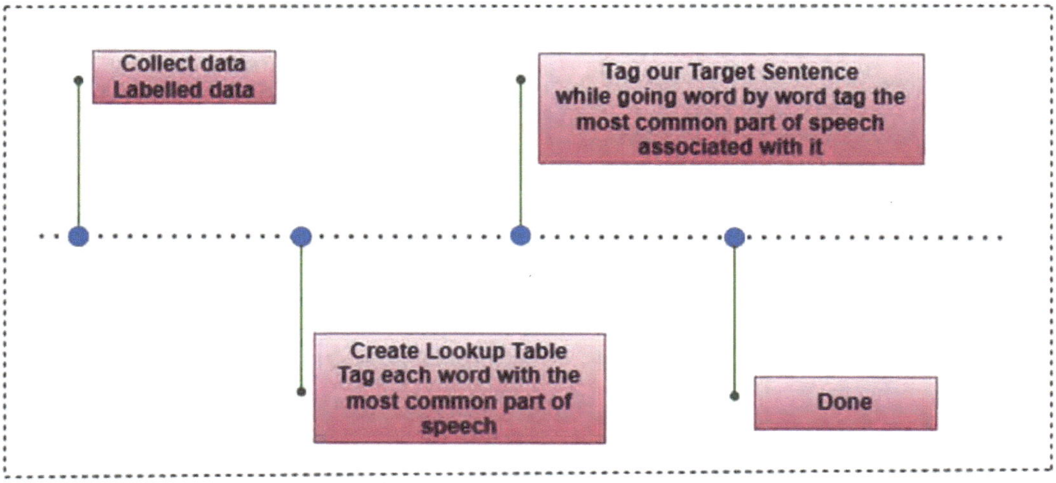

Fig. (3). Lookup table to find part of speech.

Pseudocode of Proposed Framework

1: InputDataset(ds)

2: EvaluateClassImbalance(ds):

class_counts=count occurrences of each label in labels

if class_counts are imbalanced:

new_ds=ApplySamplingTechniques(ds)

3: for each sample in new_ds do

4: Perform TextProcessing:

5: pt=PreprocessText(sample.text)

6: Apply WordEmbedding on Text:

7: wm=WordEmbedding(pt)

```
Apply FeatureAugmentation:
af=FeatureAugmentation (wm, sample. lexical_knowledge)
initialize word2vec with max_features=1000
end for
PartitionDataIntoTrainAndTestSets (length(pt)*0.8)
for each model in models do
TrainModel= (model, training_data)
pt=TestModel (model, testing_data)
CapturePerformanceMetrics (model, pt)
end for
Evaluate Model:
optimal_model=SelectOptimalModel (model, pt)
End
```

The variables and abbreviations used to design the pseudocode have been depicted in Table **1**.

Table 1. Used variables and abbreviations.

Variables	Abbreviations
ds	dataset
new_ds	resampled Dataset
pt	preprocessed_text
we	word_embedding
af	augmented_features
pt	performance_metrics

LEXICAL KNOWLEDGE RESOURCES

These resources are typically employed in Natural Language Processing (NLP) and Deep Learning applications, which involve a collection of organized sets of words and phrases to convey precise meanings. The most common use of such resources is in tasks like sentiment analysis, emotion recognition, and other language-related applications. The following are a few standard categories of lexical resources.

WordNet (WN)

WordNet is a large lexical dictionary that helps semantic analysis to identify the exact meaning of the text or word according to a particular context when that word has multiple meanings. Words are related to one another by WN in an array of ways to each other in terms of synonyms, hypernyms, hyponyms and many more. Sense disambiguation task is the main NLP application of WN.

EmoLex (EL)

EmoLex is the trademark of a specific lexical resource used particularly for emotion-related tasks. This database contains words with more than one annotation and intensity score, which further describes the degree of emotion associated with each word. This lexicon can be used by researchers and practitioners to enhance the deep learning models to interpret the emotional content in the textual data. Table **2** showcases one of the examples of the EmoLex database. This database contains a large number of emotional words and their corresponding associations, along with the intensity score outlined.

Table 2. Emolex word association and intensity score.

Word	Association	Intensity Score
Happy	Joy	0.9
	Anticipation	0.7
	Trust	0.6
Angry	Anger	0.9
	Disgust	0.8
	Fear	0.5

SentiWordNet (SWN)

It serves as a lexical record that makes use of emotional ratings to figure out a text's neutrality, positivity, and negativity. These sentiment ratings can also be used as a feature, enabling the employed classifiers in this research to understand the emotions based on the score from the entire document. The integration of scores into deep learning classifiers may also have the potential to enhance the performance of our model. Table **3** demonstrates the example of the SentiWordNet database.

Table 3. SentiWordnet word association and intensity score.

Word	Association	Intensity Score
Happy	Joy	0.9
	Contentment	0.8
	Trust	0.6
Angry	Rage	0.7
	Fury	0.8
	Irritation	0.7

EXPERIMENT AND RESULT ANALYSIS

The results and analysis from the experiment and implementation have been presented in this section. It evaluates and explores various outcomes of the employed emotion detection model, offering an in-depth assessment and its effectiveness. The performance of the proposed model is evaluated by highlighting various key metrics such as Accuracy, Precision, Recall, and F1-Score. The primary objective of the study has been to deal with multi-class classification for emotion recognition. The standard terms like True Positive (TP), False Positive (FP), True Negative (TN), and False Negative (FN) have been extended to calculate the evaluation metrics, which have been outlined earlier. Table **4** clearly explains the standard formula for the calculation of the used metrics in our study for the results and experimental analysis. Also, we have started to understand the standard terms to build a confusion matrix.

Table 4. Contingency table for emotional state and evaluation metrics calculation.

Actual	Model Prediction	Terms	Metrics	Formula
✓	✓	TP	Accuracy	(TP+TN)/(TP+FP+TN+FN)
⊠	✓	FP	Precision (P)	TP/(TP+FP)
⊠	⊠	TN	Recall (R)	TP/(TP+FN)
✓	⊠	FN	F1-Score	2*(P x R)/(P+R)

Where:
✓ -indicates the presence of Emotional State and
⊠ -indicates non-presence of Emotional State.

DATASET DESCRIPTION

The study examined 2 different datasets focused on the emotional content found on Twitter. These freely available benchmark datasets offer valuable insights and information on the wide range of emotions that users of social media networks express. Six deep learning models have been trained and tested for the computation of performance metrics on these datasets for emotion classification. The no. of features, samples, and classes present in the utilized datasets has been described in the below subsections.

Dataset 1 (SemEval 2018)

The paper discusses various conclusions by utilizing SemEval 2018 as the primary dataset, which comprises 19,400 samples with 6 distinct emotion classes. The list includes Anger, Fear, Happiness, Love, Neutral, and Sadness. Furthermore, the dataset has 2 features, Emotion and Content. Fig. (**4**) represents the distribution of different emotion classes present in the dataset.

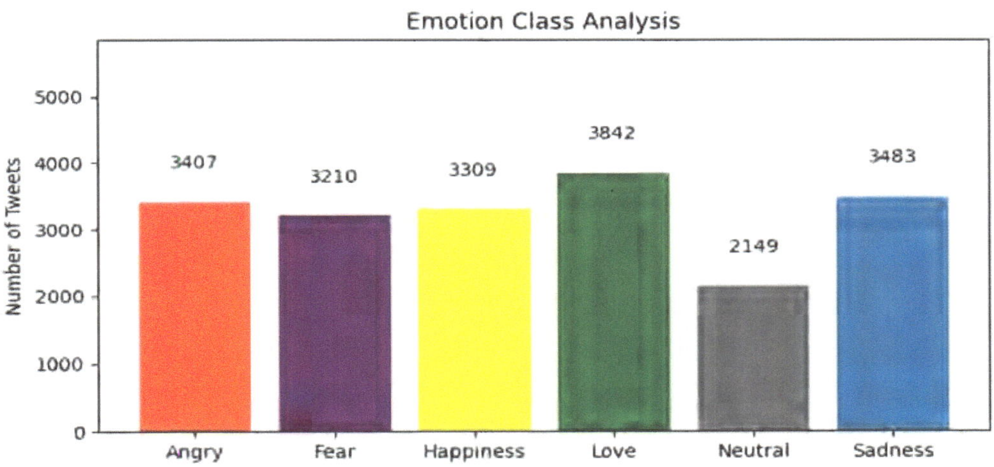

Fig. (4). Emotion class categorization for dataset 1.

Dataset 2 (DAIR-AI-Emotion)

Subsequently, the study ends with the incorporation of the Dair-AI-Emotion as the second and final dataset, having 18,000 samples, each of which has been attached with one of six different emotion classes labeled as Anger, Fear, Happiness, Sadness, Love, Neutral, and Sadness. It contains 2 features: Text and Label. The distribution of emotion classes present in the dataset has been depicted in Fig. (**5**).

PERFORMANCE AND EVALUATION METRICS

This section discusses several evaluation metrics to assess the reliability and performance of the proposed Emotion Detection Model. The primary considerations are Accuracy, Precision, Recall, and F1-score. This study has calculated different values for the above-mentioned metrics using several Deep Learning (DL) classifiers and datasets with a ratio (90% for Training and 10% for Testing). The following are the results as depicted in Tables **5** and **6**.

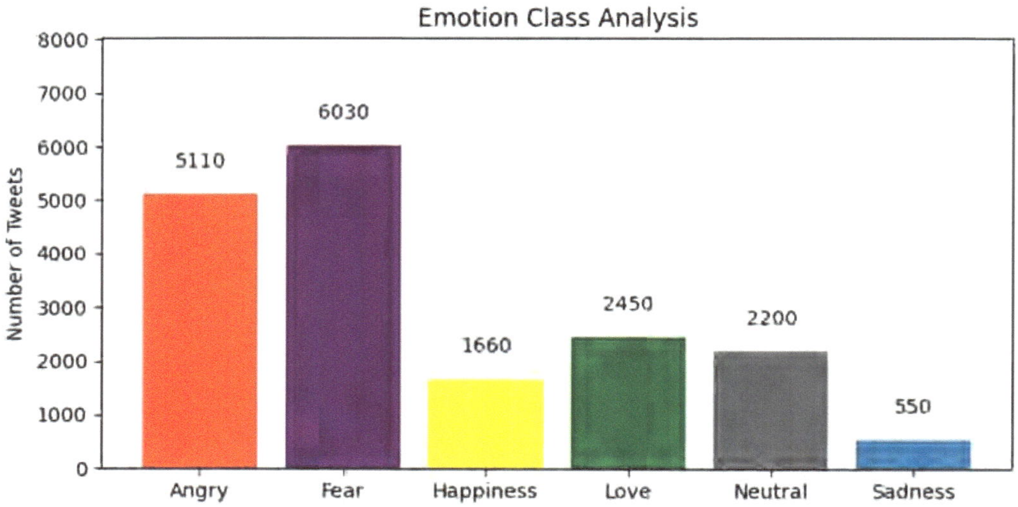

Fig. (5). Emotion class categorization for dataset 2.

Table 5. Performance Evaluation Score for Dataset 1 (SemEval 2018).

Classifiers	Accuracy (%)	Precision (%)	Recall (%)	F1-Score (%)
CNN	71	75	71	70
LSTM	72	73	72	72
BiLSTM	82	84	83	86
RNN	73	78	79	74
GRU	86	80	82	83
BERT	86	87	86	87

Table 6. Performance Evaluation Score for Dataset 2 (DAIR-AI-Emotion).

Classifiers	Accuracy (%)	Precision (%)	Recall (%)	F1-Score(%)
CNN	87	78	80	78
LSTM	89	87	89	90
BiLSTM	90	85	88	86
RNN	71	63	60	61
GRU	92	88	89	88
BERT	93	93	94	93

The above result for Dataset 1 has been represented visually in Fig. (**6**).

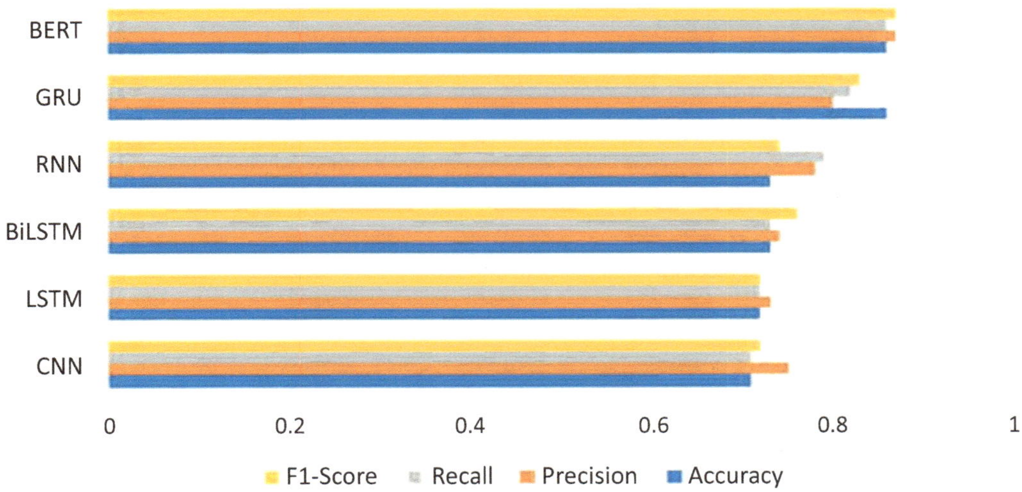

Fig. (6). Classifier performance comparison for dataset 1.

The above results for Dataset 2 have been visually represented in Fig. (**7**).

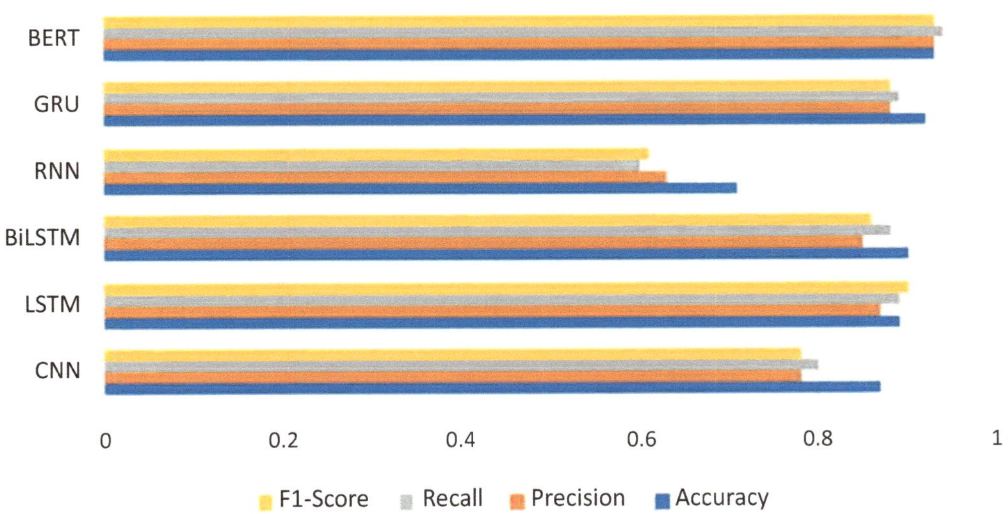

Fig. (7). Classifier performance comparison for dataset 2.

Despite using two datasets in our study for emotion classification, still, our primary emphasis is on the second dataset as it achieves superior performance when paired with the BERT model. Furthermore, the DAIR-AI-Emotion dataset has been partitioned into three ratios that can be viewed as (90% for Training and 10% for Testing), (80% for Training and 20% for Testing) and (70% for Training and 30% for Testing), respectively to facilitate a result comparison with (batch size=6 and epochs=5). In addition to this, the results of the Confusion Matrix are depicted sequentially in Figs. (**8-10**) for Dataset 2.

Fig. (8). Confusion matrix with ratio (90-10).

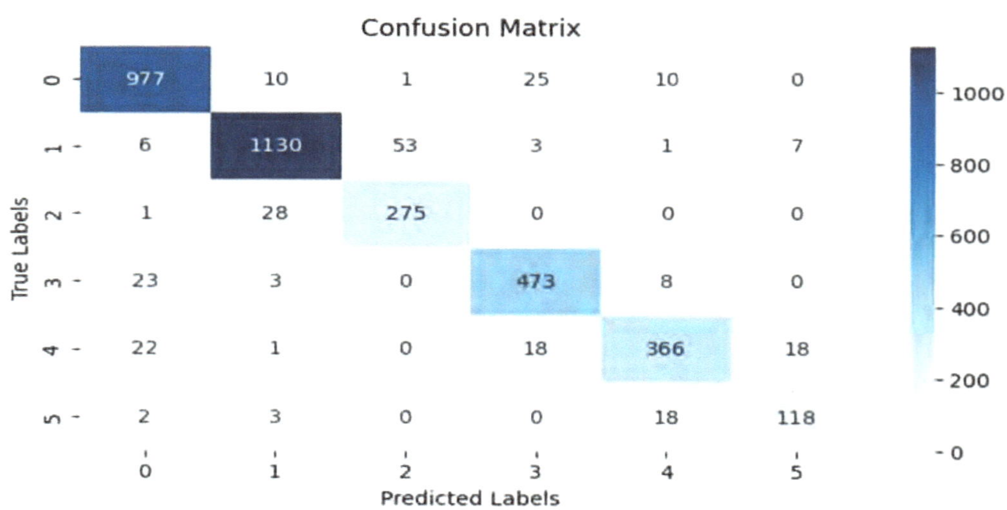

Fig. (9). Confusion matrix with ratio (80-20).

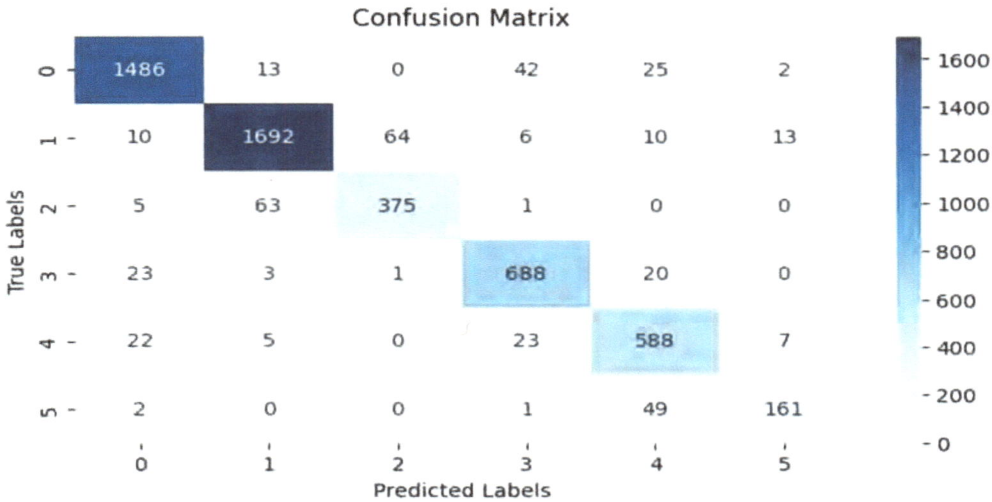

Fig. (10). Confusion matrix with ratio (70-30).

The study performs a series of experiments on the DAIR-AI-Emotion dataset, which has 6 classes with an unbalanced distribution. The proposed BERT model has promising results in all classes. The emotions Anger, Fear, Happiness, Love, Neutral, and Sadness correspond to Class 0, Class 1, Class 2, Class 3, Class 4, and Class 5, respectively. The Confusion Matrix depicted in Fig. (**8**) has provided insights with 1800 testing samples. The outputs from the figure indicate that out of 502 samples, the model has successfully predicted 487 samples for class 0 (Anger Emotion). Similarly, for Class 1(Fear Emotion), the number is 578 out of 607 samples. Furthermore, for Class 2, Class 3, Class 4, and Class 5 the number of accurate predictions is 141 out of 160, 225 out of 246, 189 out of 209, and finally 53 out of 76. Likewise, Fig. (**9**) highlights the Confusion Matrix with 3600 testing samples with 80-20% ratio. The accurate predictions for each class are as follows: 977 for Class 0, 1130 for Class 1, 275 for Class 2, 473 for Class 3, 366 for Class 4, and 118 for Class 5. Finally, Fig. (**10**) shows the Confusion Matrix with 5400 testing samples with a split ratio of (70-30). Again, the predictions for each class are 1486 for Class 0,1692 for Class 1,375 for Class 2,688 for Class 3,588 for Class 4, and 161 for Class 5. Performance Evaluation for the considered metrics is depicted in Table **7** across various ratios using the provided training and testing samples, which comprise 18,000 total samples in Dataset 2.

DEEP LEARNING (DL) MODEL COMPARISON

In this study, it has been closely examined and compared with several deep learning classifiers applied on independent datasets for emotion recognition task as illustrated in Table **8**. Due to the informal language and non-lexical nature of the contents, several key challenges have been addressed. It has also been presented with the outcomes of emotion recognition observations, which utilizes the lexical knowledge resources in detecting emotional words in sentences. The BERT classifier achieved an Accuracy (A), Precision (P), Recall (R), and F1-Score of 0.93, 0.93, 0.94, and 0.93, respectively, on the Dair-AI Emotion dataset—significantly outperforming previous work on the same task.

Table 7. Performance Evaluation based on Data Split Ratio.

Ratio	Training Samples	Testing Samples	Accuracy	Precision	Recall	F1-Score
90-10	16200	1800	0.93	0.93	0.94	0.93
80-20	14400	3600	0.92	0.92	0.92	0.91
70-30	12600	5400	0.92	0.92	0.91	0.91

Table 8. DL Classifier Comparison and Evaluation Metrics.

S. No.	Classifiers Used / Reference	Datasets Employed	Performance Metric			
			A	P	R	F1-Score
01	LSTM	SemEval-2007, ISEAR	0.68	0.67	0.65	-
02	CNN	Affect in Tweets	-	0.75	0.75	0.75
03	GPT	20newsgroups	-	0.80	0.91	0.85
04	CNN+LSTM	Twitter Glove	0.88	0.86	-	-
05	Autoencoder	JCharlies	0.76	0.74	0.72	-
06	LSTM	Emotion Classification	0.78	-	0.78	0.74
07	KNN	Aman 2007	-	0.81	0.75	0.78
08	Gaussian Supervector	USC-IEMOCAP	0.62	-	-	-
09	Context BERT	EmoWoz	0.78	0.77	0.72	-
10	**Proposed Classifier (BERT)**	**Dair-AI-Emotion**	**0.93**	**0.93**	**0.94**	**0.93**

CONCLUSION

This present study addresses the fields of emotion recognition using lexical features taken from textual data, presenting a novel contribution to the subject. The research we conducted employed two distinct datasets, each containing multiple emotion classes. We develop and evaluate six different deep learning models as stated earlier. Furthermore, the significance of preserving and blending lexical knowledge from different emotion lexicon sources aims to improve the precision of emotion recognition.

Our outcomes demonstrate that the association of lexical resources customized for a particular language increases the performance of the emotion recognition system. On the basis of these observations, it can be confidently stated that there is an imperative need to develop emotional lexicon in languages other than English. Hence, research should be carried out beyond English resources to encompass multiple languages.

Moving forward, as future work, the research will continue to focus on emotion recognition in other language resources. Furthermore, a plan shall be explored to incorporate additional linguistic information in the recognition process.

REFERENCES

[1] A.I. Middya, B. Nag, and S. Roy, "Deep learning based multimodal emotion recognition using model-level fusion of audio–visual modalities", *Knowl. Base. Syst.,* vol. 244, p. 108580, 2022. [http://dx.doi.org/10.1016/j.knosys.2022.108580]

[2] N. Alswaidan, and M.E.B. Menai, "A survey of state-of-the-art approaches for emotion recognition in text", *Knowl. Inf. Syst.,* vol. 62, no. 8, pp. 2937-2987, 2020. [http://dx.doi.org/10.1007/s10115-020-01449-0]

[3] J. Deng, and F. Ren, "A survey of textual emotion recognition and its challenges", *IEEE Trans. Affect. Comput.,* 2021.

[4] S.A. Salam, and R. Gupta, "Emotion detection and recognition from text using machine learning", *Int. J. Comput. Sci. Eng.,* vol. 6, no. 6, pp. 341-345, 2018.

[5] M.L. Barron-Estrada, R. Zatarain-Cabada, and R. Oramas-Bustillos, "Emotion Recognition for Education using Sentiment Analysis", *Research in Computing Science,* vol. 148, no. 5, pp. 71-80, 2019. [http://dx.doi.org/10.13053/rcs-148-5-8]

[6] R. CBalabantaray, M. Mohammad, and N. Sharma, "Multi-class twitter emotion classification: A new approach", *Int. J. Appl. Inf. Syst.,* vol. 4, no. 1, pp. 48-53, 2012. [http://dx.doi.org/10.5120/ijais12-450651]

[7] Karna, Mounika, D. Sujitha Juliet, and R. Catherine Joy. "Deep learning- based text emotion recognition for chatbot applications." In 2020 4th International Conference on Trends in Electronics and Informatics (ICOEI) (48184), pp. 988-993. IEEE, 2020. [http://dx.doi.org/10.1109/ICOEI48184.2020.9142879]

[8] S. Shaheen, W. El-Hajj, H. Hajj, and S. Elbassuoni, "Emotion recognition from text based on automatically generated rules", *2014 IEEE International Conference on Data Mining Workshop,* IEEE, pp. 383-392, 2014. [http://dx.doi.org/10.1109/ICDMW.2014.80]

[9] Q. Jin, C. Li, S. Chen, and H. Wu, "Speech emotion recognition with acoustic and lexical features", In: *In 2015 IEEE international conference on acoustics, speech and signal processing (ICASSP).* IEEE, 2015, pp. 4749-4753.

[10] I. Ameer, N. Bölücü, M.H.F. Siddiqui, B. Can, G. Sidorov, and A. Gelbukh, "Multi-label emotion classification in texts using transfer learning", *Expert Syst. Appl.,* vol. 213, p. 118534, 2023. [http://dx.doi.org/10.1016/j.eswa.2022.118534]

[11] S. Nandan Mohanty, S.K. Saxena, S. Satpathy, J.M. Chatterjee, Ed., *Applications of Artificial Intelligence in COVID-19. Medical Virology: From Pathogenesis to Disease Control.* Springer: Singapore. [http://dx.doi.org/10.1007/978-981-15-7317-0]

[12] S. Satpathy, S. Nandan Mohanty, J.M. Chatterjee, and A. Swain, "Comprehensive Claims of AI for Healthcare Applications-Coherence Towards COVID-19", In: *Applications of Artificial Intelligence in COVID-19. Medical Virology: From Pathogenesis to Disease Control.,* S. Nandan Mohanty, S.K. Saxena, S. Satpathy, J.M. Chatterjee, Eds., Springer: Singapore, 2021. [http://dx.doi.org/10.1007/978-981-15-7317-0_1]

[13] S. Mohapatra, S. Satpathy, and S.N. Mohanty, "A comparative knowledge base development for cancerous cell detection based on deep learning and fuzzy computer vision approach", *Multimedia Tools Appl.,* vol. 81, no. 17, pp. 24799-24814, 2022. [http://dx.doi.org/10.1007/s11042-022-12824-0]

[14] S. Satpathy, M. Mangla, N. Sharma, H. Deshmukh, and S. Mohanty, "Predicting mortality rate and associated risks in COVID-19 patients", *Spatial Information Research,* vol. 29, no. 4, pp. 455-464, 2021. [http://dx.doi.org/10.1007/s41324-021-00379-5]

CHAPTER 13

ChatCare – Enhancing Mental Health Support Using NLP

Anila Macharla[1,*], G. Kiran Kumar[1], Rishika Ponna[1], Bhasuri Polaki[1] and **Rakhee[2]**

[1] *Department of Computer Science and Engineering, Chaitanya Bharathi Institute of Technology, Hyderabad, Telangana State, India*

[2] *Faculty of Engineering, The University of the West Indies, Mona Campus, Kingston, Jamaica*

Abstract: This chapter introduces ChatCare, a chatbot-based platform designed to meet the escalating demand for accessible mental health support globally. Leveraging artificial intelligence and natural language processing algorithms, ChatCare delivers personalised emotional assistance and guidance. Notable features include sentiment analysis for tailored responses and intent recognition for meaningful interactions. Metrics such as emotion detection accuracy and intent recognition rates are employed for evaluation. With WHO reporting that nearly 1 billion people worldwide live with a mental disorder, and approximately 75% of those in low and middle-income countries do not receive treatment, the need for innovative solutions like ChatCare is evident. Moreover, a survey by Deloitte revealed that 82% of respondents believe technology can enhance mental health management, while studies indicate that 76% of users value emotional support in chatbot interactions. Against the backdrop of rising mental health challenges intensified by occurrences such as the COVID-19 pandemic, scalable and cost-effective solutions like ChatCare are crucial for promoting mental well-being globally. Through its user-friendly interface and round-the-clock availability, ChatCare aims to provide support to individuals irrespective of their geographical location or socioeconomic status.

Keywords: Mental health support, NLP, Personalized assistance, Sentiment analysis.

INTRODUCTION

In the contemporary landscape of mental health support, traditional methodologies often prove inadequate in meeting the multifaceted needs of individuals worldwide. With the escalating prevalence of mental health disorders, there arises an imperative for innovative solutions to provide effective support and

* **Corresponding author Anila Macharla:** Department of Computer Science and Engineering, Chaitanya Bharathi Institute of Technology, Hyderabad, Telangana State, India; E-mail: anilarao.m@gmail.com

Suneeta Satpathy, Sachi Nandan Mohanty & Subhendu Kumar Pani (Eds.)

intervention. Leveraging advancements in technology, particularly in Natural Language Processing (NLP) and Artificial Intelligence (AI), offers a promising avenue to enhance mental health support services. The introduction of the ChatCare initiative represents a significant stride towards addressing mental health issues. ChatCare, an intelligent platform powered by NLP and AI algorithms, aims to revolutionize mental health assistance by providing personalized and accessible support to users. Through the analysis of text-based interactions, ChatCare adeptly comprehends and responds to users' emotional states, furnishing tailored guidance and resources for coping and self-management. Employing a user-centric design ethos, ChatCare ensures inclusivity and usability for individuals from diverse backgrounds, thereby fostering trust and engagement in the mental health support process. Additionally, the collaborative synergy between experts in psychology, computer science, and user experience design strengthens ChatCare's capabilities in providing effective and empathetic care.

By seamlessly integrating state-of-the-art technologies and fostering interdisciplinary collaboration, ChatCare aims to bridge the gap in mental health care delivery, ultimately empowering individuals to navigate and enhance their mental well-being in an ever-evolving global context.

LITERATURE SURVEY

In the domain of mental health chatbots, various studies have made significant contributions to the development and evaluation of diverse approaches. Notably, a study [1] delved into the application of tokenization, featurization, intent classification, and entity extraction, attaining notable intent accuracy, particularly excelling in recognizing mental illness intent. Dhanasekar *et al.* [2] placed their emphasis on emotion recognition as a mechanism for supporting the mental health of students, achieving remarkable accuracy in discerning positive and negative emotions.

Rani and colleagues [3] harnessed Natural Language Processing (NLP), TF-IDF, N-Gram, and sentiment analysis to facilitate therapy and health monitoring. Although they introduced the Clinical Load Index as a relevant metric, specific scores were not provided. In contrast, Anandan *et al.* [4] applied a decision tree classifier and established user registration/query sessions as integral components of their approach, yet conspicuously omitted specific evaluation metrics.

Tanmay *et al.* [5] incorporated sentiment analysis and tokenization and named entity recognition, and effectively enhanced medication reminders and the quality of medical assistance. SureshKumar *et al.* [6] employed a diverse array of classification, regression, and NLP algorithms along with emotion detection and reported a great accuracy. In the pursuit of real-time emotion identification and analysis, Anastasiia *et al.* [7] harnessed artificial neural networks and text analysis, although they refrained from providing explicit metrics for their study. Crasto and co-workers [8] employed NLP and GPT alongside health questionnaires and presented training accuracy scores for various models. A study [9] made use of artificial neural networks and LSTM algorithms to predict diseases and judiciously reported accuracy, loss, and complexity metrics as measures of their model's performance. In contrast, another [10] adopted ChatterBot, Flask, and a selection of machine learning algorithms, but abstained from providing specific metrics for their evaluation. Notably, Damij *et al.* [11] scrutinized the function of AI chatbots in public services and proposed intriguing future research directions. However, they refrained from detailing the specific methodologies employed or the metrics used in their examination. Biswas and colleagues [12] employed intent classification algorithms, facial recognition techniques, and dialog flow to deliver therapy. They conscientiously reported accuracy scores for training, cross-validation, and test datasets.

Kaladevi [13] implemented NLP, support vector machines, and API integration to cater to healthcare needs. However, they chose not to furnish specific evaluation metrics for their study. Medeiros [14] leveraged IBM Watson NLU, MongoDB, and machine learning algorithms to study stress levels, although they refrained from specifying the metrics used for evaluation.

In their study, Santos [15] deftly combined NLP, supervised machine learning, and an array of diverse metrics, including precision, recall, F-score, to rigorously evaluate their chatbot's performance. Chua *et al.* [16] primarily focused on web-based server-side scripting for providing support for individuals experiencing depression, drawing conclusions from user feedback. Gupta [17] investigated information retrieval and recommendation systems in the context of mental health. However, they did not explicitly enumerate the metrics used to evaluate their research. Teles *et al.* Laijawala [18] extensively explored mobile applications and the Internet of Things (IoT) to aid those dealing with depression. Yet, they did not specify the metrics used for evaluating their approach. Siddik *et al.* [19] employed data mining methods such as Decision Tree, Random Forest, and Naïve Bayes to predict mental health issues, aiming to raise awareness among employees and employers.

Popa [20] proposed a novel deep learning chatbot framework for mental health assessment, with a keylogger module for social media tracking and a BERT-based classifier showing superior performance, effectively engaging users in motivational support and mental health screening. Another study [21] utilized NLP techniques for chatbot therapy, achieving a notable accuracy in depression symptom reduction. They also employ a dataset flagged with critical keywords to identify mental health issues and anticipate chatbots' potential as practical therapeutic tools.

A novel Deep learning model is suggested by Macharla [22] with the help of ResNet50 and SVM to diagnose Parkinson's disease. It used speech data as input and implemented their approach. The affected people with the disease usually suffer with depression, which makes them go for mental support sometimes. This model developed helps clinicians to suggest treatment in less time.

The author [23] developed a prediction model for heart disease using GridsearchCV method, it highlighted the usage of the method and how it can aid diagnosis of the disease. An LSTM-based DNN model is developed by [24 - 28] to diagnose Parkinson's using vocal data from the affected. Later comparisons were done with state-of-the-art models, it proved to be the best approach with better accuracy.

PROPOSED METHODOLOGY

Workflow

The Mental health support system revolves around four primary entities: users, chatbots, emotion detection models, and data visualization components. Upon user interaction, the chatbot engages with the user, allowing them to express their emotions and concerns freely. Advanced NLP algorithms analyze the user's text inputs in real-time, detecting underlying emotions such as sadness, happiness, or anxiety. The system then generates personalized dashboards for each user, showcasing their emotional trends over time. These dashboards include graphical representations, such as bar graphs, illustrating the distribution of emotions across selected date ranges. Additionally, they feature a curated library of mental health-related resources.

Algorithms

Algorithm 1: Chatbot

Start

Step 1: User initiates conversation with chatbot

Step 2: Chatbot receives user message

Step 3: Tokenize and preprocess the user message

Step 4: Perform intent classification using NLP techniques

If User intent is detected

Step 5: Provide appropriate responses based on detected intent

EndIf

Stop

Algorithm 2: Emotion Detection

Start

Step 1: Aggregate user conversation data

Step 2: Analyse conversation messages

Step 3: Apply deep learning models for emotion detection

Step 4: Assign emotion scores to each message

Step 5: Calculate overall emotion scores for the entire conversation

Step 6: Display emotion analysis (*e.g.*, mood is happy, angry, or sad)

Stop

Architecture

The architecture of the proposed methodology is depicted in Fig. (**1**) below. This can be understood in three phases. In phase one, an input from the user is accepted as a query and then pre-processed, which consists of tokenization of text, followed by cleaning and then removal of stop words with a stemming process. Next phase does the intent recognition using NL techniques in order to interact with the user according to the genre of queries posted. Emotion recognition and classification of the same is done by using ML models with the help of Tensorflow and Keras. The last phase revolves around the assessment of the model that is suggested in the proposed approach.

Comparative Analysis

Many researchers suggested and identified a few methods to help people seeking mental health support with Machine Learning and Deep Learning related concepts. There is a need for a simple and user-friendly application that helps users understand and self-analyze, assess their mood and get suggestions to get better support. Table **1** provides such methods developed and used by different enthusiasts. Most of them used Neural networks and their variations as they depict the human brain.

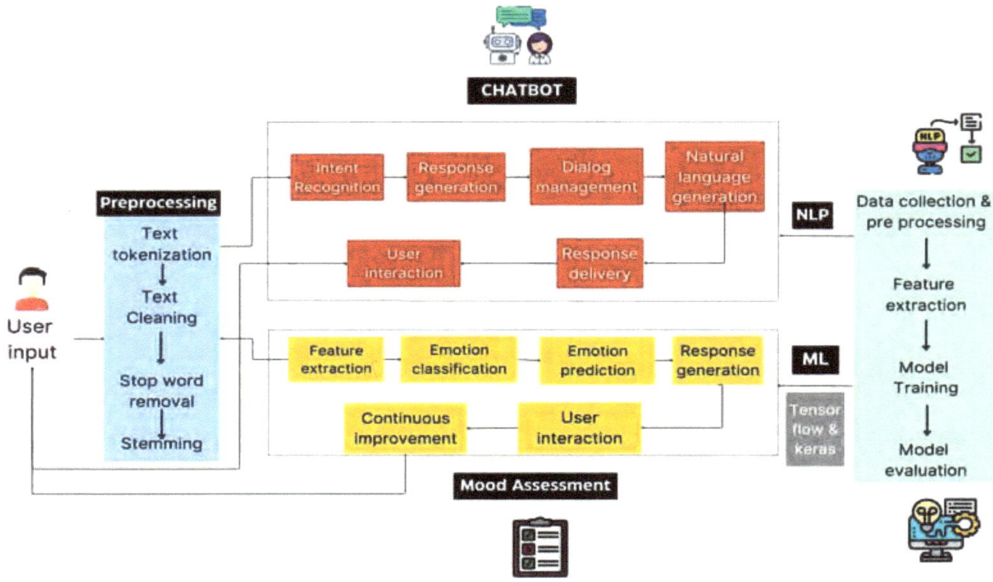

Fig. (1). Proposed flow diagram.

Table 1. Comparative analysis of various methods.

Author(s)	Methodology Used
Nayar *et al.*	Tokenization and lemmatization of text, Featurization, Intent Classification, Entity Extraction
Dhanasekar *et al.*	Intent Processing (used DialogFlow), Entity Extraction, Response Generation
Rani *et al.*	Natural Language Processing, Term Frequency-Inverse Document (TF-IDF), N-Gram, Tokenization, Sentiment Analysis, NLTK (Natural Language Toolkit), Rasa, Stop Word Removal, Flutter
Anandan *et al.*	Decision Tree Classifier, Dimensionality Reduction: Feature Selection, Feature Extraction, User Registration and Query Session
Tanmay *et al.*	Normalization, Tokenization, Named Entity Recognition, Dependency Parsing, Sentiment Analysis, and natural language generation (converting structured data into text)
SureshKumar *et al.*	Classification algorithm, Regression algorithm, Clustering algorithm, Natural Language Processing(NLP), Pattern matching algorithms, Emotion detection algorithm, RNN, data pre-processing, face detection, evaluation, Client-server architecture

(Table 1) cont.....

Author(s)	Methodology Used
Anastasiia and Korotyeva	Core methodology: Artificial neural networks- Long Short-Term Memory (LSTM), Text Processing, Text recognition, classification of categories, Emotion Identification, Emotion Analysis, User Surveys, Real-Time Analysis
Crasto *et al.*	Natural Language Processing (NLP), Generative Pre-trained Transformer (GPT), Patient Health Questionnaire (PHQ-9), WHO-5 Questionnaire, Data Extraction and Preparation Tools, Python and PyTorch, Tableau Public, Speech-to-Text and Text-to-Speech Converters
Chakraborty *et al.*	Artificial Neural Networks, like LSTM and RNN, are employed. Decision Trees and Gradient Descent are used for optimization—Tkinter for the graphical user interface, and TensorFlow for neural network implementation.
Goel *et al.*	ChatterBot, Flask, Machine learning algorithms, including deep learning techniques, are applied in combination with NLP and Naive Bayesian Classification Algorithm, N-gram Technique, Porter Algorithm, Support Vector Machine (SVM), SOAP-based Server, Multi-modal Approach.
Damij and Bhattacharya	No methodology mentioned
Kaushik	Intent Classification Algorithms such as Support Vector Machine (SVM) and Naive Bayes classifiers, Facial Recognition, SQLite as the database, Dialogflow and Java for front-end and back-end development.
Kaladevi *et al.*	Natural Language Processing (NLP), Support Vector Machine (SVM), Stanford Parser, API Integration (*e.g.*, Google API), Java, N-grams, TF-IDF (Term Frequency-Inverse Document Frequency), Cosine Similarity.
Medeiros *et al.*	IBM Watson Natural Language Understanding, MongoDB, Machine Learning Algorithms, Statistical Analysis (ANOVA, t-tests)
Santos *et al.*	Natural Language Processing (NLP), Supervised Machine Learning, Likert Scale, Rasa and Botfront (Chatbot Frameworks), Containerization (Docker and Kubernetes), HTML, CSS, and JavaScript.
Chua *et al.*	Web Server, SQL Server (for database hosting), Database Management System (DBMS) for data storage and retrieval, PHP (Hypertext Preprocessor) for server-side scripting, HTML, CSS, Bootstrap, Messenger (for chat assessments and communication),
Gupta and Basit	Information Retrieval, Artificial Intelligence, Recommendation System, Machine Learning Algorithms, Support Vector Machines, Logistic Regression, YouTube API, Sentiment Analysis
Teles *et al.*	Mobile applications, Google Play Store, Android, iOS, Internet of Things (IoT), App Store, Wearable devices
Laijawala *et al.*	Decision Tree classification, Random Forest *etc.* techniques
Siddik *et al.*	Deep learning: Conv-LSTM, BiLSTM, BERT
Raut *et al.*	Natural Language Processing (NLP), Chatbot-Based Therapy

DATASET DESCRIPTION

For building ChatCare, two datasets are utilized. The first dataset, named 'intents.json', consists of different conversation intents categorized by tags. It is represented in Fig. (**2**). Each intent includes patterns of user messages and corresponding responses. For instance, the "greeting" tag encompasses various user greetings like "Hi" or "Hello," along with corresponding chatbot responses aimed at initiating conversation and assessing the user's emotional state.

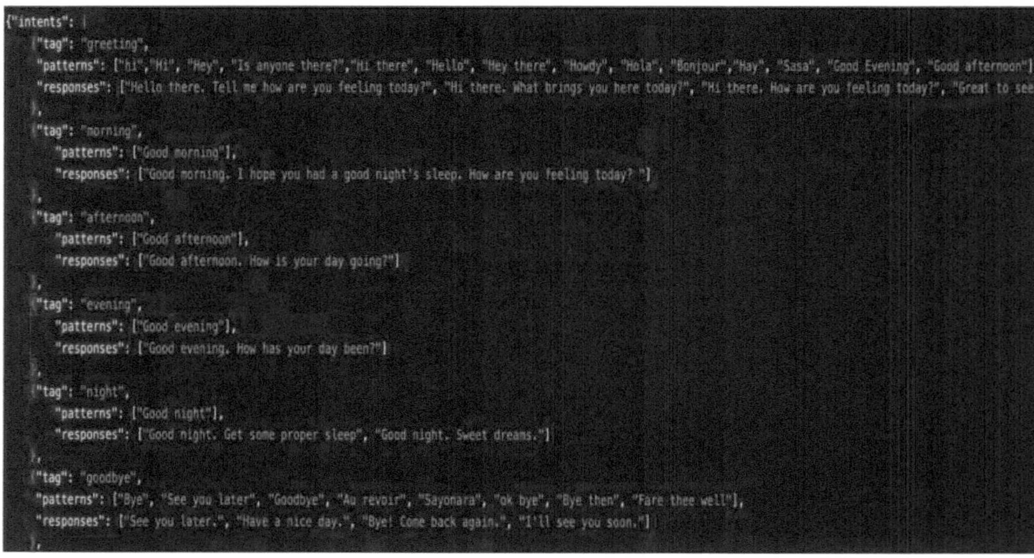

Fig. (2). Dataset for the chat box.

The second dataset, named 'data.csv', was extracted from Twitter and contains columns such as tweets, search key, and feelings and the same can be seen in Fig. (**3**). This dataset serves the purpose of emotion detection, enabling the system to analyze the emotional content of user conversations. By processing the tweets and associated feelings, the system can detect and classify various emotions expressed in the chat, facilitating further emotion analysis displayed in the dashboard.

RESULTS AND DISCUSSION

A web application is developed for the proposed methodology and is deployed with the dataset mentioned. Fig. (**4**) is the screenshot of the home page of our application.

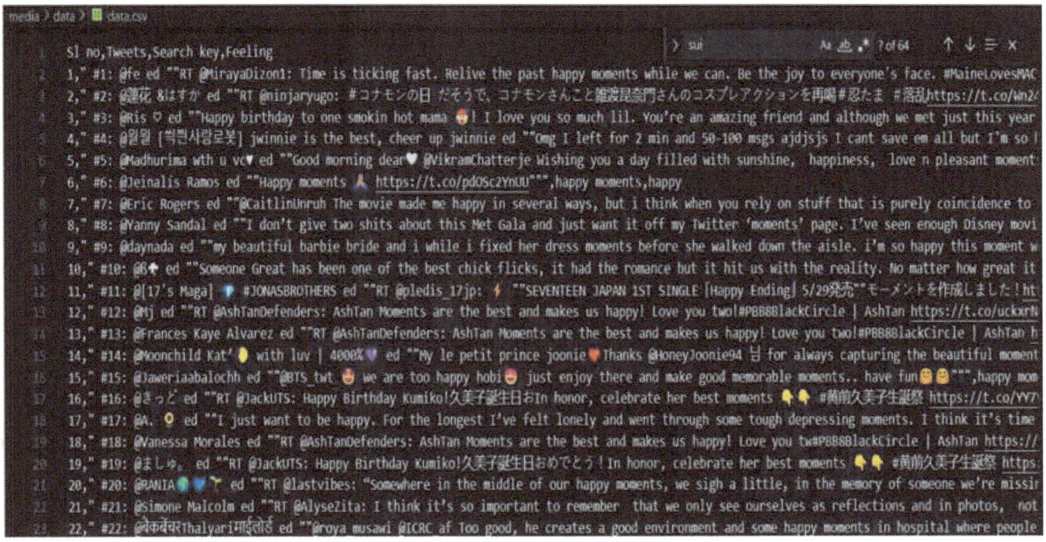

Fig. (3). Dataset for emotion detection.

Fig. (4). Home page of chatcare platform.

Fig. (**5**) indicates the registration page for the new user and login options for registered users. As the motto of this work is to provide mental health support for those in need, Fig. (**6**) is such an example from our application where the chatbot we developed understands the query and responds to the user query. Fig. (**7**) represents the personal dashboard of the registered users and from Fig. (**8**), we can see the emotional chart developed by our platform. These are the best unique features that bring more users to use the application. Our research also included suggestions of good videos to watch for mental health and well-being of users, that is presented in Fig. (**9**).

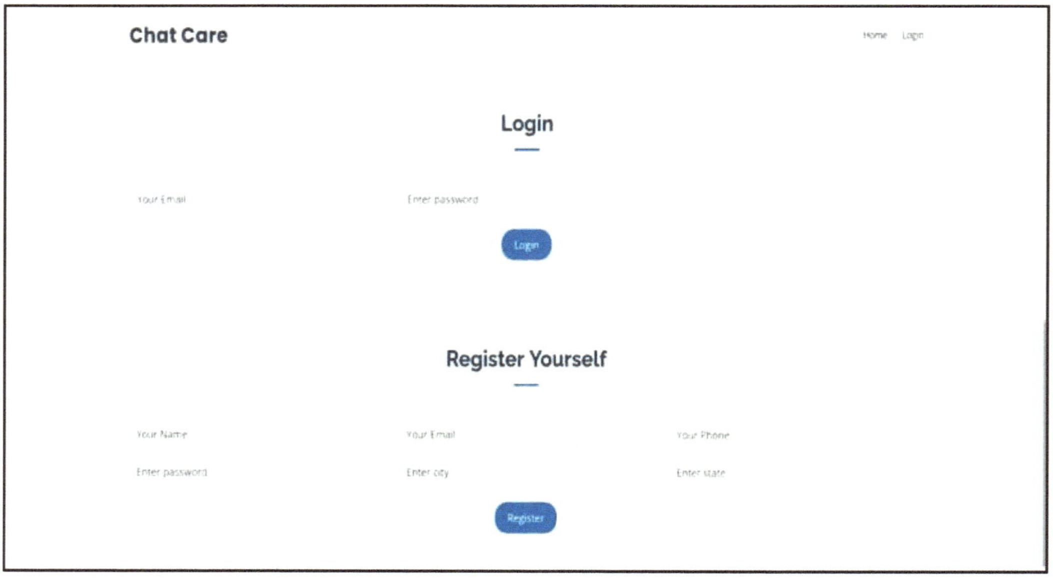

Fig. (5). Login & registration page.

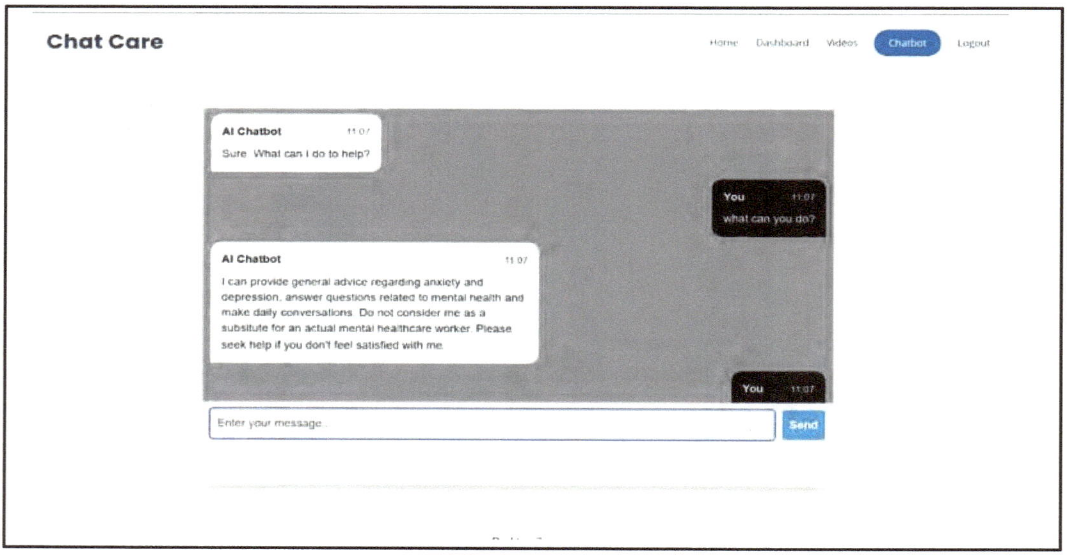

Fig. (6). The chatbot.

Fig. (7). User dashboard.

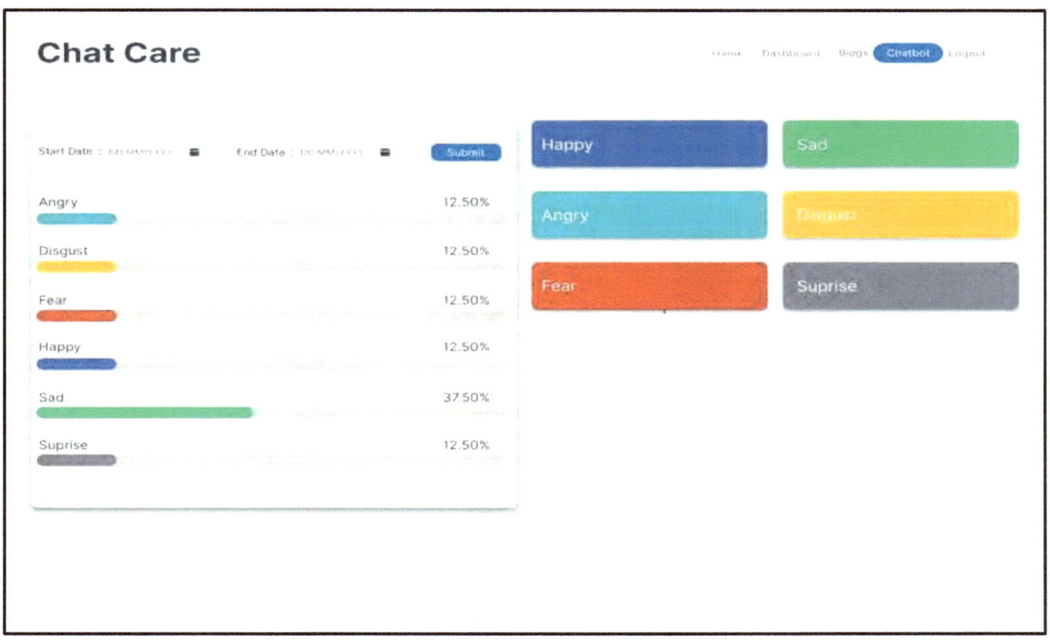

Fig. (8). Emotion analysis chart.

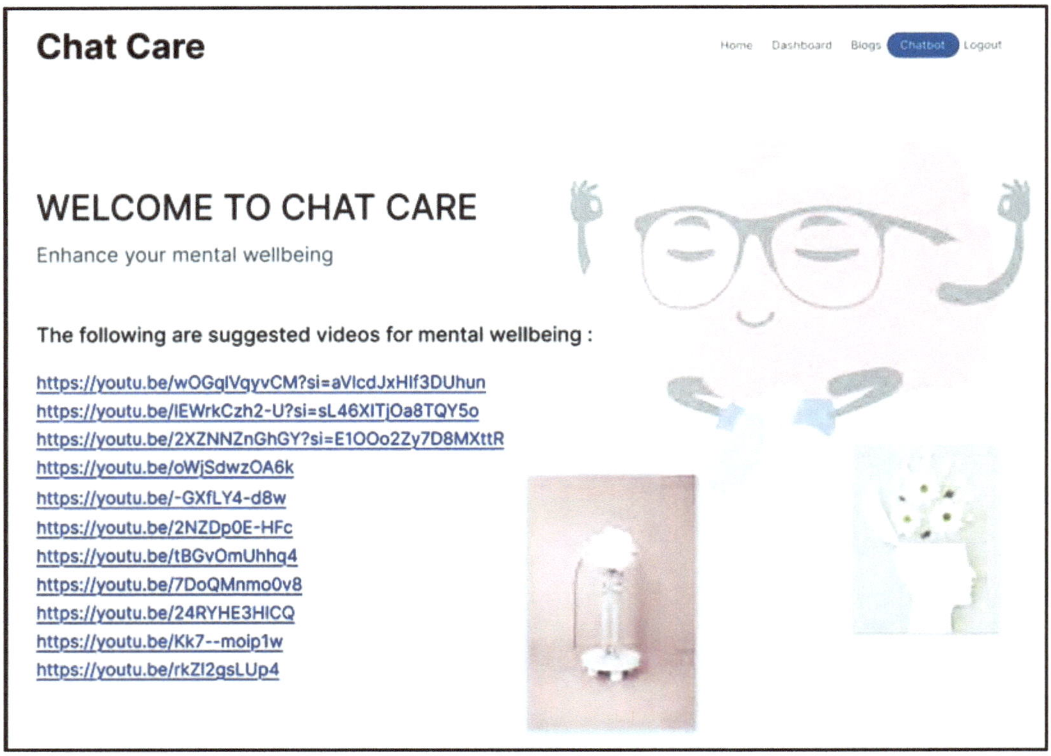

Fig. (9). Mental health-related video section.

The comparison of mental health chatbot systems showcased in Table **2**, reveals varying approaches and performance metrics. Dost-Mental Health Assistant Chatbot achieved 83.5% accuracy in intent recognition, while Depression Therapy Using Chatbot scored 84.14% on training data and 60.17% on test data. Classification Algorithms based on Mental Health Prediction reached 82.2% accuracy with the Decision Tree Algorithm.

Table 2. Performance evaluation of various models.

Evaluation Metrics	Accuracy Achieved
Intent accuracy & Confusion Matrix	83.5%
Intent Classification using Support Vector Machine and Naive Bayes classifier	84.14%
Chatbot prediction rate using Decision Tree Algorithm	82.2%
chatbot accuracy using BERT based classifier	75.3%
Emotion expression accuracy using confusion matrix	75%
ChatCare: Intent recognition	**87%**

Psyche Conversa attained 75.3% accuracy using a BERT-based classifier, and Depression Therapy Chat-Bot using NLP achieved 75% accuracy in emotion expression. ChatCare excelled with an 87% accuracy in intent recognition.

Various classifiers are implemented on the dataset and are tested, accuracy values are recorded, found that the voting classifier stands as the best approach.

The classifiers were evaluated based on their accuracy scores, with the Naive Bayes classifier performing the lowest at 66.2%. While Naive Bayes is known for its simplicity, its assumption of feature independence may limit its effectiveness. SVM and Random Forest showed moderate accuracy, leaving room for improvement. However, GBM, Decision Tree, Logistic Regression, and the Voting Classifier demonstrated high accuracy levels ranging from 90% to 92%. Among these, the Voting Classifier emerged as the best performer, leveraging the collective intelligence of diverse models to produce robust predictions as shown in Table **3**.

Table 3. Performance evaluation of various models.

Classifier	Accuracy (%)
Random Forest	89.7
Naive Bayes	66.2
SVM	78.2
GBM	90
Decision Tree	90.8
Logistic Regression	91.3
Voting Classifier	92

CONCLUSION AND FUTURE SCOPE

In conclusion, ChatCare stands at the forefront of mental health assistance, offering a multifaceted platform for managing emotional well-being. Through its integration of a conversational chatbot, emotion detection capabilities, personalized dashboards, and curated mental health resources, ChatCare provides a comprehensive solution to address users' emotional needs. By harnessing the power of natural language processing and machine learning, ChatCare facilitates meaningful interactions and empowers users to enhance their mental health proactively. Looking forward, our platform holds immense promise in filling critical gaps in mental health support, particularly for individuals lacking access to such resources. Clinicians can leverage ChatCare to augment their treatment strategies, providing personalized support to patients remotely. Moreover, the expansion of ChatCare to mobile applications will ensure broader accessibility, enabling users to access its features conveniently. Through ongoing innovation and adaptation, ChatCare is poised to make a significant and lasting impact on mental health support, offering vital assistance and guidance to individuals worldwide.

The need to provide mental health support with a variety of services that involve the development of a user-friendly website is in demand. A comparison of the various approaches with those of the proposed method is made through this work and complemented by different services that can be safely employed for a very varied number of people suffering from mental illness and looking for support. The suggested method was used to determine the types of people and provide self-analysis through the dashboard of the application developed.

REFERENCES

[1] Nayar, Atulya M., Z. Attar, S. Kachwala, T. Biswas, and S. K. Wagh. "Dost-mental health assistant chatbot." *In 2022 5th International Conference on Advances in Science and Technology (ICAST)*, pp. 252-257. IEEE, 2022. [http://dx.doi.org/10.1109/ICAST55766.2022.10039576]

[2] V. Dhanasekar, and Y. Preethi, "S. Vishali, and Praveen Joe IR. "A chatbot to promote students mental health through emotion recognition", *2021 Third International Conference on Inventive Research in Computing Applications (ICIRCA),* IEEE, pp. 1412-1416, 2021.

[3] K. Rani, H. Vishnoi, and M. Mishra, "A mental health chatbot delivering cognitive behavior therapy and remote health monitoring using NLP and AI", *2023 International Conference on Disruptive Technologies (ICDT),* IEEE, pp. 313-317, 2023. [http://dx.doi.org/10.1109/ICDT57929.2023.10150665]

[4] P. Anandan, S. Kokila, S. Elango, P. Gopinath, and P. Sudarsan, "Artificial intelligence-based chat bot for patient health care", *2022 International Conference on Computer Communication and Informatics (ICCCI),* IEEE, pp. 1-4, 2022. [http://dx.doi.org/10.1109/ICCCI54379.2022.9740912]

[5] T. Tanmay, A. Bhardwaj, and S. Sharma, "E-Health Bot to change the Face of Medicare", In: *Research, Innovation, Knowledge Management and Technology Application for Business Sustainability (INBUSH).* IEEE, 2020, pp. 49-54.

[6] Anastasiia, Maherus, and Tetyana Korotyeva. "A chatbot of a person's emotional state using a neural network." *In 2022 IEEE 17th International Conference on Computer Sciences and Information Technologies (CSIT)*, pp. 13-16. IEEE, 2022.

[7] R. Crasto, L. Dias, D. Miranda and D. Kayande. "CareBot: a mental health ChatBot." *In 2021 2nd international conference for emerging technology (INCET)* , pp. 1-5. IEEE, 2021. [http://dx.doi.org/10.1109/CSIT56902.2022.10000558]

[8] Chakraborty, Sanjay, Hrithik Paul, Sayani Ghatak, Saroj Kumar Pandey, Ankit Kumar, Kamred Udham Singh, and Mohd Asif Shah. "An AI-based medical chatbot model for infectious disease prediction." Ieee Access 10 (2022): 128469-128483. [http://dx.doi.org/10.1109/INCET51464.2021.9456326]

[9] S. Satpathy, S. Mohapatra, P.K. Swain, and B.K. Paikaray, "Securing healthcare in the cloud: a machine learning perspective", *Int. J. Internet Manuf. Serv.,* vol. 11, no. 2, pp. 114-131, 2025. [http://dx.doi.org/10.1504/IJIMS.2025.146819]

[10] Damij, Nadja, and Suman Bhattacharya. "The role of AI chatbots in mental health related public services in a (post) pandemic world: a review and future research agenda." *In 2022 IEEE Technology and Engineering Management Conference (TEMSCON EUROPE)* , pp. 152-159. IEEE, 2022. [http://dx.doi.org/10.1109/ICCES57224.2023.10192727]

[11] Biswas, Arkodeep, and Ajay Kaushik. "Depression therapy using chatbot." International Journal for Modern Trends in Science and Technology 6, no. 12 (2020): 323-327. [http://dx.doi.org/10.1109/TEMSCONEUROPE54743.2022.9801962]

[12] A. Biswas, and A. Kaushik, "Depression therapy using chatbot", *International Journal for Modern Trends in Science and Technology,* vol. 6, no. 12, pp. 323-327, 2020. [http://dx.doi.org/10.46501/IJMTST061260]

[13] R. Kaladevi, S. Saidineesha, P. Keerthi Priya, K.M. Nithiya, and S. Sai Gayatri, "Chatbot For Healthcare Using Machine Learning", *2023 International Conference on Computer Communication and Informatics (ICCCI),* pp. 1-4, 2023.

[14] L. Medeiros, T. Bosse, and C. Gerritsen, "Can a chatbot comfort humans? Studying the impact of a supportive chatbot on users' self-perceived stress", *IEEE Trans. Hum. Mach. Syst.,* vol. 52, no. 3, pp. 343-353, 2022. [http://dx.doi.org/10.1109/THMS.2021.3113643]

[15] G.A. Santos, G.G. de Andrade, G.R.S. Silva, F.C.M. Duarte, J.P.J.D. Costa, and R.T. de Sousa, "A conversation-driven approach for chatbot management", *IEEE Access,* vol. 10, pp. 8474-8486, 2022. [http://dx.doi.org/10.1109/ACCESS.2022.3143323]

[16] Gupta, Ankit, and Nada Basit. "Ranking Actionable Mental Health Advice through a Personalized Video Recommendation System." In 2022 6th International Conference on Informatics and Computational Sciences (ICICoS), pp. 140-145. IEEE, 2022. [http://dx.doi.org/10.1109/ICBIR52339.2021.9465876]

[17] Teles, Ariel, Ivan Rodrigues, Davi Viana, Francisco Silva, Luciano Coutinho, Markus Endler, and Ricardo Rabêlo. "Mobile mental health: A review of applications for depression assistance." In 2019 IEEE 32nd International Symposium on Computer-Based Medical Systems (CBMS), pp. 708-713. IEEE, 2019. [http://dx.doi.org/10.1109/ICICoS56336.2022.9930526]

[18] Laijawala, Vidit, Aadesh Aachaliya, Hardik Jatta, and Vijaya Pinjarkar. "Classification algorithms based mental health prediction using data mining." In 2020 5th international conference on communication and electronics systems (ICCES), pp. 1174-1178. IEEE, 2020. [http://dx.doi.org/10.1109/CBMS.2019.00143]

[19] Siddik, Sayed Abu Noman, B. M. Arifuzzaman, and Abul Kalam. "Psyche conversa-a deep learning based chatbot framework to detect mental health state." In 2022 10th International Conference on Information and Communication Technology (ICoICT), pp. 146-151. IEEE, 2022. [http://dx.doi.org/10.1109/ICCES48766.2020.9137856]

[20] Popa, Georgeta. "Geospatial chat bot using natural language processing." U.S. Patent 10,749,823, issued August 18, 2020.

[21] Macharla, Anila, and Gera Pradeepini. "A novel framework for the diagnosis of Parkinson's disease using transfer learning with RESNET50 and SVM classifier." Indonesian Journal of Electrical Engineering and Computer Science (IJEECS) 32, no. 2 (2023): 877-886.

[22] A. Macharla, and G. Pradeepini, "A novel framework for the diagnosis of Parkinson's disease using transfer learning with RESNET50 and SVM classifier", *Indonesian Journal of Electrical Engineering and Computer Science,* vol. 32, no. 2, pp. 877-886, 2023. [http://dx.doi.org/10.11591/ijeecs.v32.i2.pp877-886]

[23] S. Rasheed, "G. Kiran Kumar, D. Malathi Rani, MVV Prasad Kantipudi, and M. Anila. "Heart Disease Prediction Using GridSearchCV and Random Forest."", *EAI Endorsed Trans. Pervasive Health*

Technol., p. 10, 2024.

[24] S. Nandan Mohanty, S.K. Saxena, S. Satpathy, J.M. Chatterjee, Ed., *Applications of Artificial Intelligence in COVID-19. Medical Virology: From Pathogenesis to Disease Control.* Springer: Singapore.
[http://dx.doi.org/10.1007/978-981-15-7317-0]

[25] S. Satpathy, S. Nandan Mohanty, J.M. Chatterjee, and A. Swain, "Comprehensive Claims of AI for Healthcare Applications-Coherence Towards COVID-19", In: *Applications of Artificial Intelligence in COVID-19. Medical Virology: From Pathogenesis to Disease Control.,* S. Nandan Mohanty, S.K. Saxena, S. Satpathy, J.M. Chatterjee, Eds., Springer: Singapore, 2021.
[http://dx.doi.org/10.1007/978-981-15-7317-0_1]

[26] S. Mohapatra, S. Satpathy, and S.N. Mohanty, "A comparative knowledge base development for cancerous cell detection based on deep learning and fuzzy computer vision approach", *Multimedia Tools Appl.,* vol. 81, no. 17, pp. 24799-24814, 2022.
[http://dx.doi.org/10.1007/s11042-022-12824-0]

[27] S. Satpathy, M. Mangla, N. Sharma, H. Deshmukh, and S. Mohanty, "Predicting mortality rate and associated risks in COVID-19 patients", *Spatial Information Research,* vol. 29, no. 4, pp. 455-464, 2021.
[http://dx.doi.org/10.1007/s41324-021-00379-5]

[28] M. Anila, G. Kiran Kumar, and D. Malathi Rani, "MVV Prasad Kantipudi, and D. Jayaram. "An LSTM based DNN Model for Neurological Disease Prediction Using Voice Characteristics."", *EAI Endorsed Trans. Pervasive Health Technol.,* p. 10, 2024.

CHAPTER 14

Electronic Health Records Using Blockchain

Srinivasa Rao Dhanikonda[1], **B. Srinivasulu**[2,*], **C. H. Ravi**[3], **Ponnuru Sowjanya**[4] and **N. Subhash Chandra**[5]

[1] *Department of AI & DS, Faculty of Science and Technology (IcfaiTech), ICFAI Foundation for Higher Education (Deemed to Be University), Hyderabad, Telangana, India*

[2] *Department of IT, BVRIT Hyderabad College of Engineering for Women, Hyderabad, Telangana, India*

[3] *Department Of CSE, CMRIT, Hyderabad, Telangana, India*

[4] *School of Technology, GITAM (Deemed to be University), Hyderabad, Telangana, India*

[5] *CVR College of Engineering, Hyderabad, Telangana, India*

Abstract: The healthcare industry is one area where blockchain technology is being used to ensure security and reliability. Blockchain is like a big list where each new transaction is added to the end of the list, and each transaction is connected to the one before it. This creates a secure and unchangeable record of all transactions. To ensure everyone agrees on the new transactions being added, blockchain uses methods that involve everyone checking each other's work. Patient information is very sensitive, so it is essential to keep it safe. In the past, each hospital stored patient information separately, making sharing information difficult and risky. By using blockchain, it can share information more easily without worrying about being changed or stolen. This research aims to combine blockchain and healthcare to make sharing information more secure and efficient.

Keywords: Blocks, Breaches, Cmmutability, Consensus, Dependability, Diagnosis, Healthcare, Hash functions, Methods, Research, Security, Tampering, Transactions.

INTRODUCTION

Blockchain technology's varied applications accelerate scientific research in various disciplines critical to societal advancement. The blockchain, a distributed record of interconnected blocks protected by cryptographic procedures, is at the vanguard of this disruptive wave. Researchers are looking into how it will affect finance, healthcare, supply chain management, and energy to increase openness,

* **Corresponding author B. Srinivasulu:** Department of IT, BVRIT Hyderabad College of Engineering for Women, Hyderabad, Telangana, India; E-mail: srinivas.nivas47@gmail.com

security, and efficiency. The technology's decentralized structure also bodes well for decentralized identity management and intellectual property protection applications. As blockchain evolves, its cryptographic processes and distributed consensus mechanisms assure user protection while maintaining ledger consistency, resulting in a dependable and tamper-resistant record. This continuing scientific investigation into blockchain's possible applications defines a future characterized by innovation and interconnection across multiple sectors.

In the healthcare sector, electronic health records, or EHRs, are essential because they store private data that is needed for diagnosis and treatment. Big data in healthcare today involves a lot of information. Thus, it must be shared with many parties, including researchers, insurance companies, patients, families, and healthcare providers. For efficient healthcare administration, this complex web of data interchange is essential [1, 2].

Patients may need access to their medical history at different times and frequently need to see other healthcare providers for several consultations. Coordinating and gaining access to this data in a centralized system can provide formidable obstacles. Blockchain technology offers a solution using a shared and immutable ledger to address these problems. This ledger guarantees an accurate and transparent record.

Healthcare data coordination and accessibility are improved by utilizing blockchain. The technology's decentralized and secure architecture facilitates the easy exchange of information between authorized parties, such as patients, healthcare providers, and other pertinent entities. Transparency in documenting each participant's contributions to the patient's medical journey promotes accuracy and confidence in medical records.

Blockchain technology makes electronic health records more efficient by offering a decentralized, impenetrable system that makes safe and transparent data transmission amongst the various parties in the healthcare ecosystem possible. This innovation can enhance healthcare service effectiveness, guarantee data integrity, and improve patient care.

Electronic Health Records Objectives are:

- Isolation: Only authorized individuals can obtain information.
- Information should be accurate. It's unchangeable by unauthorized individuals.
- Validity: Identity checks should be done before granting access to sensitive information.
- Anonymity: Users should be completely anonymous or appear to be someone else.

LITERATURE WORK

Blockchain can improve the interactivity and reliability of EHRs by allowing people to own their medical records, according to a report [3]. The blockchain only stores metadata, while the health cloud keeps all other data. When a function is called, a node is created with the help of a patient using the key. The key is necessary to retrieve this information.

This section discusses several earlier works on blockchain-based electronic medical record systems. Blockchain can improve the interactivity and reliability of EHRs by allowing people to own their medical records, according to a report [4]. The blockchain will only store metadata [5], while the health cloud will keep all other data. When a function is called, a node is created with the help of a patient using the key. The key is necessary to retrieve this information.

Another study [6] claims that most earlier works relied on public blockchains like Ethereum and were therefore expensive to compute. Public blockchain privacy concerns may also be a weak point. Thus, Hyperledger, a permissioned blockchain framework, is employed in this work to assure a higher level of anonymity and to lower the computational cost associated with the consensus mechanism. By guaranteeing that no peers are anonymous and that only a small number of authorized peers participate in the consensus mechanism for the Hyperledger framework, the system created following the methods outlined in this paper can avoid the escape of sensitive medical information.

In a study [7], they used various encryption methods to encrypt data and Hyperledger fabric in their paper. The system lacks scalability and the database's capacity to hold data is limited.

In 2017, Al Omar, A., Rahman, M.S., Basu, A., and Kiyomoto, S [8]. published a paper introducing Medibchain, a blockchain-based platform for managing healthcare data while protecting patient privacy. The research, which was presented at the International Conference on Security, Privacy, and Anonymity in Computation, Communication, and Storage, focuses on using blockchain technology to address privacy concerns in the healthcare industry. Medibchain is suggested as a way to protect patient privacy, secure sensitive healthcare data, and facilitate safe data exchange throughout the healthcare system. The article highlights Medibchain's function in boosting the security and confidentiality of medical data while offering insights into the system's conception and execution. The proceedings can be cited by using these specific details: Al Omar [9], A., Kiyomoto, S. (2017), Basu, A., and Rahman, M.S. "Medibchain: a privacy-preserving blockchain technology"

In this paper [10], they tried to use multi-party communication to eradicate data leak-age, but during the process of decryption, data leakage is still possible.

In this paper [11], the authors suggested the encryption of data using private keys, but data tampering without the user's awareness may be caused by duplicate copies of the data.

Nguyen [12] integrates smart contracts with IPFS to increase accessibility by gaining access to restricted data sharing and decentralised data storage. Which, in turn, makes it possible for private data to be compromised.

This suggested framework is built on IoT, blockchain, data archiving, and the gathering of patient wearable devices. It also uses LPWAN and the cloud [13]. The drawbacks set here are that Multi-jurisdictional operations are a part of cross-border health information sharing, which could reduce blockchain's benefits.

The light-edge IoT device identification protocol was first presented by Arani *et al.* [14]. It has a three-layer architecture that consists of cloud service providers, a network-edge trust center, and an IoT device layer. The results show that the suggested protocol is superior to competing options in terms of time costs, information exchange costs, and assault resistance.

A blockchain-based architecture framework has been designed to secure control over how different health service providers share a patient's personal health-related data. The framework that is being provided enables the least amount of data risk by developing a mechanism for data control. According to the author, by using the framework offered, the health care provider may ensure more robust data management security than is now achievable with the system in place.

A blockchain was utilized in a 2015 study [15] to safeguard user privacy. The authors developed a system that transforms a blockchain into an automatic access-control manager that does not require faith in a third party, with the goal of guaranteeing that users own and control their data. Since 2016, a number of studies have focused on using blockchain technology to manage medical data.

A study [16] developed a medical transaction control system with distributed ledger capabilities that is constructed with sensors and other networked devices. The suggested framework steers clear of issues with the conventional IoT strategy, like expensive server hardware and single points of failure.

In their paper [17], Nguyen, D.C., Pathirana, P.N., Ding, M., and Seneviratne, A. investigate how blockchain technology can be used to securely share Electronic Health Records (EHRs) in cloud-based mobile e-health systems. The authors concentrate on using blockchain technology in the context of mobile cloud platforms to improve the security and privacy of EHRs. With a focus on blockchain's potential advantages in protecting sensitive health data, the research attempts to provide insights on how to use it for secure data exchange in e-health systems.

Patient data is presently stored in Iryo, a Slovenia-based enterprise, an open-sourced OpenEHR platform. Patientory, Atlanbased - Using an app, users can maintain a patient profile to record their medical history. Patients may easily keep track of doctor appointments, hospital bills, insurance, personal medical information, vaccines, and prescription drugs with the help of this program. EHRs are available in Slovenia, Atlanta, UAE, Estonia, Netherlands, UK, Singapore, which use Cloud and Guard time. The major disadvantage of the existing system is cross-border jurisdiction.

PROPOSED SYSTEM

The primary technology used in this project is the local Ethereum Blockchain, specifically Ganache, which is a public blockchain. This decentralized, blockchain-based platform ensures privacy and immutability, resulting in reduced costs for sharing information. Patient data is also stored in a local database on the AMPPS server, providing a backup in case of technical issues or tampering.

An electronic health record (EHR) system is an ideal solution for ensuring secure data storage, ease of access, and data sharing while addressing privacy concerns. To address these issues, a web-based decentralized application is being developed, utilizing a blockchain-based EHR system.

In the above image (Fig. **1**), the website can be accessed through either mobile or desktop devices. In blockchain, every function is considered a transaction. Hospitals are connected through a peer-to-peer network, and transactions are initiated and verified based on specific conditions, such as whether the gas value is sufficient or not, as well as general verifications that occur during execution.

Once all conditions are met, the transaction is considered verified, and a new block is created for the data. This block is then appended to the existing chain, and upon completion, this message “Record is added successfully” confirms the transaction's verification.

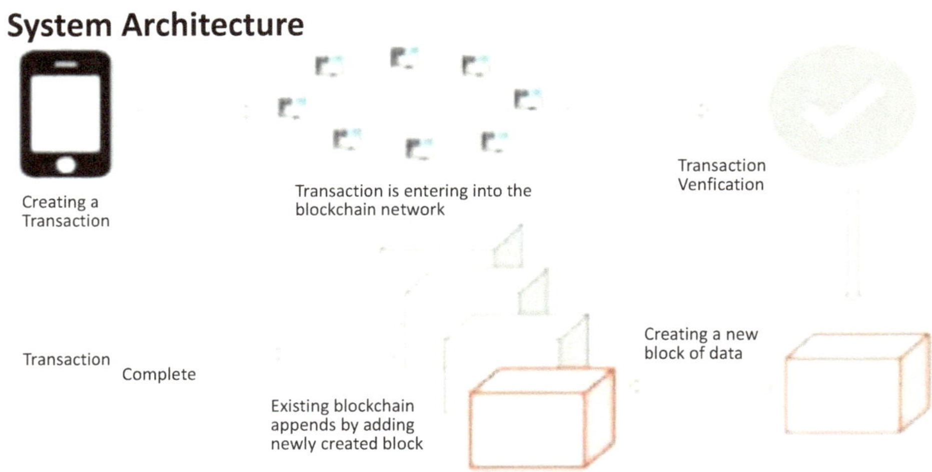

Fig. (1). System architecture.

Problem Definition

A happy and content existence is built on good health. The scholars are interested in the social relevance of public healthcare. Private information pertaining to patient diagnosis and care is included in extremely sensitive healthcare data. This data must be securely stored and shielded from unauthorised access because they are kept in the format of electronic health records (EHR). With the conventional system, each hospital stores these data independently, making data interchange more time-consuming and increasing the likelihood of data loss or harm. Also, the patient has little control over his medical data. Moreover, access to health data is severely constrained due to the possibility of sensitive data leakage. The privacy and security difficulties with current EHR systems can be resolved with a blockchain-based EMR system because it will guarantee the security of data storage, make it simple to access, use, and share data, and address privacy issues. In this project, a web-based decentralised application will be created in an effort to address such issues using a blockchain-based EMR system.

IMPLEMENTATION

This framework starts with doctors adding patient information using a unique identifier called patient ID, which can be the patient's phone number for easy implementation. After entering the patient ID, doctors provide other details such as their name, hospital name, and other relevant information. Upon submitting the data, we move to the fetch page, where we retrieve records of the patient. Here, we type in the patient ID to retrieve all the records associated with them on the blockchain. These records are stored in a MySQL database on the local computer and are hyperlinked. Clicking on a record opens a new page that shows the data stored in that record. As we are using an app server, the website is hosted on a local computer using the link 'http://localhost/EHR.'

The framework includes two modules: the patient module and the doctor module. The doctor module is an upload page where doctors enter patient data, while the patient module uses the patient ID to fetch data, serving as an example of a fetch page.

This algorithm1 offers a condensed summary of the main functions and exchanges in the blockchain framework for Electronic Health Records (EHRs) that is described. The framework presumes the website will be hosted locally on an AMPPS server. The patient ID is a unique identification code for obtaining and accessing particular patient records. On the local PC, the data are kept in a MySQL database. The framework supports essential functions like physician data submission and patient data retrieval.

Algorithm1: EHR Blockchain Framework
Doctor Module:
1. Start
2. Collect patient information from the doctor.
- Input: Patient ID (unique identifier, *e.g.*, phone number), name, hospital name, and other relevant details.
3. Submit the data to the system.
- Action: Store the data in a MySQL database on the local computer.
4. End
Patient Module:
1. Start
2. Access the fetch page.
- Action: Open the website hosted locally (http://localhost/EHR).
3. Enter the patient ID.
- Input: Patient ID to retrieve the associated records.
4. Retrieve records from the blockchain.
- Action: Query the MySQL database to fetch patient records associated with the provided ID.
5. Display hyperlinked records.
- Output: Show hyperlinked records on the fetch page.
6. Click on a record.
- Action: Open a new page to display the data stored in the selected record.
7. End

RESULTS

This is the website's homepage (Fig. **2**); all pages, like about, upload, and fetch, are accessed through it. It is hosted using Apache from the Ampps software.

Fig. (2). Home page.

Fig. (**3**) shows us various fields of information that need to be entered about the patient, such as patient ID, patient name, doctor name, hospital name, vital statistics, symptoms, treatment, date, next checkup date, and any extra notes that doctors usually need to know.

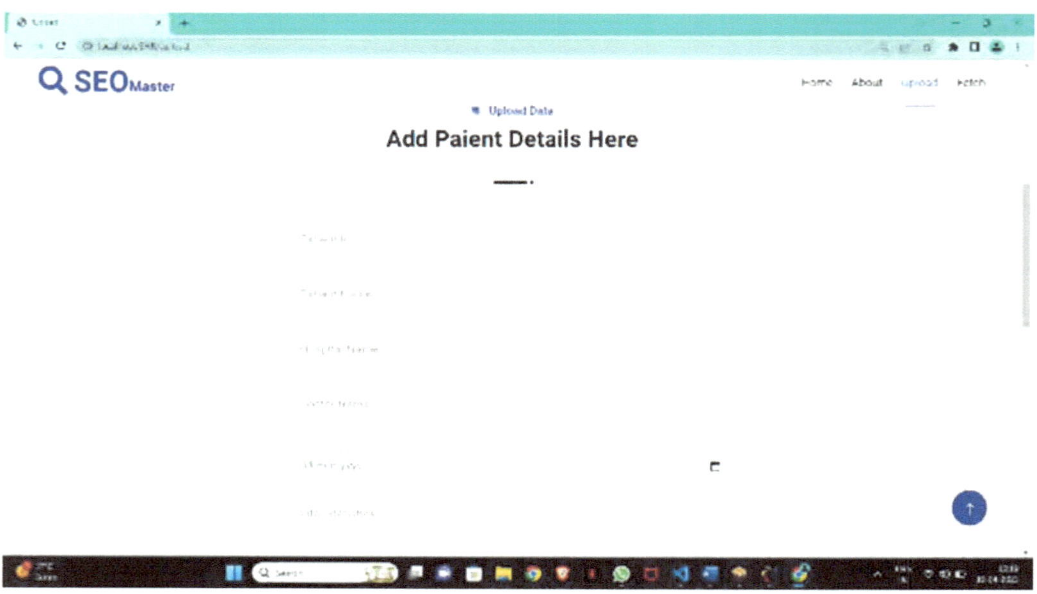

Fig. (3). Add patient details.

The above image, Fig. (**4**), shows us all the records associated with the given patient ID.

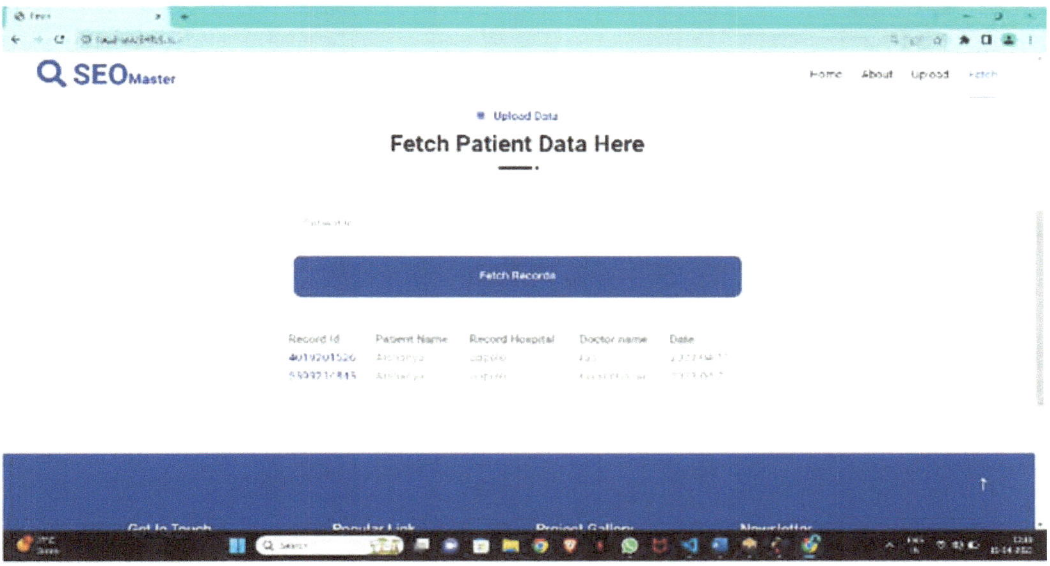

Fig. (4). Fetching data.

CONCLUSION AND FUTURE WORK

Patients can actively monitor their medical records in this proposed work, as well as concerns about privacy and security. *via* patient ID, the patient controls access to their health history and can revoke that permission at any time to better comprehend and evaluate; the doctor can access the patient’s health records if they are rapidly updated on the blockchain network. The doctor can append the patient's present status to the blockchain if any prescriptions are written and medical procedures are performed. The flaws in the paper can be mainly divided into three groups: To begin with, the system is not scalable. Secondly, participation is restricted to medical professionals and patients. Lastly, there are significant limitations on the database's capacity to store data. A few essential diagnostic details are considered, all of which are text-based.

The forthcoming paper has two parts. The given solution is a prototype designed to demonstrate how blockchain operates when integrated with EHR. At the first level, there is more participation, similar to that of an insurance company or hospital. The system must scale up, and independent database upkeep must be accomplished before the authentic prescription in a picture or other format may be stored. For instance, there will still be a demand for a hybrid or layered blockchain.

REFERENCES

[1] King Yip, "BlockChain and alternative payment models," White paper, 2016.

[2] A. Ackerman, A. Chang, N. Diakun-Thibault, L. Forni, F. Landa, J. Mayo, and R. van Riezen, "Blockchain and health IT: Algorithms, privacy and data," Project PharmOrchard of MIT's Experimental Learning, MIT FinTech: Future Commerce, White Paper, August 2016.

[3] H. Li, L. Zhu, M. Shen, F. Gao, X. Tao, and S. Liu, "Blockchain-based data preservation system for medical data," J. Med. Syst., vol. 42, no. 8, p. 141, Jun. 2018, doi: 10.1007/s10916-018-0997-3.

[4] Gupta, N., Jha, A., & Roy, P. (2016). Adopting blockchain technology for electronic health record interoperability. Cognizant Technol. Solutions, Teaneck, NJ, USA, White Paper.

[5] S. Amofa, E. B. Sifah, K. O.-B. Obour Agyekum, S. Abla, Q. Xia, J. C. Gee, and J. Gao, "A blockchain-based architecture framework for secure sharing of personal health data," in Proc. 2018 IEEE 20th Int. Conf. e-Health Networking, Applications and Services (Healthcom), 2018, pp. 1–6, doi: 10.1109/HealthCom.2018.8531160.
[http://dx.doi.org/10.1109/HealthCom.2018.8531160]

[6] A. Azaria, A. Ekblaw, T. Vieira and A. Lippman, "MedRec: Using Blockchain for Medical Data Access and Permission Management," 2016 2nd International Conference on Open and Big Data (OBD), Vienna, 2016, pp. 25-30.
[http://dx.doi.org/10.1109/OBD.2016.11]

[7] A. Al Omar, M.S. Rahman, A. Basu, and S. Kiyomoto, "Medibchain: a blockchain based privacy preserving platform for healthcare data", *International Conference on Security, Privacy and Anonymity in Computation, Communication and Storage,* Springer, pp. 534-543, 2017.
[http://dx.doi.org/10.1007/978-3-319-72395-2_49]

[8] G. Zyskind, O. Nathan, and A. Pentland, "Decentralizing privacy: Using blockchain to protect personal data," in 2015 IEEE Security and Privacy Workshops (SPW), San Jose, CA, USA, 2015, pp. 180–184, doi: 10.1109/SPW.2015.27..
[http://dx.doi.org/10.1109/SPW.2015.27]

[9] X. Zheng, R. R. Mukkamala, R. Vatrapu, and J. Ordieres-Mere, "Blockchain-based personal health data sharing system using cloud storage," in 2018 IEEE 20th International Conference on e-Health Networking, Applications and Services (Healthcom), 2018, pp. 1–6.
[http://dx.doi.org/10.1109/HealthCom.2018.8531125]

[10] X. Yue, H. Wang, D. Jin, M. Li, and W. Jiang, "Healthcare data gateways: found healthcare intelligence on blockchain with novel privacy risk control", *J. Med. Syst.,* vol. 40, no. 10, p. 218, 2016.
[http://dx.doi.org/10.1007/s10916-016-0574-6] [PMID: 27565509]

[11] W. Gao, W. G. Hatcher and W. Yu, "A Survey of Blockchain: Techniques, Applications, and Challenges," 2018 *27th International Conference on Computer Communication and Networks (ICCCN)* , Hangzhou, 2018, pp. 1-11.
[http://dx.doi.org/10.1109/ICCCN.2018.8487348]

[12] A. Shahidinejad, M. Ghobaei-Arani, A. Souri, M. Shojafar, and S. Kumari, "Light-Edge: A Lightweight Authentication Protocol for IoT Devices in an Edge-Cloud Environment", *IEEE Consum. Electron. Mag.,* vol. 11, no. 2, pp. 57-63, 2022.
[http://dx.doi.org/10.1109/MCE.2021.3053543]

[13] J. D. Vyas, M. Han, L. Li, S. Pouriyeh, and J. S. He, "Integrating Blockchain Technology into Healthcare," in Proceedings of the 2020 ACM Southeast Conference (ACM SE '20), Association for Computing Machinery, New York, NY, USA, 2020, pp. 197–203.
[http://dx.doi.org/10.1145/3374135.3385280]

[14] M. Usman and U. Qamar, "Secure Electronic Medical Records Storage and Sharing Using Blockchain Technology," Procedia Computer Science, vol. 174, pp. 321–327, 2020, ISSN 1877-0509.
[http://dx.doi.org/10.1016/j.procs.2020.06.093]

[15] T. Dey, "HealthSense: A medical use case of Internet of Things and blockchain", *International Conference on Intelligent Sustainable Systems (ICISS),* pp. 486-491, 2017. [http://dx.doi.org/10.1109/ISS1.2017.8389459]

[16] D. C. Nguyen, P. N. Pathirana, M. Ding, and A. Seneviratne, "Blockchain for Secure EHRs Sharing of Mobile Cloud Based E-Health Systems," IEEE Access, vol. 7, pp. 66792–66806, 2019. [http://dx.doi.org/10.1109/ACCESS.2019.2917555]

[17] M. T. Quasim, A. A. E. Radwan, G. M. M. Alshmrani, and M. Meraj, "A Blockchain Framework for Secure Electronic Health Records in Healthcare Industry," in Proceedings of the 2020 Conference on 09–10 October 2020, 2020.

CHAPTER 15

Employing Blockchain for a Tamperproof Organ Donation Model in the Post-Supply Chain

S. Nonita[1], **Monika Mangla**[2,*], **Y. Gokul**[3] and **R. Manik**[4]

[1] *Department of Information Technology, Indira Gandhi Delhi Technical University for Women, Delhi, India*

[2] *Department of Information Technology, Dwarkadas J Sanghvi College of Engineering, Navi Mumbai, India*

[3] *School of Computer Science and Engineering, VIT-AP University, Amaravati, India*

[4] *Department of Computer Science & Engineering, Lovely Professional University, Jalandhar, India*

Abstract: The administration of organs and tissues is tedious and may lead to black-marketing activity, particularly given high demand and low supply. This mishandling and mismanagement of organs may be alleviated by using technological advancements. Employment of technological advancements in organ donation leads to enhanced security and a significant reduction in intermediaries in organ acquisition. Thus, it enables the achievement of greater openness amongst donors and procurers. However, structures that keep all their data in a centralized database are more vulnerable to manipulation and illegal access. These risks of manipulation and illicit access may be compacted if the system is distributive in structure or sustains a distributive ledger/database, *i.e.*, each network maintains a replica of the record. For the same, blockchain technology may significantly contribute to organ donation to overcome the associated challenges and constraints and meet the required levels of dependability. The current manuscript outlines OrDo (organ donation), a blockchain-based organ donation system that creates a transparent platform between organ donors and receivers. Furthermore, the proposed system uses Decentralized Identifiers (DIDs) and Verifiable Credentials (VCs) to maintain a layer of trust. To explain the possession requirement, DIDs, and VCs are used to provide a distinct identity and credentials. A trustworthy consensus algorithm is also used to establish a connection between the patient and the donor to achieve encrypted, consistent, reliable, and justified communication, which can be employed in various real-world scenarios.

Keywords: Blockchain, Consensus algorithm, Decentralized identifiers, Organ donation, Verifiable credentials.

* **Corresponding author Monika Mangla:** Department of Information Technology, Dwarkadas J Sanghvi College of Engineering, Navi Mumbai, India; E-mail: manglamona@gmail.com

Suneeta Satpathy, Sachi Nandan Mohanty & Subhendu Kumar Pani (Eds.)

INTRODUCTION

Organ Donation is one of the most significant advancements in healthcare. Still, it involves many complications from a medical point of view and other aspects like legal, ethical, and personal choices, *etc* [1]. Organ transplants started in the 1950s and revolutionized medicine expeditiously, saving numerous lives and comforting people with end-stage organ failure (ESOF) [2]. Organ donation can be done at any age, but the organ must be healthy, and its donation may not pose a life risk to the donor. Although this is a noble cause, supply chain management is very complex, with high demand and low supply [3]. Owing to the significant risks involved and the high discrepancy between supply and demand, there is an imperative need to streamline the organ donation process so that the donor and receiver can come on a single platform and see a transparent view of the process. Blockchain has numerous applications in the healthcare domain and is equally proficient in the organ donation sector. Blockchain innovation could improve the procedure by providing a more dependable and straightforward approach to checking compatibility between contributors and beneficiaries, facilitating accessibility with the request, and providing a secure way to store data and key health information [4].

When one refers to the word "Blockchain" in the context of healthcare, this refers to the ledger-like system that runs on the Internet of Things (IoT) to track medical records and to keep the flow of information secure [5]. For instance, using a blockchain for the organ donation program helps reduce fraud and identity theft and helps protect donors' private information from outside access. The term comes from how the ledger is formed, where each transaction is recorded in a block, referred to as a blockchain, on which other transactions can be grouped. Healthcare companies use the ledger to identify and track their transactions [6]. It also helps to reduce fraud by ensuring that only legitimate transactions will be included in the blockchain and that the transaction is only valid if it is part of the ledger and not forged. This ledger can be used to track the receiver's medical records. It can be used to prevent identity theft [7]. In its place, a smart contract, a set of instructions, will be executed on individual nodes, and a consensus is made to decide whether a particular transaction is acceptable [8].

In today's era, when decentralization is gaining immense popularity, relying on central authorities must not be acceptable. In the current timeline, unique identifiers, somewhere or the other, are heavily dependent upon central authorities. The solution to all this problem can be DIDs, which are defined as unique identifiers that enable decentralized digital identity, and the Uniform Resource Name is employed for encoding them. Three basic entities for controlling DIDs are the controller, requesting party, and subject. The DID controller is mainly responsible for making alterations to the already generated DID. For instance, if there is any update in the identity record of the participants, DID controllers take care of the required changes. DID request party is the entity that needs to create a unique identity on the network, effectively for which DID is generated in the consortium. DID subject refers to the entities uniquely identified by the DIDs. Here, entities refer to assets, clients, or organizations. The structure of the DID fundamental entity is demonstrated in Fig. (**1**).

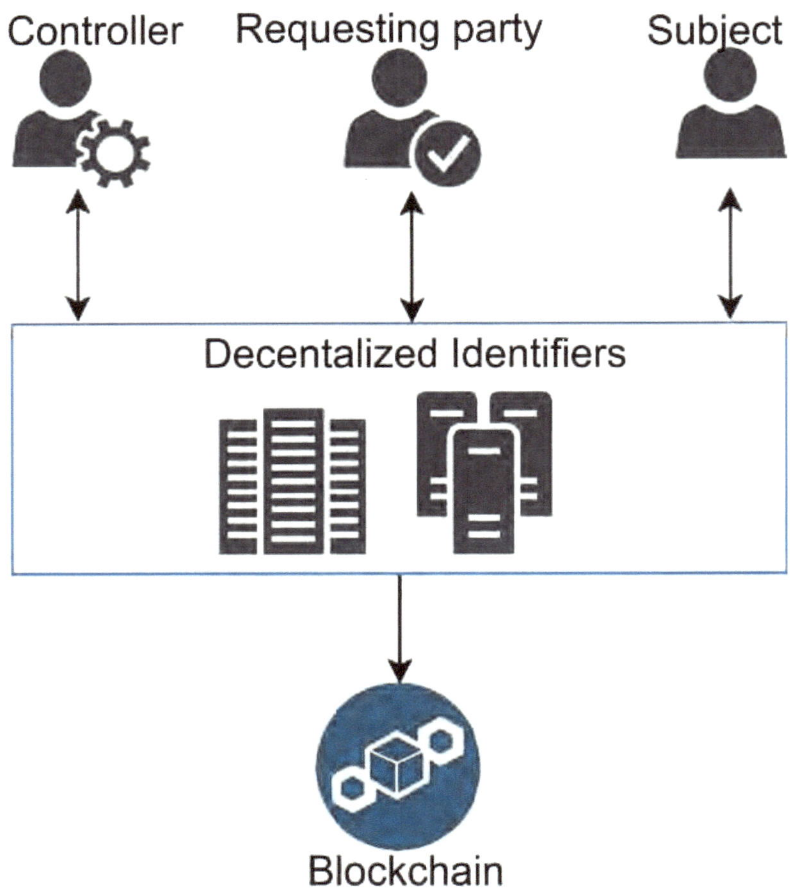

Fig. (1). Basic action performing entities.

DIDs are correlated to the DID Document, which stores the basic information about DID. Apart from this, it also contains an attestation protocol, public key, and service endpoints. In a general sense, the DID Document contains specific attributes that give a unique identity to the network participant. A few of the attributes are as follows:

Id: It is the attribute representing the Decentralised ID of the entity.

Public Key: In blockchain-based identities, two keys are associated with a particular DID: public and private. This attribute contains the details of the public key, *i.e.*, the value of the public key and the type of public key.

Created: The date and time stamp of the new DID generation are stored in this attribute.

Signature: It contains the signatures of the creator of the new DID.

The usage of DIDs empowers the digitally signed VCs. Every entity in the real world needs claims as an essential part of day-to-day activities. More importantly, digital claims prove their importance as more and more entities are moving on the Internet. For example, in healthcare, suppose a person is traveling to another city where they fall ill and need to provide their medical prescription, so they will present their digital credentials to prove their medical background. Since these credentials must be shared with the right person and the amount of information shared must be checked, they need the correct address and keys to be mapped with before sharing the credentials. DIDs can prove their worth in that case. This is the fundamental logic behind the collaboration of DIDs and VCs. VCs are authentic, trustworthy, and tamper-resistant credentials, using which the issuer can propose claims related to the subject. Now, the issuer can propose allegations about the subject by signing the corresponding to the specific DID. VC, generally, contains a field named ID, which DID can fill, dereferenced document, where DID Document can be passed.

This research aims to present a blockchain-based prototype called Organ Donation, *i.e.*, OrDo. OrDo is the secure, trustable, and justifiable interaction model that acts as the bridge between the donor and patient, where the hospital helps the patient and the donor by building a direct link among them. The proposed model contains two layers: the Blockchain layer, which will act as a back-end security layer, and the End-User Interactive layer, which comprises an innovative user interface that makes it easy to port from one module to another. The proposed model targets eradicating mismanagement in organ donation and provides an easy platform for the willing donor to come forward and contribute to this noble cause.

The manuscript is structured as follows: Related work is demonstrated in section II. The framework is proposed in section III, and the demonstration of the proposed work is detailed in section IV. Results are discussed in section V, and finally, the conclusion and future scope are given in section VI.

RELATED WORK

A comprehensive literature review helps the authors understand various blockchain applications influencing the medical domain. Despite the significant relevance of blockchain in the medical domain, it lacks intense research and rigorous efforts to streamline the donation process and bring transparency to the entire chain [9]. The authors aim to include the findings of various researchers. Clinical well-being represents an assortment of difficulties for public institutions all around the globe. The medical industry has unique security and privacy constraints as legal complications are equally involved, along with life complications [10]. The main concerns regarding security and privacy are patient data confidentiality, which requires authentication, interoperability, information sharing, and clinical records exchange [11]. Medical care frameworks seek to achieve long-haul maintainability and improved medical care results. Blockchain has been perceived as an idyllic solution in the healthcare industry, and numerous use cases have evolved recently, specifically secure health records management, drug fabrication, enhanced clinical preliminary administration, and communal well-being observation (*e.g.*, immunizations and epidemics control and prevention) [12]. Researchers are working on several use cases of blockchain in the healthcare industry, *viz.* identity management for the patient, information recovery, data consent usage, and others [13].

The research on blockchain applications in the healthcare industry has provided solutions to the impending problems related to centralized databases: privacy, security, confidentiality, and tamperproof effective communication. However, among several blockchain applications, organ donation is the least explored, and the study shows that the organ donation blockchain network is either region-specific [14, 15] or organ-specific [16]. Data integrity issues in the network demand critical attention [17]. The identifiers to identify the entities in the network are strongly reliant upon central authorities, assigning the task to local organizations, who further assign it to another organization for completing identifiers. This chain works well for small systems, but from a long-term perspective, the identifier's integrity is questionable, as the survival of central authorities justifies the survival of the identifier. Hence, applying a new type of DID is imperative, making the IDs persistent, resolvable, and, most importantly, cryptographically verifiable. The manuscript proposes an organ donation framework based on DIDs to provide a secure, trustworthy, and justifiable interaction model for organ donation. The subsequent section demonstrates the proposed organ donation framework. The system architecture, which includes the various layers and the modules that encompass them, is discussed in the following section.

PROPOSED FRAMEWORK

The section demonstrates the proposed framework for OrDO. The proposed architecture comprises two layers: blockchain and end-user interaction. Further, the end-user interactive layer consists of three modules, *i.e.*, the hospital's consortium management module, the patient module, and the organ donor management module. The proposed system is the use case of Hyperledger Fabric, where a network or consortium is prepared for independent hospitals adhering to specific endorsement policies. One example of endorsement policy may be the requirement of the signature of all the existing participants for any public transaction, such as adding any new hospital to the consortium. On the other hand, there is no need to maintain integrity for any private transactions. A detailed description of the proposed framework is given below:

System Architecture

As mentioned earlier, OrDo comprises the blockchain and End-user interactive layers. The blockchain layer is based on the Permissioned Network, DIDs & VCs issuing and management layer where the Permissioned network is built by preparing a decentralized network using Hyperledger Fabric.

Blockchain Layer

This layer is the most critical layer for the proposed model, as this layer acts as the primary solution to the objective. This layer has three main components: a decentralized network, which is built on hyper ledger fabric; contract (known as chain code in hyper ledger fabric); and DIDs & VCs creation and management.

Smart Contracts play a vital role in performing transactions in the blockchain network. Smart Contracts are the only way to interact with blockchain databases. State databases, or state ledgers of blockchain, allow the creation, Reading, Updating, and Deleting (CRUD) operations to be performed. Smart contract functions are generally of two types: invoke and query functions. Using the invoke functions, create, update, and delete operations can be performed on the state ledger of the blockchain, and query functions allow one to read or view the state ledger. Since invoke functions alter the state ledger, transactions are recorded for invoke functions only in fabric-based blockchain networks. Technically, smart contract functions can be the intermediatory layer between the API and the blockchain ledger and act as the communication link.

The proposed system's initial network is built by adding three hospital nodes as the organizations. Adding an odd number of organizations in the decentralized network where the endorsement policy is set on MAJORITY is always recommended. All the organizations are added to a single channel on which a single smart contract (chain code) is deployed to carry transactions and CRUD operations in the blockchain ledger or world state database. When the Fabric network is prepared, the developer defines the endorsement policies, and depending upon them, invocation is carried out in the network. In a blockchain network, a transaction means changing the state of the asset, and changing the state of the asset is done by invoking the function of the chain code (smart contract). Furthermore, calling any intelligent contract function (either the invoke or query function) depends on the endorsement policy and is performed by the organization's peers. In this particular case, the donor and patient are the clients who will be registered with the organization. Now, suppose the client carries out any transaction. In that case, the gateway must interact with the blockchain, which can only be made using the organization's peer, and the organization's peers will work according to the endorsement policy defined while preparing the network.

There are two data types for the proposed framework, *i.e.*, On-Chain data and Off-Chain data. On-chain data is stored on Blockchain, and Off-chain is stored on other storage platforms, locally or in the cloud. For the proposed model, On-Chain data constitutes the information about DIDs and VCs, *i.e.*, their issuing date and time, credential schemas, *etc.* In the fabric-based blockchain consortium, network certificates can be generated by implementing the Fabric Certification Authority (CA) concept. Fabric CA is the sub-node of the primary organizational nodes. While creating the network, the template of each certificate is prepared, and access is given to the Fabric CA component. So when any client or node requests certificate generation, it is processed through smart contracts considering the type of client, *i.e.*, Donor or patient. Here, implementing a smart contract ensures that the request is generated from legitimate users only. All this information is stored in Blockchain using different cryptographic techniques, which is defined as a Hyperledger Ursa package. Moreover, in the proposed model, each hospital organization will maintain its donor ledger using the Organ Donor module; this ledger is prepared and maintained on Blockchain to provide secure and efficient ledger management. Storing and matching the DID of each participant available in the consortium will be part of the off-chain data for the proposed model. Fig. (**2**) gives the pictorial representation of the Blockchain layer implemented for the proposed model. The Hyperledger Aries project can be employed to implement this layer as it provides predefined modules for implementing the concept of DIDs and VC. The Chain employed for the proposed model is the test chain of the Sovrin Foundation.

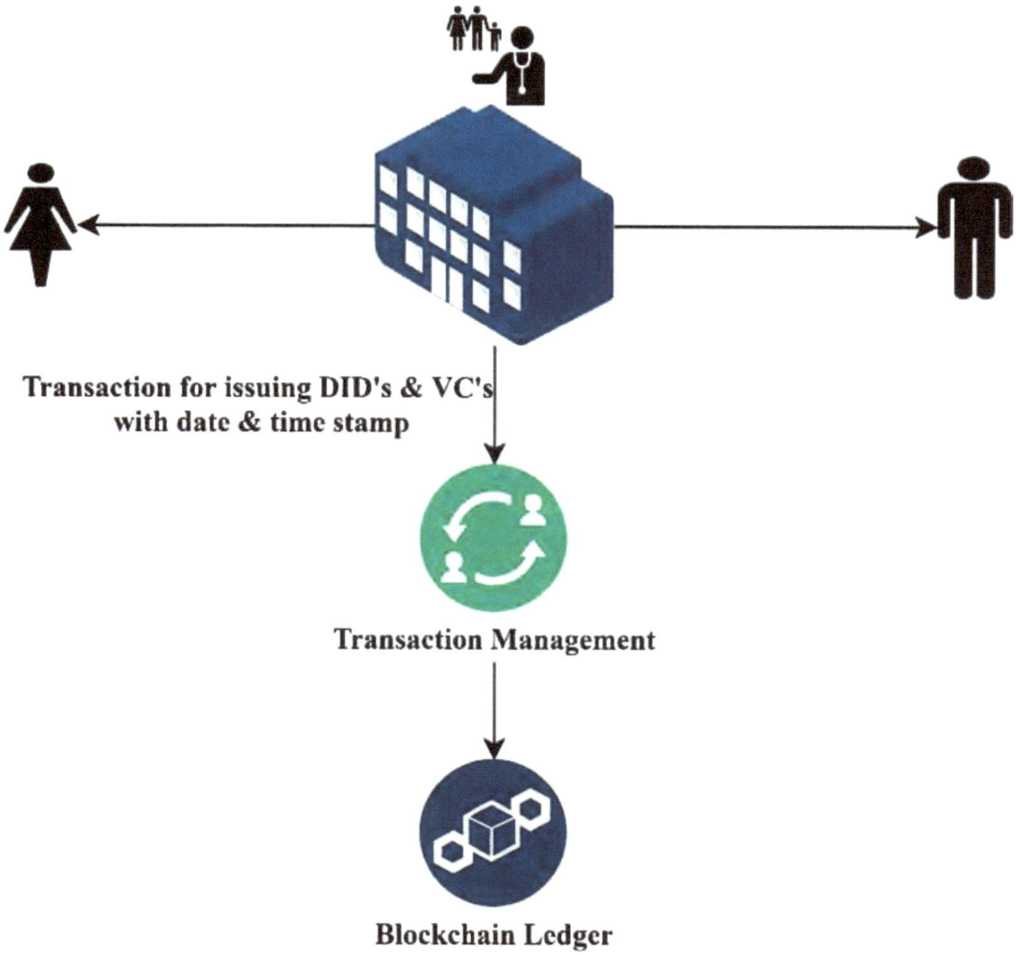

Fig. (2). Blockchain layer.

The proposed system is flexible enough to add new nodes, *i.e.*, new hospitals, to the proposed network. The addition of the new hospital to the network will depend on how the other existing nodes react when the new hospital requests to update the genesis block of the blockchain network. To add a new node to the consortium, new node configurations need to be updated in the genesis block, *i.e.*, the very first block of the network, which contains the details of network configurations and participants. Once the update request has been made using the Fabric SDK, changes will be committed only when the existing organizations sign that transaction. So, suppose any organization has a problem adding a new hospital node to the consortium. In that case, they may not sign the transaction, which will deny the whole update of network configurations, and the request to be part of the consortium made by the new node will be rejected. Furthermore, network configurations play a significant role here. If configurations are set while developing the network, such as all participants must sign the transaction to add a new node, *i.e.*, setting it up for ALL, then the request must be signed by each existing node. In contrast, there is another option, *i.e.*, MAJORITY, which will ask for each node's vote. The final decision will be made according to the majority response made by the existing nodes.

In the blockchain network, whenever there is a CRUD operation on any asset, the transaction corresponds to that operation, which is stored in the chain of blocks. The number of transactions stored in the blocks is defined in the network configurations. Furthermore, the user can verify the performed operations on the asset by comparing them with the transaction. The transaction corresponding to the state change of any asset can be retrieved using the intelligent contract function, which will be counted as the query function, *i.e.*, there will be no separate transaction. For instance, if the hospital staff want to track the operations performed by the patients, they will directly pass the DID of the patient in the hospital module, and correspondingly, they view the list of transactions corresponding to the past operations the patients have performed. Similarly, suppose the patient wants to see the donor's past operations. In that case, they need to add the patient's DID, which will act as an argument for the intelligent contract function, and the list of transactions will be displayed in front of the patient.

End-User Interactive Layer

This layer will directly interact with the end-user and perform fundamental functionalities of the proposed model. This layer comprises three modules, *i.e.*, the Hospital consortium management module, the patient module, and the organ donor management module.

Hospital Consortium Module

This module handles all the existing hospitals in the network. One of the prime aims of blockchain is to prepare immutable transactional logs to achieve transparency. In addition to openness, a fabric-based blockchain network helps the organization maintain integrity. The confidential data about a particular organization will be recorded and stored in the form of blocks. Thus, transparency while maintaining confidentiality can be achieved in the proposed model. This module will handle all the unique De-centralized identities issued to the hospital node during network/consortium registration. Moreover, this module will add a new organization to the network. The addition of new organizations will depend upon the results of the voting consensus organized among the consortium's existing organizations. Once the majority of existing organizations/nodes of the consortium agree to add a new organization/node to the consortium, the system will generate a unique DID for that organization and add its VC to its wallet. This VC will act as the authentication entity for the organization among the consortium participants.

Organ Donor Module

This module's main aim is to take care of the donor ledger, which each hospital maintains. Each hospital will create a DID for the registered donor. Corresponding to that, DID, VC, which carries all the necessary information, will be issued by the hospital and added to his wallet. Parallel to that, this module will update the individual hospital's ledger of donors with the necessary information and medical details. Fig. (**3**). gives an insight into the donor module. Once the hospital recognizes the patient through the patient module, the ledger of this module will be prompted to the patient.

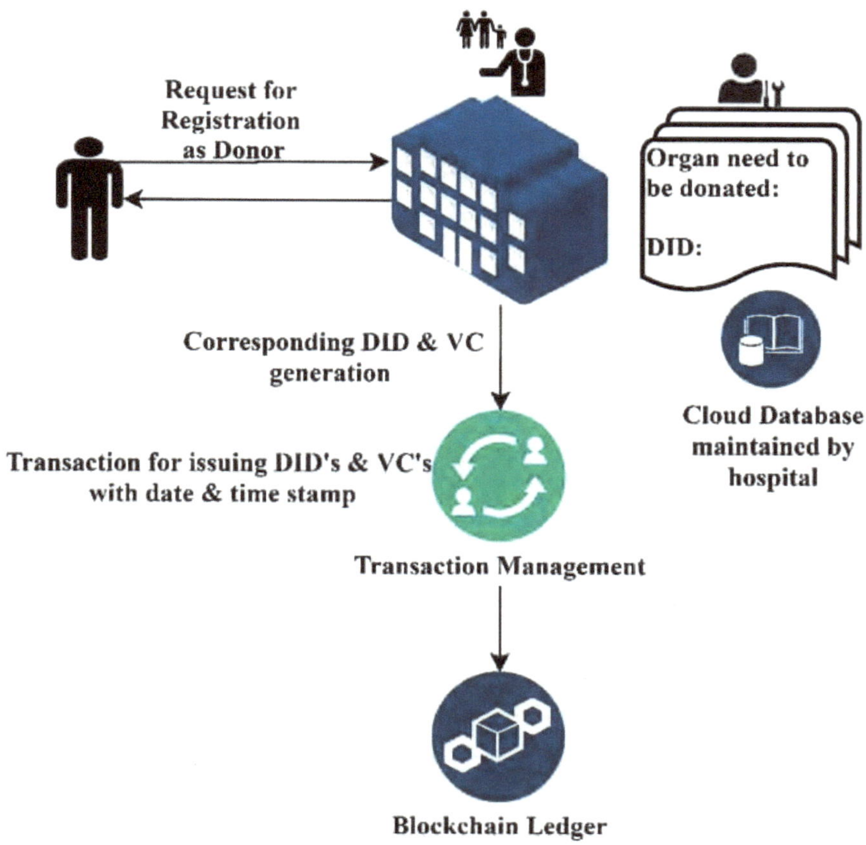

Fig. (3). Organ donor module.

Patient Module

This module works around the portal's management of patients. This module will build a connection between the hospital and the patient. After verifying patients' medical records, the hospital will create a DID for the patient, generate a VC, carry all the patient's medical history, and add that VC to the patient's wallet. Once the VC is issued to the patient, the system will prompt the list of donors of the respective hospital, depending on the request acceptance/rejection of ledger visibility.

In the proposed system, only one channel exists among the participating organizations (hospitals). A single smart contract (chain code) is deployed in that channel for ledger communication. Once the patient identifies the donor, the invite and VC will be sent to the donor using the intelligent contract function. Since they are part of the same channel, sending requests will be easier. If the patient requests multiple donors, it may also be performed efficiently as it does not require the creation of a separate channel for each donor, thus providing a cost-effective solution.

Basic Flow

The proposed model is cleaved into three different modules, *i.e.*, Hospital Consortium, Organ Donor module, and Patient module, and each module has its flow and functionality. For the proposed model, the basic flow is cited in Fig. (**4**). The basic flow of all modules is elaborated below:

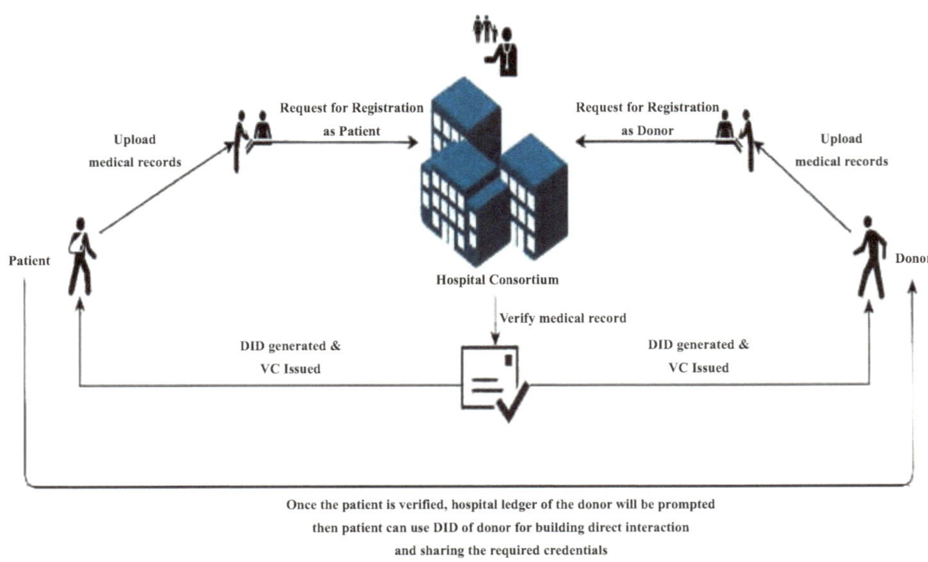

Fig. (4). Basic flow of the proposed model.

Function of Hospital Consortium

This module will register the new hospital organizations and generate their corresponding DID. Once the DID is generated, the module will create the organization's web-based wallet, which will use its local storage to store all its private data, such as its necessary identity certificates. Moreover, this consortium will define a particular schema for verifying the new Donor or Patient participant. The Verification Schema is just like empty data fields, which patients and donors will use to gather required information about each other. Verification schema helps the patient and the donor share the information necessary for the interaction. The Verification Schema will remain the same for all the organizations, *i.e.*, all hospitals participating in the consortium. Moreover, after registering a new hospital organization, each hospital's donor ledger will be initiated by this module, which will be editable by the hospital authorities. Transactions for managing the ledger will be stored on the Blockchain to keep a tamper-resistant record.

Function of Organ Donor Module

This module is mainly meant for donors who want to donate their organs. They will get registered to the network through the registered hospital organization and upload their medical records for verification. Once the hospital management confirms the basic medical requirements for organ donation, they will generate the DID using this module and issue VCs depending on the schema defined in the consortium. The issued VC will be uploaded to the donor's web-based wallet. This VC will be accepted by any hospital organization in the consortium to verify the donor's essential medical credentials. Moreover, these VCs will also implement the zero-knowledge proof concept, *i.e.*, without sharing medical records, the donor can prove their medical state. Moreover, when hospital authorities generate DID, corresponding DID matching with necessary revealable credentials will be stored in the cloud for easy search of the donor by the hospital authorities. When DID is generated, and VC is issued, their transactional records are stored on Blockchain to maintain a tamper-resistant and justifiable ledger for all the donors.

Function of Patient Module

This module works for patients wanting to receive an organ. The patient will get registered to the network as a participant using this module. The patient will send their requirement to the hospital organization using this module. Hospital authorities will verify the requirement depending on the medical records uploaded. Once the hospital authorities successfully verify the medical records, they will create the patient's DID and issue the VC to the patient, depending on the schema the consortium decided. Once the DID is generated, the patient's essential revealable credential is matched with the DID and stored on cloud storage. Parallel to cloud storage, issuing DID, and VC transactions are stored on the Blockchain to maintain a justifiable ledger of all patients. Once the VC is issued to the patient, a list of all the donors will be sent with the donor's necessary details. Once the patient identifies the required donor, the patient will send the invite request to the donor and share their VC to share medical information with the donor if needed. This way, the proposed model will develop safe, secure, trustworthy, and justifiable interaction between the patient and donor.

DEMONSTRATION

The proposed model is implemented using a different cryptographic algorithm, which provides security to the critical documents uploaded by each consortium participant. Moreover, the concept of DID makes the model more reliable and trustworthy. Also, the Zero Knowledge Proof concept prepares a reliable model for sharing each participant's personal information without sharing the documents. Since all the transactions performed on the portal are stored on the Blockchain, there will be a tamper-resistant, justifiable, and reliable place for organ donation. The detailed flowchart of the entire system is demonstrated in Fig. (**5**).

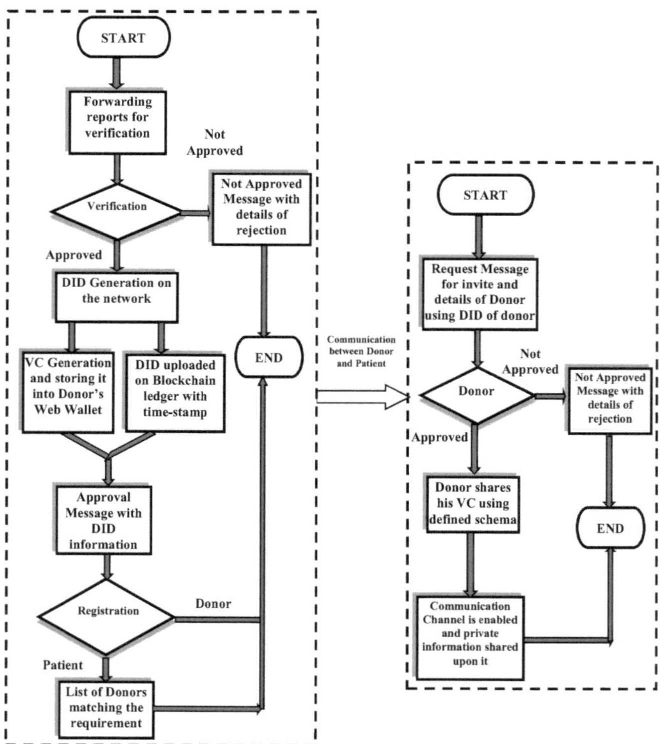

Fig. (5). Flowchart of the system demonstrating the communication between various modules.

Implementation of Decentralized Identities and Verifiable Concept

The main objective of proposing this model is to build a secure, trustworthy, and direct bridge between the organ donor and receiver without the intervention of the middleman. It is essential to recognize the end-user securely for sharing personal details with them. To counter this situation, this prototype generates a DID for each end-user. It stores its issuing information on the Blockchain, which gives a sense of security to the end-user and recognizes the second person more effectively and quickly. Using this prototype, hospital authorities issue VCs for each patient and donor, which contain all the required information about the person and their medical details. In this way, each patient and donor must have verified personal credentials that each participant in the consortium accepts. Suppose anyone among the network participants wants to view details of any other participant, *i.e.*, patient or donor. In that case, they will retrieve the schema from the hospital authority, which is typical for all hospital organizations, and the credential request will be sent to the other participant. Depending upon the acceptance or rejection of a request, fields mentioned in the schema will be updated with participants' credentials and forwarded to the end person automatically. In this way, the participant did not share their actual document but proved their authenticity and shared the required details. The algorithm for the same is given below:

Algorithm for DID generation: Issue Credential/VC Generation

it

1. Generate *Pair(Public Key, Private Key)* Locally.
2. Generate *PrivateKey(Privatekey-> Name)*
3. Get *PublicKey(PublicKey-> PrivateKey)*

stration

1. *Registration request -> public key, private key*, i.e., generating a based key for the network.
2. *Json(private and public key)*
3. *Web Signature (Public key)*
4. Header of the (*web signature-> Public id, proof of work, operation*).
5. Sign the web key using the private key.
6. Receive Registration API.
7. Pass the Web-based key to the network operator.
8. The operator will renovate the network configuration and add the key to the network depending upon the proof of work.
9. In return, the network operator will return the DID.
10. Fetch the schema from the organization ledger.
11. Search the world state of schema defined for credential definition using smart contract or chaincode.
12. Pass the schema to the peer.
13. Fill in the required attributes from the schema with all the required details.
14. Hospital Peer will fill the schema attribute depending upon the participant's details.
15. Finally, the Peer will generate a credential card.
16. Sign the Credential with the public key of the Credential Owner.
17. Finally, the peer will sign the credential with the participant's public key.

l Result

1. DID Generated
2. Peer issue the credential

Implementation of Zero Knowledge Proof Concept

Sharing actual documents with an unknown person may result in the misuse of documents. If somehow information is shared without actually sharing the original document with genuine authenticity, it may become easy for the end-user. The solution for this situation lies in the concept of zero-knowledge-proof implementation. In the proposed model, if any participant wants to retrieve information about another participant, the participant will request credential sharing using the hospital organization's schema in the consortium. Initially, they will retrieve the consortium schema stored in Blockchain; then, they will request credential sharing, where the schema will be offered to the other participant in the request. Depending upon the request acceptance and rejection, the schema's fields will be updated automatically by the system with the required credentials and forwarded to the request sender. This way, there will be zero involvement in document sharing, but authentic information sharing will be employed as the universal schema.

Algorithm for Zero-Knowledge Proof

1. Received the authorized credential from the organization.
   a. The algorithm of credential request and further issue credentials will return the authorized credential to the participant.
   b. The credential card is stored in the architecture web wallet.
2. Verifier Request to prove the authenticity of another participant.
   a. Patient will fetch the donor from the ledger prompted after network registration.
   b. The patient will request an Appropriate donor's identity using a user-defined function and intelligent contract/chain code function, i.e., requestIdentity().
3. The provider and Hospital organization put the limited details in the schema required for proving authenticity.
   a. The requested donor will receive the verification message.
   b. The requested donor will call the user-defined function to prove identity(), which will,
      i. Fetch the required schema to provide limited details on the Blockchain.
      ii. The Donor's details will fill in the required attributes of the schema.
   c. Providing detailed credentials will be prepared and returned to the donor.
4. Prepared data is shared with the verifier.
   a. Providing detail cards will be shared with the patient to verify and prove the donor's realness.

Transparency in Record Keeping and Record Retrieval

Transparency in record management in the medical field is crucial. The proposed model eradicates the problem of misuse of medical documents as the medical records are uploaded on the portal at the time of registration only. Further, asymmetric encryption is employed for the record onboarding process, where required medical documents are uploaded using the hospital's public key and can only be decrypted by the hospital's private key. So, this provides end-to-end encryption of documents. Other than that, there is no document movement in the proposed model. Additionally, all the required information is retrieved or transferred using VCs. Furthermore, the flow of personal details among the participants depends on the schema defined by the hospital, which is common among all hospital organizations.

RESULTS & DISCUSSION

The section presents the analysis of the OrDo with the existing systems and the performance analysis of the proposed framework concerning two parameters: Batch time out and transaction rate. A detailed discussion of the analysis is given as under:

Performance Analysis

The performance of the proposed OrDo system is analyzed in terms of Batch time out and transaction rate using Hyperledger Explorer. Batch time out is the waiting time (in seconds) for ordering services to generate a new block. The Transaction Rate is the period between registering new transactions in the system.

In the Blockchain system, time is mainly consumed for transferring and updating the ledger of blocks at each node [18]. There are two main measures to be considered: block generation and transaction rate. Transaction rate remains almost identical for generating and registering new transactions, whereas Block generation is measured by Batch timeout. Batch Time-out affects the block generation but has the least effect on block propagation. The batch-out time is illustrated in Fig. (**6**), which indicates that each newly generated block takes almost a similar amount of time for propagation among the different peers. Mainly, it revolves around block propagation among the nodes. Configuration blocks take the least time for block transference, read operation blocks take almost the same time for propagation, and write operation blocks take the maximum time for propagation in the chain.

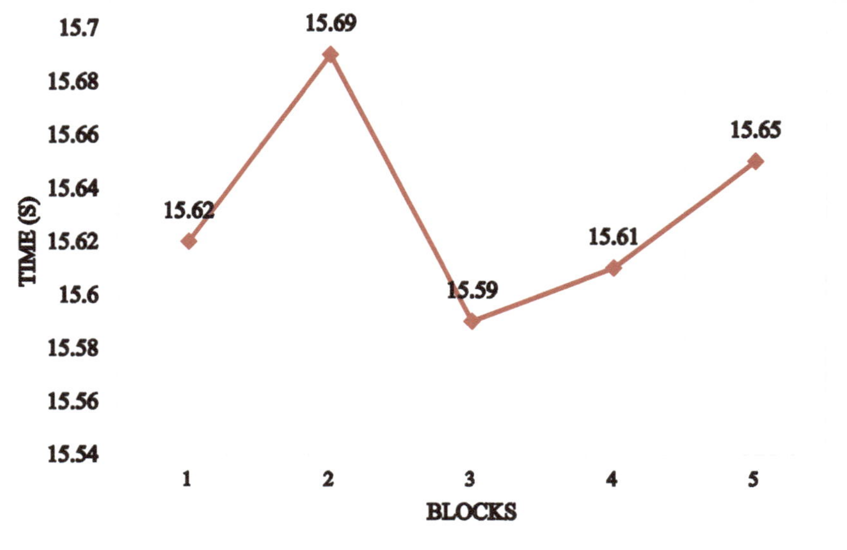

Fig. (6). Analysis of batch time out of proposed ordo.

Moreover, the propagation time of blocks containing write operations (basically generated when the invoke function is called) increases linearly with the increase in newly generated blocks, as shown in Fig. (7).

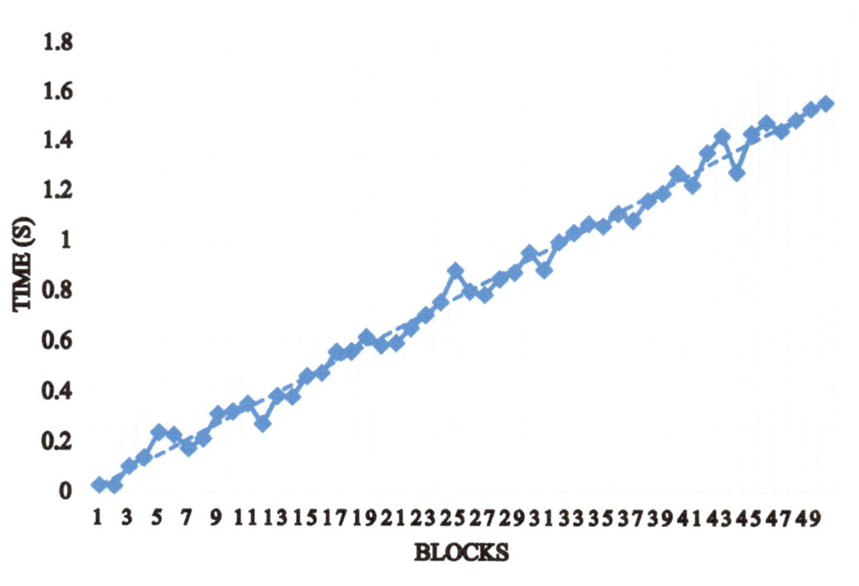

Fig. (7). Analysis of propagation time for write operation blocks.

Comparative Analysis

The section compares the proposed OrDo concerning the existing centralized technologies, considering ease of auditing, single-source dependence, and cross-border relations. Further, Table **1** analyzes the proposed OrDo framework with existing blockchain technologies, namely Ethereum and Hyperledger framework [19, 20].

Table 1. Comparative analysis of the proposed framework with existing methodologies.

Points	Hyperledger	Ethereum	Proposed OrDo Framework
Purpose	Improve the exchange of inter-industrial knowledge by planning the same contact platform for them.	To promote the development of decentralized software to achieve transparency	Decentralized, but also targets to build communication channels among the different users.
Accessibility	Restricted individuals from numerous organizations may come together to exchange essential information. In addition, there are small members, and only approved members can do mining.	Ethereum is known as a mining friendly. Decentralized Blockchain Networks and the general public can participate and bring forward the ideas concerned.	A permissioned network of donors and receivers, where any authorized public can be brought on the same channel.
Confidentiality	Only authorized individuals are visible in the transactions existing on the network. Less transparent relative to the public network.	Each transaction will be public. Transfers are encrypted to preserve the identity of the participants.	Reveals limited information about the donors to limited participants.

The proposed decentralized system is also compared with the prevalent centralized system, as demonstrated in Table **2** [21, 22]. The decentralized system provides immutable logs with proper time stamps to easily audit transactions. Further, being based on a distributed network with Self Sovereign Identity (SSI), each participant has their own identity and participation in the functioning of the system in the decentralized system [23, 24]. Additionally, the proposed decentralized model eliminates geographical borders and puts donors and receivers in the same channel for enhanced information sharing.

Table 2. Comparative analysis of proposed or do with the prevailing centralized system.

Points	Prevailing Centralized System	Proposed Decentralized System
Ease in Auditing	Complex and time-consuming, a Lack of ordered logs management.	Ease in the auditing process. Provides immutable logs with proper time stamps for transactions for easy auditing.

(Table 2) cont.....

Points	Prevailing Centralized System	Proposed Decentralized System
Single Source Dependence	Receivers need to rely on one particular organization, the organization's stakeholders, which limits the exploring options and decreases the stakeholder's trust.	It uses a distributed network with Self-Sovereign Identity (SSI) implementation for DID and VCs, alleviating a single authority. Each participant has their own identity and participation.
Cross-Border Relation	A centralized system always relies on legitimate government authority to define protocols. Moreover, it limits the system to the geographical boundaries of the particular region.	Eliminates geographical borders and brings donors and receivers to the same channel for enhanced information sharing.

Thus, it is evident from the comparative analysis illustrated in Tables **1** and **2** that the proposed framework yields promising results and has the potential for widespread deployment in the real world. The proposed model is not limited to the healthcare domain; it can be effectively employed in any scenario demanding any critical resource management system.

CONCLUSION AND FUTURE SCOPE

Organ transplants and organ donations are significant tasks and should be adequately governed as they have a massive impact on the lives of those concerned. Further, it may also invite black marketing of the organs without central governance. Hence, the proposed system proposes a framework that develops a reliable, encrypted, and truthful ledger of organ donation based on Blockchain Technology. The proposed framework's prime objective is to bring transparency to verified organ donors and procurers in the organ donation chain. It suggests the concept of decentralization, which the consortium of verified Hospitals can efficiently govern. The Proposed model is mainly bifurcated into three different modules, *i.e.*, Hospital Module, Donor Module, and Patient Module, each having its significance and functionality. DIDs and VCs' concept is introduced to each module to provide a unique identifier for each individual and their corresponding justifiable credentials. Moreover, the proposed model comprises the Zero-Knowledge Proof concept, which helps individuals submit only relevant details. The most substantial positive aspect of the proposed model is a transparent log that maintains information about the organ acting as an asset. Once the chain is active, it generates lots of data that needs to be appropriately managed, as it contains the personal information of donors and patients. The results have been motivating; thus, the proposed model can be implemented in real-world applications involving critical resource management. The proposed work can be further extended in the direction of using appropriate indexing techniques or any other feasible way-out model enabling better data management.

REFERENCES

[1] G. Moloney, and I. Walker, "Talking about transplants: Social representations and the dialectical, dilemmatic nature of organ donation and transplantation", *Br. J. Soc. Psychol.,* vol. 41, no. 2, pp. 299-320, 2002.
[http://dx.doi.org/10.1348/014466602760060264] [PMID: 12133230]

[2] T. McGhin, K.K.R. Choo, C.Z. Liu, and D. He, "Blockchain in healthcare applications: Research challenges and opportunities", *J. Netw. Comput. Appl.,* vol. 135, pp. 62-75, 2019.
[http://dx.doi.org/10.1016/j.jnca.2019.02.027]

[3] A. Ríos, A. López-Navas, A. López-López, F.J. Gómez, J. Iriarte, R. Herruzo, G. Blanco, F.J. Llorca, A. Asúnsolo, P. Sánchez, P.R. Gutiérrez, A. Fernández, M.T. de Jesús, L. Martínez Alarcón, A. Lana, L. Fuentes, J.R. Hernández, J. Virseda, J. Yélamos, J.A. Bondía, A.M. Hernández, M.A. Ayala, P. Ramírez, and P. Parrilla, "A Multicentre and stratified study of the attitude of medical students towards organ donation in Spain", *Ethn. Health,* vol. 24, no. 4, pp. 443-461, 2019.
[http://dx.doi.org/10.1080/13557858.2017.1346183] [PMID: 28665141]

[4] L. Ismail, H. Materwala, and S. Zeadally, "Lightweight blockchain for healthcare", *IEEE Access,* vol. 7, pp. 149935-149951, 2019.
[http://dx.doi.org/10.1109/ACCESS.2019.2947613]

[5] C. Gan, A. Saini, Q. Zhu, Y. Xiang, and Z. Zhang, "Blockchain-based access control scheme with incentive mechanism for eHealth systems: patient as supervisor", *Multimedia Tools Appl.,* pp. 1-17, 2020.
[http://dx.doi.org/10.1007/s11042-020-09322-6]

[6] T.T. Kuo, H. Zavaleta Rojas, and L. Ohno-Machado, "Comparison of blockchain platforms: a systematic review and healthcare examples", *J. Am. Med. Inform. Assoc.,* vol. 26, no. 5, pp. 462-478, 2019.
[http://dx.doi.org/10.1093/jamia/ocy185] [PMID: 30907419]

[7] B.K. Pramanik, A.Z.M.S. Rahman, and M. Li, "Blockchain-based reward point exchange systems", *Multimedia Tools Appl.,* vol. 79, no. 15-16, pp. 9785-9798, 2020.
[http://dx.doi.org/10.1007/s11042-019-08341-2]

[8] L. Soltanisehat, R. Alizadeh, H. Hao, and K.K.R. Choo, "Technical, Temporal, and Spatial Research Challenges and Opportunities in Blockchain-Based Healthcare: A Systematic Literature Review", *IEEE Trans. Eng. Manage.,* pp. 1-16, 2020.
[http://dx.doi.org/10.1109/TEM.2020.3013507]

[9] Singh, B., Sharma, K. P. & Sharma, N. (2020). *Blockchain Applications, Opportunities, Challenges and Risks: A Survey Proceedings of the International Conference on Innovative Computing & Communications (ICICC) 2020.*
[http://dx.doi.org/10.2139/ssrn.3565930]

[10] I. Abu-elezz, A. Hassan, A. Nazeemudeen, M. Househ, and A. Abd-alrazaq, "The benefits and threats of blockchain technology in healthcare: A scoping review", *Int. J. Med. Inform.,* vol. 142, p. 104246, 2020.
[http://dx.doi.org/10.1016/j.ijmedinf.2020.104246] [PMID: 32828033]

[11] Y.L. Gao, X.B. Chen, G. Xu, W. Liu, M.X. Dong, and X. Liu, "A new blockchain-based personal privacy protection scheme", *Multimedia Tools Appl.,* pp. 1-14, 2020.
[http://dx.doi.org/10.1007/s11042-020-09867-6]

[12] X. Yue, H. Wang, D. Jin, M. Li, and W. Jiang, "Healthcare data gateways: found healthcare intelligence on blockchain with novel privacy risk control", *J. Med. Syst.,* vol. 40, no. 10, p. 218, 2016.
[http://dx.doi.org/10.1007/s10916-016-0574-6] [PMID: 27565509]

[13] Q. Xu, C. Jin, M.F.B.M. Rasid, B. Veeravalli, and K.M.M. Aung, "Blockchain-based decentralized content trust for docker images", *Multimedia Tools Appl.,* vol. 77, no. 14, pp. 18223-18248, 2018.
[http://dx.doi.org/10.1007/s11042-017-5224-6]

[14] S. Morande, and M. Marzullo, "Application of Artificial Intelligence and Blockchain in healthcare management - donor organ transplant system", *Annals of Management and Organization Research,* vol. 1, no. 1, pp. 25-38, 2020.
[http://dx.doi.org/10.35912/amor.v1i1.261]

[15] L.A. Dajim, S.A. Al-Farras, B.S. Al-Shahrani, A.A. Al-Zuraib, and R. Merlin Mathew, "Organ Donation Decentralized Application Using Blockchain Technology", *2nd International Conference on Computer Applications & Information Security (ICCAIS),* pp. 1-4, 2019.Riyadh, Saudi Arabia.
[http://dx.doi.org/10.1109/CAIS.2019.8769459]

[16] Y. Chen, Y. Li, J.D. Kalbfleisch, Y. Zhou, A. Leichtman, and P.X. Song, "Graph-based optimization algorithm and software on kidney exchanges", *IEEE Trans. Biomed. Eng.,* vol. 59, no. 7, pp. 1985-1991, 2012.
[http://dx.doi.org/10.1109/TBME.2012.2195663] [PMID: 22542649]

[17] Chavez, N., Kendzierskyj, S., Jahankhani, H., & Hosseinian, A. (2020). Securing Transparency and Governance of Organ Supply Chain Through Blockchain. In Policing in the Era of AI and Smart Societies Springer Cham:97-118.
[http://dx.doi.org/10.1007/978-3-030-50613-1_4]

[18] Y. Ren, F. Zhu, K. Zhu, P.K. Sharma, and J. Wang, "Blockchain-based trust establishment mechanism in the internet of multimedia things", *Multimedia Tools Appl.,* pp. 1-24, 2020.
[http://dx.doi.org/10.1007/s11042-020-09578-y]

[19] S. Yadav, N. Sharma, M. Mangla, and A. Mahajan, "Blockchain and ipfs based framework for secure student document record keeping", *J. Educ. Multimed. Hypermedia,* vol. 30, no. 2, pp. 165-181, 2021.

[20] W., Hu, J., Zhu, T., Ren, Y. and Choo, K.K.R. (2020). A flexible method to defend against computationally resourceful miners in blockchain proof of work. Information Sciences 507:161-171.
[http://dx.doi.org/10.1016/j.ins.2019.08.031]

[21] S. Yadav, M. Mangla, N. Sharma, and A. Mahajan, "Blockchain framework for smart contract and distributive ledger for entity marketplace", *International Journal of Blockchains and Cryptocurrencies,* vol. 3, no. 2, pp. 95-111, 2022.
[http://dx.doi.org/10.1504/IJBC.2022.124004]

[22] H. Yang, J. Hwang, D. Paik, "The Identification of Blur Extent from Space-variant Motion Blurred Image," Journal of Internet Computing and Services, vol. 8, no. 6, pp. 169-0, 2007.
[http://dx.doi.org/10.3745/JIPS.01.0024]

[23] J. Cheng, L. Xie, X. Tang, N. Xiong, and B. Liu, "A survey of security threats and defense on Blockchain", *Multimedia Tools Appl.,* pp. 1-30, 2020.
[http://dx.doi.org/10.1007/s11042-020-09368-6]

[24] N. Sharma, M. Mangla, and S.N. Mohanty, "A Markov decision process-based secure consensus framework for leveraging blockchain technology in IoT applications", *Int. J. Electron. Bus.,* vol. 17, no. 3, pp. 270-282, 2022.
[http://dx.doi.org/10.1504/IJEB.2022.124329]

CHAPTER 16

Brain Tumor Classification Using Deep Learning Techniques

Jyotismita Talukdar[1,*] and **Zahid Mansur**[2]

[1] *Tezpur University, Assam, India*

Abstract: Ensuring timely detection and precise categorization of brain tumors plays a vital role in the effective diagnosis and management of neuro-oncological conditions. Traditional manual approaches to brain tumor segmentation in imaging are frequently labor-intensive and prone to subjectivity, underscoring the need for a transition to automated methodologies. This research introduces a sophisticated deep learning architecture aimed at improving the accuracy and speed of brain tumor analysis. There are combinations of several classification algorithms, like binary classification (CNN algorithm, VGG19 algorithm) and multi-class classification (EfficientNetB0 algorithm). VGG19, with its 19 layers containing 16 convolution layers, 3 fully connected layers, and several max-pooling layers, achieves an accuracy of 86%. For multiclass classification, differentiating among various types of brain tumors, EfficientNetB0 is used. This model employs a compound scaling method to balance network depth, width, and resolution, resulting in higher accuracy and efficiency. The images are labelled using one-hot encoding: 0 for 'Glioma Tumor', 1 for 'No Tumor', 2 for 'Meningioma Tumor, and 3 for 'Pituitary Tumor'. EfficientNetB0 outperforms VGG19, achieving an accuracy of 97%, and demonstrates higher efficiency in computational power usage. Finally, we developed a user-friendly interface allowing users to upload MRI images and predict the presence and type of brain tumor using the trained models. This integration of advanced segmentation and classification techniques highlights the potential for improved performance in automated brain tumor diagnostics.

Keywords: Brain tumor, Binary classification, Classification, CNN, EfficientNetB0, VGG19, Multiclass classification.

INTRODUCTION

Brain tumors are abnormal growths of brain cells in the brain that can be either benign (non-cancerous) or malignant (cancerous) [1, 2]. Its presence can disrupt normal brain functions, leading to a range of neurological symptoms. Early and

* **Corresponding author Jyotismita Talukdar:** Tezpur University, Assam, India; E-mail: jyoti4@tezu.ernet.in

precise detection of these tumors is crucial for effective medical intervention [3, 4, 5]. Timely and accurate identification of tumor regions is essential for determining the type, size, and location of the tumor, as well as its impact on surrounding brain structures [6, 7].

Brain tumor classification is crucial in medical imaging for accurate diagnosis and personalized treatment plans [8]. Traditional methods, relying on manual analysis by radiologists, can be time-consuming and inconsistent. The introduction of deep learning algorithms, especially VGG19 and EfficientNetB0, has revolutionized this field by providing more precise and efficient solutions [9].

VGG19 is a deep convolutional neural network, with 19 layers, excels in recognizing intricate patterns in medical images, making it ideal for binary classification tasks like detecting tumors [10]. EfficientNetB0 uses a novel compounding scaling method to balance network width, depth and resolution, achieving high accuracy with fewer computational resources [11, 12]. This makes it suitable for multiclass classification, distinguishing between various brain tumor types. Using VGG19 and EfficientNetB0 for brain tumor classification offers significant advantages, such as enhanced diagnostic accuracy and efficiency, reducing radiologists' workload, and minimizing errors [13]. These models' advanced feature extraction capabilities allow them to handle the complexity of brain tumor imaging better than traditional algorithms and other machine learning techniques [14, 15]. Their integration into medical imaging represents a major advancement, providing substantial benefits over conventional methods [16].

WORKFLOW

The detailed workflow of the proposed classification of brain tumor using MRI images is shown below in Fig. (**1**).

BINARY CLASSIFICATION

A total of 700 MRI images of brain tumors have been collected from Kaggle. The dataset is segmented with the help of Threshold-Based segmentation and its result is used for Binary classification. Our study starts with the data preparation process for image classification where images are labeled as either "no" or "yes" based on presence or absence of brain tumor from the MRI images as shown in Figs. (**2** and **3**).

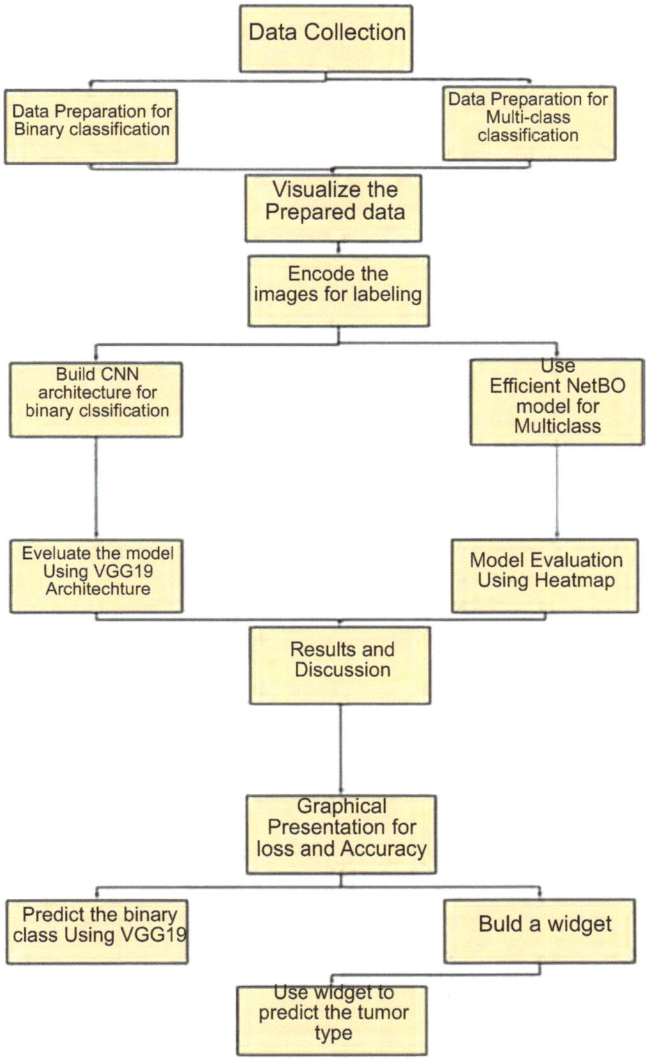

Fig. (1). Detailed workflow for MRI image classification.

Initially, the code lists all images in the specified directories for "no" and "yes" categories. Each image is read using OpenCV, which is converted to a PIL Image object, resized to 240x240 pixels, and then appended to the 'dataset' list as a NumPy array as label 0 and label 1. Finally, the lists are converted to NumPy arrays, and the shapes of the 'dataset' and 'label' arrays are printed to verify the data preparation. This ensures that the images are uniformly sized and correctly labeled for subsequent machine-learning tasks.

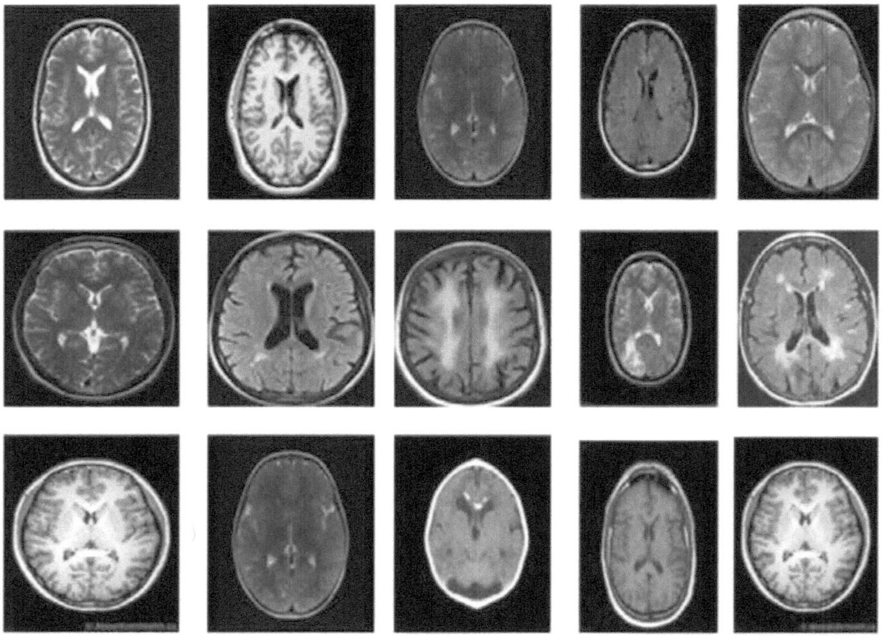

Fig. (2). MRI dataset with no tumor.

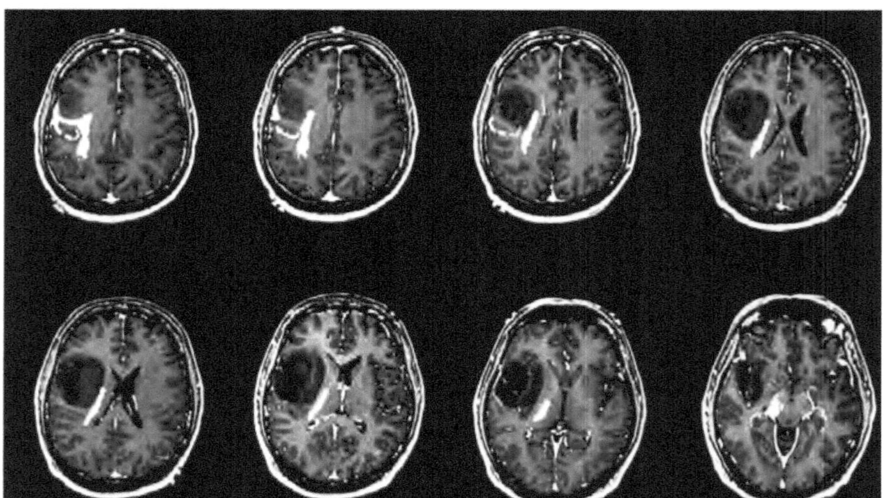

Fig. (3). MRI dataset with tumor.

CNN Architecture

A Convolutional Neural Network (CNN) is a deep learning model particularly effective for analyzing visual data. It is widely used in image and video recognition, as well as in other applications like medical image analysis and self-driving cars [17]. CNNs work automatically and adaptively learning spatial features from input images. They consist of several layers, each performing a specific function. The primary layers are convolutional layers, which apply a filter set to the input to produce feature maps that capture local patterns such as edges and textures [18, 19]. Pooling layers follow these layers. Additional layers, including fully connected layers, further process the features for the final task, such as classification or detection. Through backpropagation, CNNs iteratively adjust their filters and weights during training to minimize prediction errors, thereby improving their accuracy and effectiveness in recognizing patterns in new data, which down sample the feature maps to reduce their dimensions and computation requirements, making the model more manageable and less prone to overfitting [20, 21]. Through backpropagation, CNNs iteratively adjust their filters and weights during training to minimize prediction errors, thereby improving their accuracy and effectiveness in recognizing patterns in new data [22]. In our study, the model is a neural network designed for the task of image classification [23]. It begins with a convolutional layer that uses 32 filters and a 3x3 kernel to detect features in the input images, which are 240x240 pixels with 3 color channels. This is followed by a max-pooling layer, which reduces the image size. Another similar convolutional and max-pooling layer pair further extracts and condenses image features. The data is then flattened into a 1D array. A dense layer with 64 units and ReLU activation is included, using regularization techniques to prevent overfitting by applying penalties on large weights. A dropout layer is added next, randomly ignoring half the units during training to improve model generalization [24]. Finally, a dense layer with a single unit and sigmoid activation produces an output between 0 and 1, indicating the classification result. This architecture is designed to effectively process and classify images with measures to enhance model performance and robustness [25].

VGG19 Model

VGG19 is a convolutional neural network (CNN) architecture renowned for its deep stack of 19 layers, including 16 convolutional and 3 fully connected layers, primarily utilized for the task of image classification. Created by the Visual Geometry Group at the University of Oxford, VGG19 functions by incrementally extracting hierarchical features from input images by using convolutional layers, which apply learnable filters to detect edges, textures, and shapes. Max-pooling layers interspersed within the network downsample feature maps, reducing computational complexity while retaining crucial information. Following convolutional processing, the features are flattened and passed through fully connected layers to make predictions, typically employing softmax for multi-class classification or sigmoid for binary classification [26, 27]. VGG19's architecture depth facilitates learning intricate image patterns, making it highly effective for various computer vision tasks.

In our study, we use an architecture built upon the pre-trained VGG19 model, which excels at extracting intricate features from images due to its deep convolutional layers. Initially, the VGG19 model is loaded with pre-trained weights from ImageNet, ensuring it possesses a strong foundation for image recognition tasks. The existing layers are then frozen to retain their learned patterns, preventing them from being altered during training. Additional layers are then appended to the model: a flattening layer to convert the output into a 1D array, followed by a dense layer with 64 units and ReLU activation. This added layer helps in discerning more intricate patterns in the image features. Regularization techniques are applied to the dense layer to overcome overfitting. Finally, a dense layer with a single unit and sigmoid activation is added for binary classification, producing a probability score that indicates the likelihood of the image belonging to a particular class. This comprehensive architecture combines the strengths of the pre-trained VGG19 model with additional layers tailored to enhance its performance in specific classification tasks [28].

MULTICLASS CLASSIFICATION

For multi-class classification, we have collected the MRI dataset from Kaggle of 3200 images. The images contain Glioma, Meningioma, Pituitary and, and no-tumor data. The dataset is segmented with the help of Threshold-Based segmentation and used for multi-class classification. In this study, we initialize empty lists 'X_train' and 'y_train' to store image data and corresponding labels, respectively. The variable 'image_size' is set to 150, representing the dimensions to which all images will be resized. Then iterates through each category in both the training and testing datasets. For each image in each category, it reads the image using OpenCV, resizes it to the specified dimensions, and appends the resized image data to 'X_train' while appending the corresponding label to 'y_train'. Once all images from both training and testing sets are processed, 'X_train' and 'y_train' are converted to numpy arrays. We efficiently collect the image data and their corresponding labels, preparing them for use in training a machine-learning model for brain tumor classification. It ensures that the images are uniformly sized and properly labeled, which are crucial steps in preparing data for machine learning tasks. Additionally, the use of 'tqdm' allows for progress bars, providing visual feedback on the processing of images, particularly useful when dealing with large datasets. Overall, this algorithm operates as a foundational step in a machine learning pipeline for brain tumor classification, effectively organizing and preparing the data for subsequent model training and evaluation. Fig. (**4**) represents each type of brain tumor.

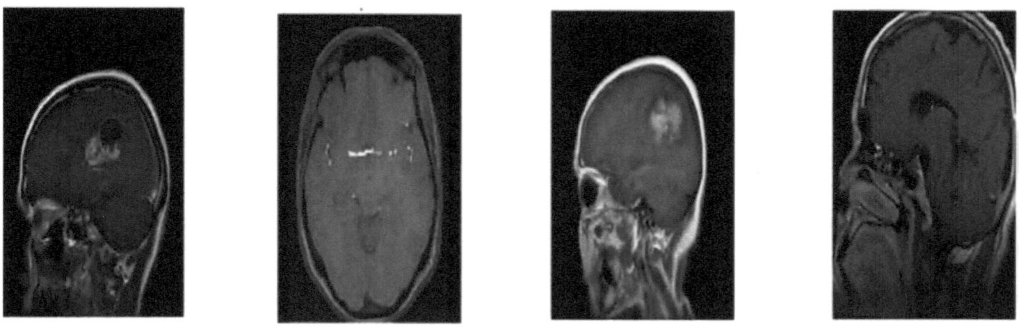

Glioma tumor **No tumor** **Meningioma tumor** **Pituitary tumor**

Fig. (4). Image visualization for multiclass classification.

One-Hot-Encoding for Labelling the Images

Essentially, a data preprocessing step aimed at converting categorical labels into a numerical format called "one-hot encoding", which is commonly used in machine learning tasks. Initially, it creates empty lists, 'y_train_new' and 'y_test_new', to hold the transformed labels. Then, it loops through each label in the training and testing datasets, searching for its index within the list of tumor types. For instance, if the label is 'no_tumor', its index might be 1 in the list of tumor types. We have recorded this index in the new list. After processing all the labels, the next step involves converting these indices into a special numerical format known as one-hot encoding. This conversion is done using TensorFlow's 'to_categorical()' function, which converts each index into a binary vector, whereas only one element is 1 and the rest are 0s, corresponding to the position of the label within the list of tumor types. This transformation is crucial because it allows the machine learning model to interpret the labels as numerical data, making it simple for the model to understand and learn from them during the training process. By converting the labels into a format that the model can easily process, this preprocessing step lays the groundwork for building an accurate brain tumor classification model.

EfficientNetB0 Model

EfficientNetB0 represents the base model in the EfficientNet series, known for its outstanding performance and efficiency in image classification. This architecture is pre-trained on the large and diverse ImageNet dataset, enabling it to grasp a wide range of visual concepts and patterns. The code snippet provided showcases the instantiation and customization of EfficientNetB0 for a specific classification task [29].

Upon loading the EfficientNetB0 model pre-trained on ImageNet, custom layers are added to tailor the model. First, a global average pooling layer is appended, enabling the model to aggregate spatial information from the feature maps.

Next, a dropout layer with a 0.5 dropout rate is added to help prevent over- fitting by randomly tanning off neurons during training. Finally, a dense layer with a softmax activation function is added to produce probabilities for each class, effectively transforming the model into a multi-class classifier.

The model is subsequently compiled with categorical cross-entropy as the loss function, the Adam optimizer for gradient descent, and accuracy as the metric for evaluation. This setup optimizes the model's ability to minimize prediction errors and maximize classification accuracy.

To facilitate model training and monitoring, various callbacks are defined. These include Tensor Board for visualizing training metrics, Model Check point for saving the best performing model based on validation accuracy, and ReduceLROnPlateau for dynamically adjusting the learning rate to enhance convergence.

During training, the model is provided with training data (X_train) with their corresponding labels (y_train), with a portion reserved for validation (10% in this case). The training process spans 12 epochs, with a batch size of 32 samples per iteration. Throughout training, the defined callbacks are invoked to log progress, save the best model, and adjust the learning rate as needed.

In essence, EfficientNetB0 serves as a powerful backbone for image classification tasks, offering a potent combination of accuracy and efficiency. Through customization and fine-tuning, it can be adapted to various domains and achieve state-of-the-art performance with minimal computational resources [30].

PERFORMANCE EVALUATION

The classification report offers a detailed assessment of the model's performance on the test data. In this report, each class (denoted by labels 0, 1, 2, and 3) is individually assessed in terms of precision, recall, and F1-score.

- ***Precision:*** Precision evaluates the ratio of true positive predictions out of all positive predictions made by the model. For class 0, the precision is 0.97, indicating that 97% of the instances predicted as class 0 were correct. Similar high precision scores are observed for classes 1, 2, and 3, ranging from 0.96 to 0.98.

- ***Recall:*** Recall, also referred to as sensitivity, assesses the ratio of correctly predicted positive instances to all actual positive instances in the dataset. High recall scores, ranging from 0.96 to 1.00, indicate that the model effectively identified the majority of instances belonging to each class.
- ***F1-score:*** The F1-score, which is the harmonic mean of precision and recall, offers a balanced assessment between these two metrics. It considers both false positives and false negatives in its calculation. The F1-scores for all classes range from 0.96 to 0.99, indicating robust performance across all classes.
- ***Support:*** Support refers to the count of instances in the test data associated with each class. For instance, class 0 has 97 instances, class 1 has 48 instances, class 2 has 94 instances, and class 3 has 89 instances.

Furthermore, the report includes overall metrics:

- ***Accuracy:*** This reflects the average of precision, recall, and F1-score computed separately for each class. The macro average for these metrics is 0.97, indicating a consistently high performance across all classes.
- ***Macro Average:*** This denotes the average precision, recall, and F1-score computed separately for each class. The macro average for precision, recall, and F1-score is 0.97, suggesting a well-balanced performance across all classes.
- ***Weighted Average:*** These metrics reflect the average precision, recall, and F1-score calculated for each class separately. A macro average of 0.97 across these metrics signifies consistently high performance across all classes.

Overall, the high precision, recall, and F1-score, along with a high accuracy rate, suggest that the model has successfully learned to classify the instances across different classes with robust performance. The results are given below as shown in Table **1**:

Table 1. Accuracy metrics.

-	Precision	Recall	F1-score
0	0.97	0.96	0.96
1	0.96	0.96	0.96
2	0.97	0.96	0.96
3	0.98	1.00	0.99

Heatmap Presentation for Confusion Matrix

We have created a heatmap visualization of the confusion matrix for a classification task, using the 'seaborn' library's 'heatmap' function as shown in Fig. (**5**). The confusion matrix, computed from the true labels ('y_test_new') and predicted labels ('pred'), is displayed with annotations to show the count of predictions for each class. The heatmap employs a sequential colormap ('YlGnBu') that transitions from yellow to green to blue, enhancing the visual distinction of values. The plot includes labels on both axes, corresponding to the class labels, and features a title above the heatmap for context.

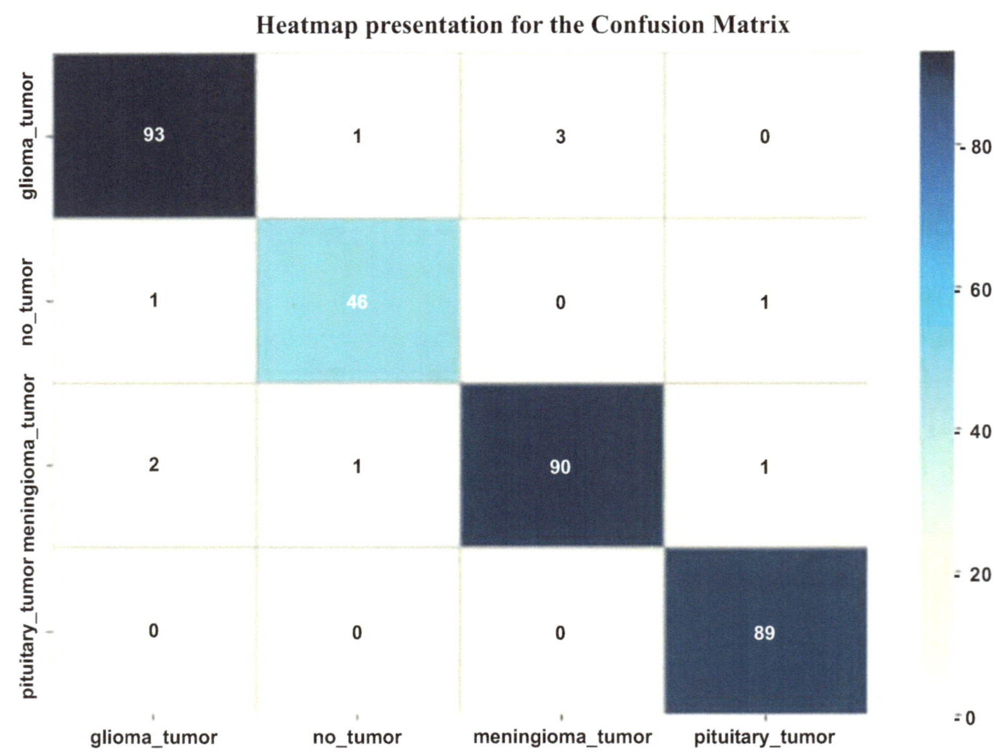

Fig. (5). Heatmap presentation for confusion matrix.

RESULTS AND DISCUSSION

A custom CNN model was built, trained, and evaluated for binary classification, achieving high accuracy on the test dataset. The architecture includes two convolutional layers with 32 filters each, using ReLU activation, followed by max-pooling layers to reduce spatial dimensions and control overfitting. After flattening, a Dense layer with 64 units, incorporating L1 and L2 regularization, and a dropout layer with a rate of 0.5, was added to prevent overfitting. The final Dense layer uses a sigmoid activation function, suitable for binary classification. The model was compiled with the Adam optimizer and binary cross-entropy loss. Class weights were applied during training to address class imbalance. Achieved accuracy highlights the model's effectiveness, with convolutional layers for feature extraction and regularization techniques enhancing performance. This CNN model is a practical approach for image classification, balancing complexity and efficiency.

For binary classification, we first build a CNN architecture and calculate the loss using Binary cross entropy and Adam optimization algorithm to evaluate the accuracy. The dataset was divided into training and testing sets with an 80:20 ratio. We train the model up to 30 epochs and calculate the loss and accuracy for training and validation. After compiling we get 0.6 training loss and 0.8 validation loss, and 80% accuracy. The graphical representation of training and validation is shown below in Figs. (**6** and **7**), respectively.

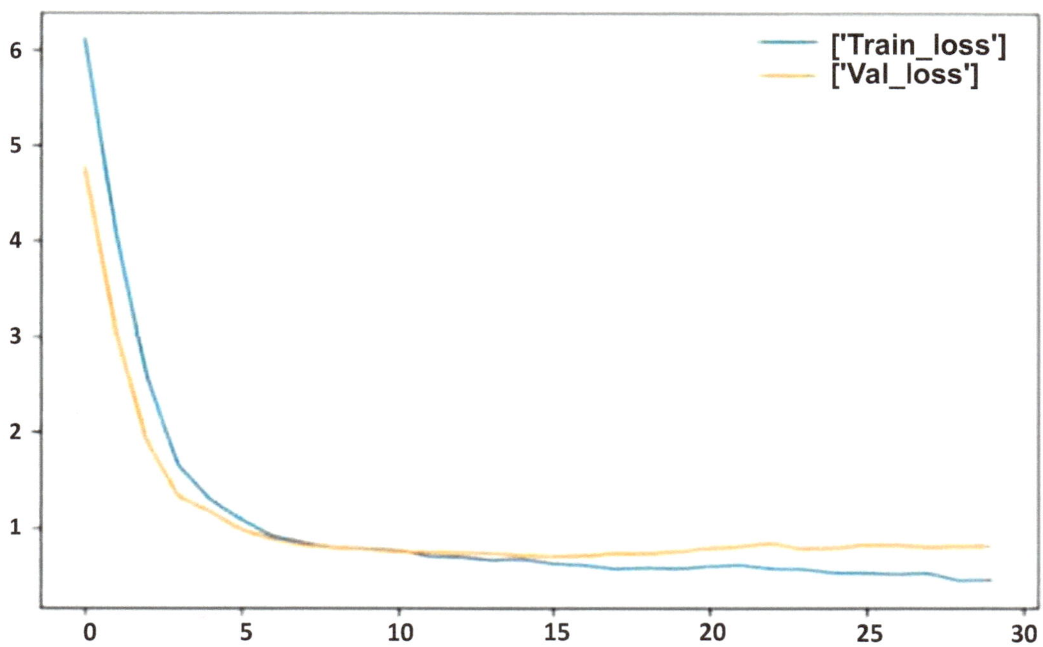

Fig. (6). Loss during training and validation for CNN.

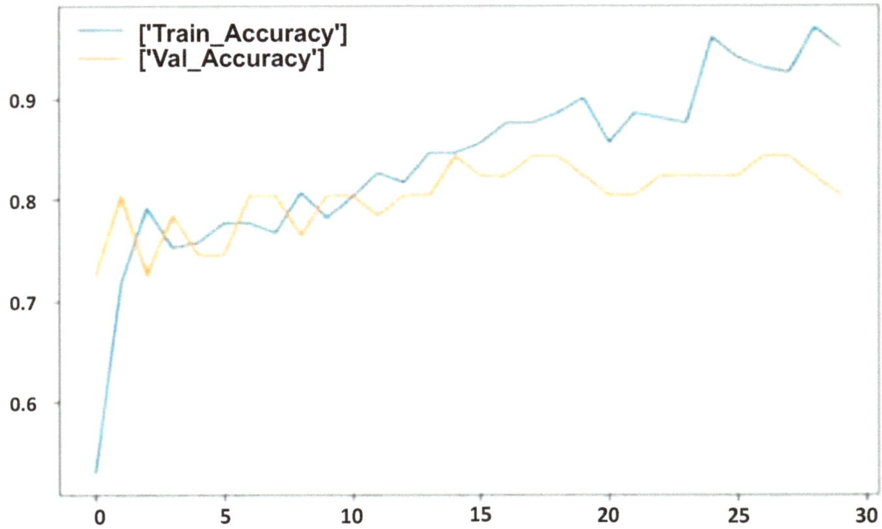

Fig. (7). Accuracy during training and validation for CNN.

Binary Classification Using the VGG19 Model

The VGG19 model was modified and evaluated on the test dataset. The modifications involved resizing the input images to 240x240 pixels and using the pre-trained VGG19 model with ImageNet weights, while freezing its existing layers to retain the learned features. Additional layers were added: a Flatten layer to convert the 2D output of VGG19 to a 1D vector, a Dense layer with 64 units and ReLU activation incorporating L1 and L2 regularization to prevent overfitting, and a final Dense layer with a sigmoid activation function for binary classification.

Freezing the pre-trained layers allowed the model to leverage pre-learned features, making training faster and requiring less data. The additional Dense layers helped in fine-tuning the model to the specific classification task. The achieved accuracy indicates that this approach effectively adapted the VGG19 model to perform well on the given dataset, combining the power of pre-trained features with new task-specific layers [31, 32]. This method is efficient and practical, especially when dealing with limited computational resources and data. For binary classification, we have built a VGG19 and calculated the loss using Binary cross-entropy and using the Adam optimization algorithm to evaluate the accuracy. We train the model up to 30 epochs and calculate the loss and accuracy for training and validation. After compiling, we obtain 0.57 training loss, 0.72 validation loss, and 86% accuracy. The graphical representation of training and validation is shown below in Figs. (**8** and **9**):

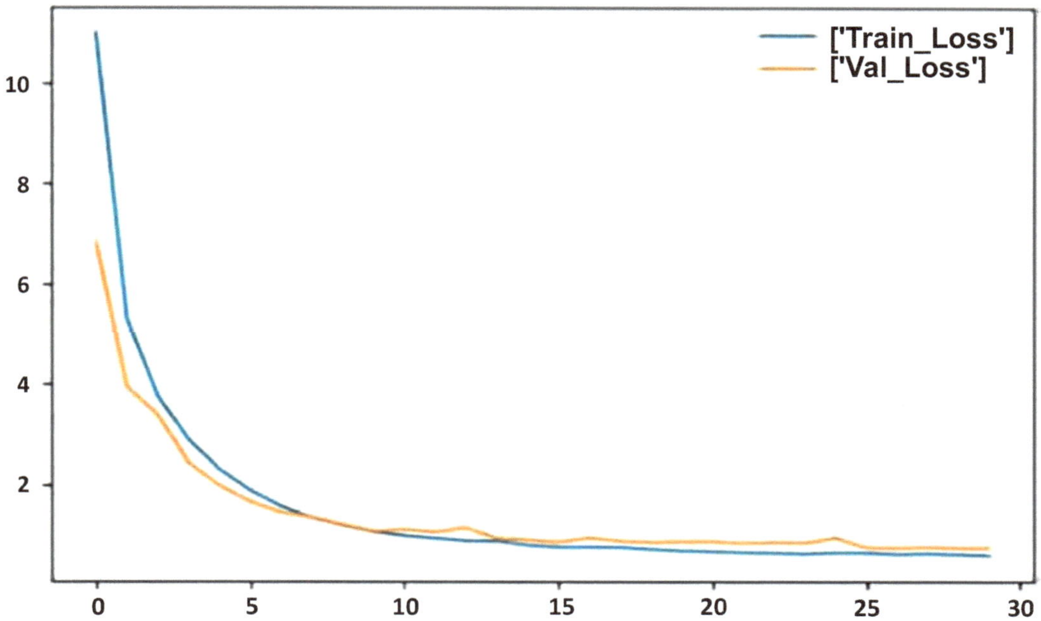

Fig. (8). Loss during training and validation for VGG19.

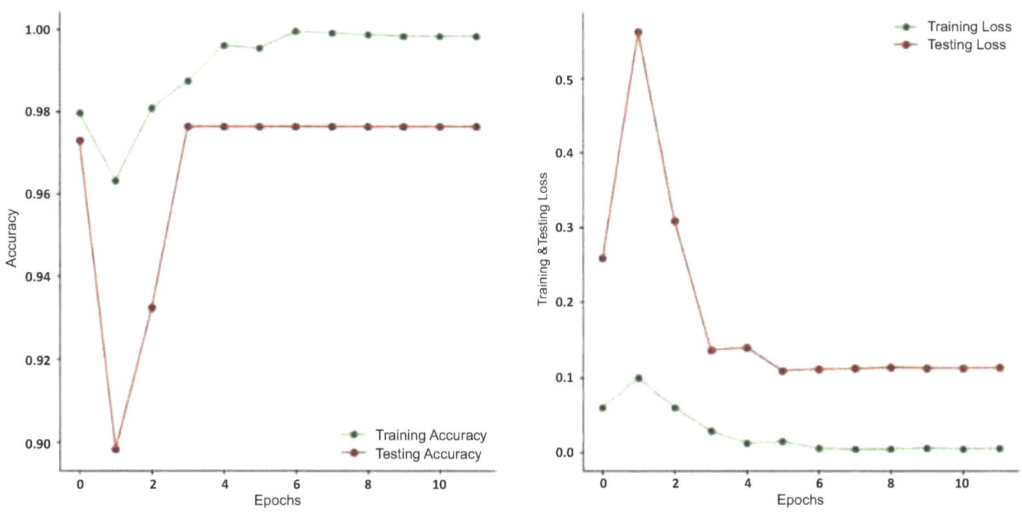

Fig. (9). Accuracy during training and validation for VGG19.

Once the model has been trained, predictions can be made on the test data using VGG19. It can successfully do the prediction part for binary classification. The prediction result is shown below in Fig. (**10**):

```
[ ] print(y_pred)

[1,0,1,1,1,0,1,1,0,0,1,1,1,1,1,0,1,0,0,1,1,
```

Fig. (10). Prediction result with binary classification.

Multiclass Classification Using EfficientNetB0 Model

The EfficientNetB0 architecture, a highly efficient convolutional neural network (CNN) pretrained on ImageNet for image classification tasks. The model is configured with the categorical cross-entropy loss function and the Adam optimizer, which work together to reduce the classification error throughout the training process. Additionally, metrics such as accuracy are tracked during training to assess model performance [33]. The algorithm implements callbacks including TensorBoard for visualizing training metrics, ModelCheckpoint to save the best model based on validation accuracy, and ReduceLROnPlateau to dynamically adjust the learning rate based on validation accuracy improvements [34].

Throughout the training phase, the model is trained on the training dataset, with 10% set aside for validation. A batch size of 32 is utilized, and the training process continues for a total of 12 epochs. These training settings aim to optimize the model's performance while preventing overfitting. The resulting 'history' object contains training metrics such as loss and accuracy over epochs, providing insights into the model's learning process and performance trends. Overall, this code orchestrates a comprehensive training pipeline for EfficientNetB0, aiming to develop an accurate and robust image classification model.

After completing the epochs, we get a 0.068 Training loss and 0.11 testing loss. Also, we get 97% accuracy in training and 89% accuracy in Testing. The results of loss and accuracy with respect to epoch are shown graphically in Fig. (**11**):

After completing the epochs, our model is ready for prediction. Therefore, we save the model for prediction and build a widget for a better user experience. We simply click the upload button to upload any MRI image with any dimension or size, just the image belonging to a category like Glioma, Meningioma, Pituitary, and No-tumor, then click the predict button and the widget can predict the tumor type accurately as shown in Fig. (**12**) below.

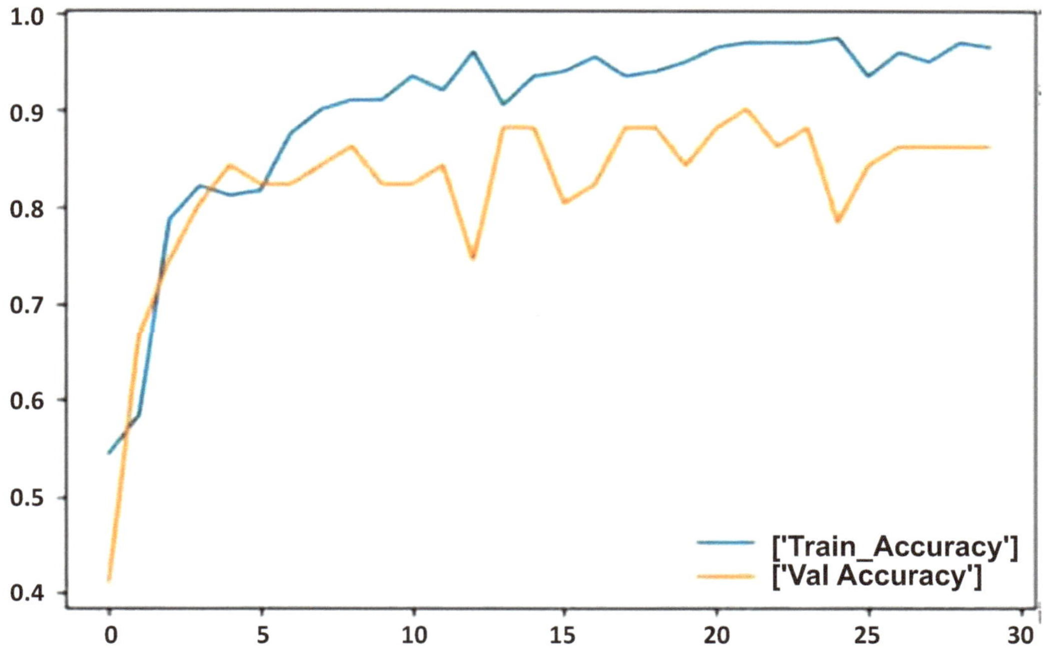

Fig. (11). Epochs *vs*. training and validation accuracy/Loss.

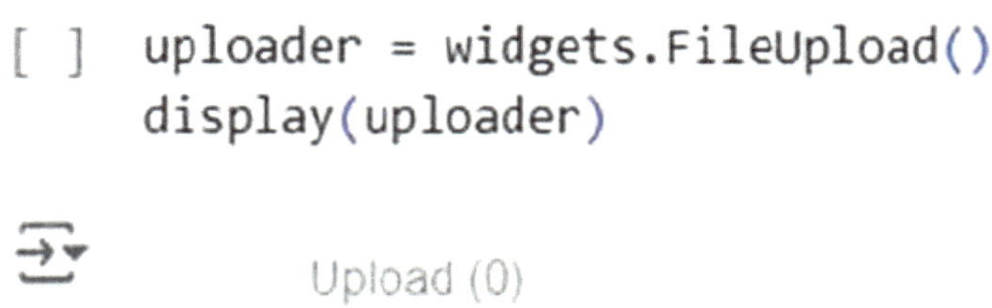

Fig. (12). Upload the MRI image by clicking the upload button.

CONCLUSION

Brain tumor classification from MRI images is a critical task in medical diagnostics, aimed at accurately identifying and categorizing tumors to inform treatment decisions. The complexity and variability in tumor appearance necessitate the use of advanced computational methods to assist radiologists in making accurate diagnoses. Conventional approaches frequently depend on manual interpretation, which may consume significant time and be prone to human inaccuracies. Hence, utilizing deep learning methods, notably Convolutional Neural Networks (CNNs), has gained popularity for their capacity to autonomously learn and extract complex features from MRI images. This results in enhanced classification accuracy and efficiency.

The application of Convolutional Neural Networks (CNNs) for brain tumor classification has demonstrated encouraging outcomes, utilizing sophisticated architectures such as VGG19 and EfficientNetB0 to enhance diagnostic precision. In the beginning, a conventional CNN model underwent training, attaining a training loss of 0.6 and a validation loss of 0.8, along with an accuracy rate of 80%.These metrics indicate a moderate performance, suggesting the model could differentiate brain tumor images from non-tumorous ones to a fair extent, but there was room for enhancement in terms of both loss and accuracy, pointing to the necessity for more sophisticated architectures or further tuning.

To address these limitations, the VGG19 and EfficientNetB0 models were employed. VGG19, recognized for its depth and proficient feature extraction, obtained a training loss of 0.57 and a validation loss of 0.72, showcasing an enhanced accuracy of 86%. This improvement highlights VGG19's ability to generalize better on unseen data due to its depth and complexity. However, the most significant advancement was observed with EfficientNetB0, which recorded a remarkably low training loss of 0.068 and a testing loss of 0.11, demonstrating a high efficiency in learning and generalization. Furthermore, EfficientNetB0 achieved 97% accuracy in training and 89% in testing, underscoring its superior performance in brain tumor classification tasks. This dramatic enhancement in both loss metrics and accuracy emphasizes EfficientNetB0's optimized architecture, balancing depth and computational efficiency, making it a highly effective tool for medical image classification tasks.

REFERENCES

[1] G. Litjens, T. Kooi, B.E. Bejnordi, A.A.A. Setio, F. Ciompi, M. Ghafoorian, J.A.W.M. van der Laak, B. van Ginneken, and C.I. Sánchez, "A survey on deep learning in medical image analysis", *Med. Image Anal.,* vol. 42, pp. 60-88, 2017. [http://dx.doi.org/10.1016/j.media.2017.07.005] [PMID: 28778026]

[2] S. Pereira, A. Pinto, V. Alves, and C.A. Silva, "Brain tumor segmentation using convolutional neural networks in MRI images", *IEEE Trans. Med. Imaging,* vol. 35, no. 5, pp. 1240-1251, 2016. [http://dx.doi.org/10.1109/TMI.2016.2538465] [PMID: 26960222]

[3] G.S. Tandel, A. Balestrieri, T. Jujaray, N.N. Khanna, L. Saba, and J.S. Suri, "Multiclass magnetic resonance imaging brain tumor classification using artificial intelligence paradigm", *Comput. Biol. Med.,* vol. 122, p. 103804, 2020. [http://dx.doi.org/10.1016/j.compbiomed.2020.103804] [PMID: 32658726]

[4] W. Ayadi, W. Elhamzi, I. Charfi, and M. Atri, "Deep CNN for brain tumor classification", *Neural Process. Lett.,* vol. 53, no. 1, pp. 671-700, 2021. [http://dx.doi.org/10.1007/s11063-020-10398-2]

[5] T. Sadad, A. Rehman, A. Munir, T. Saba, U. Tariq, N. Ayesha, and R. Abbasi, "Brain tumor detection and multi-classification using advanced deep learning techniques", *Microsc. Res. Tech.,* vol. 84, no. 6, pp. 1296-1308, 2021. [http://dx.doi.org/10.1002/jemt.23688] [PMID: 33400339]

[6] B.N. Subudhi, V. Thangaraj, E. Sankaralingam, and A. Ghosh, "Tumor or abnormality identification from magnetic resonance images using statistical region fusion based segmentation", *Magn. Reson.*

Imaging, vol. 34, no. 9, pp. 1292-1304, 2016.
[http://dx.doi.org/10.1016/j.mri.2016.07.002] [PMID: 27477599]

[7] A. Pashaei, H. Sajedi and N. Jazayeri, "Brain Tumor Classification *via* Convolutional Neural Network and Extreme Learning Machines," 2018 8th International Conference on Computer and Knowledge Engineering (ICCKE), Mashhad, Iran, 2018, pp. 314-319.
[http://dx.doi.org/10.1109/ICCKE.2018.8566571]

[8] A.M. Sarhan, "Brain tumor classification in magnetic resonance images using deep learning and Wavelet Transform", *J. Biomed. Sci. Eng.,* vol. 13, no. 6, pp. 102-112, 2020.
[http://dx.doi.org/10.4236/jbise.2020.136010]

[9] N. Abiwinanda, M. Hanif, S.T. Hesaputra, A. Handayani, and T.R. Mengko, "Brain tumor classification using convolutional neural network", *IFMBE Proc.,* vol. 68, pp. 183-189, 2019.
[http://dx.doi.org/10.1007/978-981-10-9035-6_33]

[10] B. Badjie, and E. Deniz Ülker, "A deep transfer learning-based architecture for brain tumor classification using MR images", *Inf. Technol. Control,* vol. 51, no. 2, pp. 332-344, 2022.
[http://dx.doi.org/10.5755/j01.itc.51.2.30835]

[11] S. Deepak, and P.M. Ameer, "Brain tumor classification using deep CNN features via transfer learning", *Comput. Biol. Med.,* vol. 111, p. 103345, 2019.
[http://dx.doi.org/10.1016/j.compbiomed.2019.103345] [PMID: 31279167]

[12] K. S. Gill, A. Sharma, V. Anand, and R. Gupta, “Brain Tumor Detection using VGG19 model on Adadelta and SGD Optimizer,” 2022 6th International Conference on Electronics, Communication and Aerospace Technology, Dec. 2022.
[http://dx.doi.org/10.1109/ICECA55336.2022.10009496]

[13] V. Goutham, A. Sameerunnisa, S. Babu, and T. B. Prakash, “Brain Tumor Classification using EfficientNet-B0 Model,” 2022 2nd International Conference on Advance Computing and Innovative Technologies in Engineering (ICACITE), Apr. 2022.
[http://dx.doi.org/10.1109/ICACITE53722.2022.9823526]

[14] A. E. Minarno, M. H. C. Mandiri, Y. Munarko, and H. Hariyady, “Convolutional Neural Network with Hyperparameter Tuning for Brain Tumor Classification,” Kinetik Game Technology Information System Computer Network Computing Electronics and Control, May 2021.
[http://dx.doi.org/10.22219/kinetik.v6i2.1219]

[15] F. Bashir-Gonbadi, and H. Khotanlou, "Brain tumor classification using deep convolutional autoencoder-based neural network: multi-task approach", *Multimedia Tools Appl.,* vol. 80, no. 13, pp. 19909-19929, 2021.
[http://dx.doi.org/10.1007/s11042-021-10637-1]

[16] Y. Bhanothu, A. Kamalakannan and G. Rajamanickam, "Detection and Classification of Brain Tumor in MRI Images using Deep Convolutional Network," 2020 6th International Conference on Advanced Computing and Communication Systems (ICACCS), Coimbatore, India, 2020, pp. 248-252.
[http://dx.doi.org/10.1109/ICACCS48705.2020.9074375]

[17] J. Kang, Z. Ullah, and J. Gwak, "MRI-Based Brain Tumor Classification Using Ensemble of Deep Features and Machine Learning Classifiers", *Sensors (Basel),* vol. 21, no. 6, p. 2222, 2021.
[http://dx.doi.org/10.3390/s21062222] [PMID: 33810176]

[18] A. Tiwari, S. Srivastava, and M. Pant, "Brain tumor segmentation and classification from magnetic resonance images: Review of selected methods from 2014 to 2019", *Pattern Recognit. Lett.,* vol. 131, pp. 244-260, 2020.
[http://dx.doi.org/10.1016/j.patrec.2019.11.020]

[19] A.K. Mandle, S.P. Sahu, and G.P. Gupta, "CNN-Based deep learning technique for the brain tumor identification and classification in MRI images", *Int. J. Softw. Sci. Comput. Intell.,* vol. 14, no. 1, pp. 1-20, 2022.
[http://dx.doi.org/10.4018/IJSSCI.304438]

[20] A.M. Alhassan, and W.M.N.W. Zainon, "Brain tumor classification in magnetic resonance image using hard swish-based RELU activation function-convolutional neural network", *Neural Comput. Appl.,* vol. 33, no. 15, pp. 9075-9087, 2021. [http://dx.doi.org/10.1007/s00521-020-05671-3]

[21] M. Rizwan, A. Shabbir, A.R. Javed, M. Shabbir, T. Baker, and D. Al-Jumeily Obe, "Brain tumor and glioma grade classification using Gaussian convolutional neural network", *IEEE Access,* vol. 10, pp. 29731-29740, 2022. [http://dx.doi.org/10.1109/ACCESS.2022.3153108]

[22] A.H. Khan, S. Abbas, M.A. Khan, U. Farooq, W.A. Khan, S.Y. Siddiqui, and A. Ahmad, "Intelligent model for brain tumor identification using Deep learning", *Appl. Comput. Intell. Soft Comput.,* vol. 2022, pp. 1-10, 2022. [http://dx.doi.org/10.1155/2022/8104054]

[23] K. Maheswari, A. Balamurugan, P. Malathi, and S. Ramkumar, "Hybrid clustering algorithm for an efficient brain tumor segmentation", *Mater. Today Proc.,* vol. 37, pp. 3002-3006, 2021. [http://dx.doi.org/10.1016/j.matpr.2020.08.718]

[24] A. A. Abbood, Q. M. Shallal, and M. A. Fadhel, "Automated brain tumor classification using various deep learning models: a comparative study," Indonesian Journal of Electrical Engineering and Computer Science, vol. 22, no. 1, p. 252, Apr. 2021. [http://dx.doi.org/10.11591/ijeecs.v22.i1.pp252-259]

[25] P. Cheng, M. Han, A. Li, and F. Zhang, "STC□IDS: Spatial-temporal correlation feature analyzing based intrusion detection system for intelligent connected vehicles", *Int. J. Intell. Syst.,* vol. 37, no. 11, pp. 9532-9561, 2022. [http://dx.doi.org/10.1002/int.23012]

[26] S. Bagyaraj, R. Tamilselvi, P.B. Mohamed Gani, and D. Sabarinathan, "Brain tumour cell segmentation and detection using deep learning networks", *IET Image Process.,* vol. 15, no. 10, pp. 2363-2371, 2021. [http://dx.doi.org/10.1049/ipr2.12219]

[27] Y. Guan, M. Aamir, Z. Rahman, A. Ali, W.A. Abro, Z.A. Dayo, M.S. Bhutta, and Z. Hu, "A framework for efficient brain tumor classification using MRI images", *Math. Biosci. Eng.,* vol. 18, no. 5, pp. 5790-5815, 2021. [http://dx.doi.org/10.3934/mbe.2021292] [PMID: 34517512]

[28] S.A. El-Feshawy, W. Saad, M. Shokair, and M. Dessouky, "Brain Tumour Classification Based on Deep Convolutional Neural Networks", *2021 International Conference on Electronic Engineering (ICEEM),* pp. 1-5, 2021.Menouf, Egypt. [http://dx.doi.org/10.1109/ICEEM52022.2021.9480637]

[29] V. Rajinikanth, S. Kadry, and Y. Nam, "Convolutional-Neural-Network assisted segmentation and SVM classification of brain tumor in clinical MRI slices", *Inf. Technol. Control,* vol. 50, no. 2, pp. 342-356, 2021. [http://dx.doi.org/10.5755/j01.itc.50.2.28087]

[30] M. Sajjad, S. Khan, K. Muhammad, W. Wu, A. Ullah, and S.W. Baik, "Multi-grade brain tumor classification using deep CNN with extensive data augmentation", *J. Comput. Sci.,* vol. 30, pp. 174-182, 2019. [http://dx.doi.org/10.1016/j.jocs.2018.12.003]

[31] B. Pydala, T. P. Kumar, and K. K. Baseer, "Smart_Eye: A Navigation and Obstacle Detection for Visually Impaired People through Smart App," Journal of Applied Engineering and Technological Science (JAETS), vol. 4, no. 2, pp. 992–1011, Jun. 2023. [http://dx.doi.org/10.37385/jaets.v4i2.2013]

[32] M.M. Rahman, M.M. Islam, S. Ahmmed, and S.A. Khan, "Obstacle and Fall Detection to Guide the Visually Impaired People with Real Time Monitoring", *SN Computer Science,* vol. 1, no. 4, p. 219,

2020.
[http://dx.doi.org/10.1007/s42979-020-00231-x]

[33] G. Presti, D. Ahmetovic, M. Ducci, C. Bernareggi, L. Ludovico, A. Barate, F. Avanzini, and S. Mascetti, *WatchOut: Obstacle Sonification for People with Visual Impairment or Blindness. United States of America.* Association for Computing Machinery: New York, United States, 2019.
[http://dx.doi.org/10.1145/3308561.3353779]

[34] M. Shimakawa, K. Matsushita, I. Taguchi, C. Okuma, and K. Kiyota, *Smartphone Apps of Obstacle Detection for Visually Impaired and its Evaluation.* Association for Computing Machinery: New York, United States of America, 2019.
[http://dx.doi.org/10.1145/3325291.3325381]

SUBJECT INDEX

D

E

F

G

H

I

K

L

M

N

O

P

R

S

T

V

X

Z

www.ingramcontent.com/pod-product-compliance
Ingram Content Group UK Ltd.
Pitfield, Milton Keynes, MK11 3LW, UK
UKHW060011051026
15591UKWH00002B/147